THE BIOTECHNOLOGY DEBATE

LIBRARY OF ETHICS AND APPLIED PHILOSOPHY

VOLUME 29

Editor in Chief

Marcus Düwell, *Utrecht University, Utrecht, NL*

Editorial Board

Deryck Beyleveld, *Durham University, Durham, U.K.*
David Copp, *University of Florida, USA*
Nancy Fraser, *New School for Social Research, New York, USA*
Martin van Hees, *Groningen University, Netherlands*
Thomas Hill, *University of North Carolina, Chapel Hill, USA*
Samuel Kerstein, *University of Maryland, College Park, USA*
Will Kymlicka, *Queens University, Ontario, Canada*
Philippe Van Parijs, *Louvaine-la-Neuve (Belgium) en Harvard, USA*
Qui Renzong, *Chinese Academy of Social Sciences, China*
Peter Schaber, *Ethikzentrum, University of Zürich, Switzerland*
Thomas Schmidt, *Humboldt University, Berlin, Germany*

For further volumes:
http://www.springer.com/series/6230

THE BIOTECHNOLOGY DEBATE

Democracy in the Face of Intractable Disagreement

by

BERNICE BOVENKERK
Utrecht University, Netherlands

Bernice Bovenkerk
Ethics Institute
Philosophy Department
Utrecht University
Janskerkhof 13a
3512 BL Utrecht
Netherlands
b.bovenkerk@uu.nl

ISSN 1387-6678
ISBN 978-94-007-2690-1   e-ISBN 978-94-007-2691-8
DOI 10.1007/978-94-007-2691-8
Springer Dordrecht Heidelberg London New York

Library of Congress Control Number: 2011941906

© Springer Science+Business Media B.V. 2012
No part of this work may be reproduced, stored in a retrieval system, or transmitted in any form or by
any means, electronic, mechanical, photocopying, microfilming, recording or otherwise, without written
permission from the Publisher, with the exception of any material supplied specifically for the purpose
of being entered and executed on a computer system, for exclusive use by the purchaser of the work.

Printed on acid-free paper

Springer is part of Springer Science+Business Media (www.springer.com)

To my parents,
Frank & Louise
and my children,
Siena & Inuk

Foreword

Democratic theory took a deliberative turn in the 1990s. Deliberative democracy has since taken many of its own turns, beginning with an empirical turn that focused on the deliberative dynamics of mini-publics, followed by an institutional turn that looked at the way different local designs influence deliberation in mini-publics. This book builds upon all of these developments while adding a number of further twists and turns, all of which considerably enrich and refine our understanding of both the virtues and limits of deliberative democratic theory and practice.

Most deliberative democrats recognise that some conflicts are intractable, but they have not yet fully grappled with the question as to how liberal democratic governments should best handle them. Bernice Bovenkerk sets out to explore the role that deliberation might play in liberal democracies that purport to be neutral but must make regulatory decisions in response to problems that are the subject of intractable disagreement. Through her comparative study of biotechnology ethics committees and consensus conferences in Australia and the Netherlands, she shows that disagreement comes in different shapes and sizes, and that therefore government responses must be calibrated to suit the particular confluence of disagreement presented by different problems. She shows that the debate over modern biotechnology is a quintessential case of intractable conflict because it encompasses disagreement on virtually all conceivable levels – factual, scientific, definitional, interest-based, value-based, moral and metaphysical. The upshot is that conflict permeates both expert and lay debates and therefore ought not and cannot be contained by being relegated to expert committees. She argues convincingly that the best strategy for governments in response to conflicts of this kind is to enable rather than foreclose a vigorous debate in the public sphere, and that politicisation needs to take place within and across scientific and normative discussions.

Bernice Bovenkerk also delves deeply into the aims of deliberative democracy and exposes some key tensions, particularly between the goal of consensus, on the one hand, and inclusiveness and quality of debate, on the other. One of the acute insights to emerge from her study is that deliberation can serve different purposes vis-à-vis different types of problem. In the case of deeply unstructured problems, like the modern biotechnology debate, she argues that the aim of inclusion is more appropriate than the aim of consensus; it is fairer from a pluralist standpoint since there is much less of a risk of suppression of dissensus.

This book adds a much needed comparative dimension to the study of deliberative democracy and highlights the importance of political culture and broader institutional settings in shaping the capacity and propensity of citizens to engage in deliberation and the degree to which governments are prepared to relinquish authority to deliberative mini-publics. Bernice Bovenkerk writes with great authority, since she has not only plumbed the depths of deliberative democratic theory and practice as a scholar but also has first-hand knowledge of the practical workings of expert committees, having worked as a scientific assistant at the Dutch Committee for Animal Biotechnology. She has also lived in both of the worlds that she has studied and therefore understands the idiosyncrasies of both political cultures.

Bernice Bovenkerk demonstrates with considerable force that there can be no grand theory of deliberative democracy because there is no common political culture of deliberation. She concludes that 'Tailoring deliberative democracy to specific contexts may be the next big challenge for deliberative democrats'. In my view, this book has already admirably responded to this challenge and is a must-read for anyone interested in democracy, deliberation and/or intractable conflict.

Melbourne, Australia   Robyn Eckersley

Preface

Disagreement is part of the human condition. We could call it a paradoxical force; for social and psychological reasons we try to avoid disagreement, while on a cognitive level we need it; the confrontation with other points of view can lead us to either reconsider or strengthen our own views and disagreement is a catalyst for change and progress. Even if we simply want to accept or even celebrate difference and disagreement, on a political level we still have to deal with it. This book addresses the question of how governments should deal with the existence of intractable disagreement about novel technologies. The debate about the genetic modification of animals and plants is examined as a paradigmatic case for intractable disagreement in today's pluralistic societies. This examination reveals that the disagreements in this debate are multi-faceted and multi-dimensional and can often be traced to fundamental disagreements about values or worldviews. How can governments acknowledge divergence of opinions and stay neutral between them for as long as possible while still making regulatory decisions? Two solutions to this problem of neutrality are explored: political liberalism and deliberation. The central argument of this book is that certain deliberative theories and practices, developed in the book, can deal better with the existence of intractable disagreement about novel technologies than general political liberal theories and liberal democratic practices. This argument is advanced both on a theoretical and on a practical level; the latter through the use of case-studies, concerning two different deliberative practices – consensus conferences and biotechnology ethics committees – in two different countries – the Netherlands and Australia. This double comparative analysis serves to highlight the relevance of broader institutional factors in accounting for the success or failure of such deliberative experiments. The aim of this empirical analysis is twofold: first, to find out to what extent deliberative fora conform to the ideals of deliberative democracy; and second, to examine to what extent these ideals may need to be revised in the light of practice. In this respect, this work seeks to offer a more refined normative deliberative theory grounded in a more refined sociological understanding of deliberative practice.

It is argued that different types of problem – either structured, unstructured, or moderately structured – call for different responses and that particularly the goal of deliberation needs to be matched to the right policy problem. The empirical analysis, furthermore, suggests that political culture is very influential for the way public deliberation is regarded and for its success. We cannot expect a grand theory of deliberative democracy that is equally applicable to each country. While political liberalism is criticised it is not argued that the liberal framework needs to be abandoned, but rather that, firstly, more emphasis needs to be put on its deliberative elements, and secondly, that successful deliberation calls for careful attention to the political culture of countries. Particular challenges exist for countries that can be characterised as antagonistic, but where consensus democracies have elitist tendencies, these need to be dealt with as well. The aims, form, and scope of the deliberation should, then, be chosen carefully, dependent on the specific context. Tailoring deliberative democracy to specific contexts may be the next big challenge for deliberative democrats.

This book was developed from my doctoral dissertation. My initial interest in the topic of environmental politics was aroused when I was an exchange student at the University of Melbourne and Robyn Eckersley introduced me to the debate between atomists and holists. I have been hooked to the problem of disagreement about environmental decision-making ever since. I cannot even begin to express my gratitude to Robyn for taking me on as a Ph.D. student and inspiring me to think outside of the box. Her dedication, meticulous correction work, and unfailing enthusiasm were indispensable in the realisation of my thesis. I feel privileged to have worked so closely with one of the leading deliberative democrats of our time. The department of Political Science at the University of Melbourne supported me throughout that period. This research is based on a comparative analysis between the Netherlands and Australia, and I have been fortunate enough to be able to travel between these two countries. This research could never have been undertaken without generous grants from the University of Melbourne and the Government of Australia. In Australia I was also stimulated to write my thesis and to stay critical by Richard Hindmarsh. I would like to thank Richard and also Carolyn Hendriks for their helpful comments.

The foundations of this book were further laid when I was working as scientific assistant at the Dutch Committee for Animal Biotechnology. While I have learned a lot from the interesting discussions between the members of this committee, being so close to the fire also enabled me to critically reflect on the committee-system. I would like to thank in particular Egbert Schroten and Ronno Tramper for giving me this opportunity and for our stimulating discussions.

In the Netherlands I was generously welcomed at the Ethics Institute of the University of Utrecht. I want to thank all my colleagues at the Institute for their kind support and for providing critical and constructive feedback. First and foremost I am indebted to Frans W. A. Brom, without whom this book would not exist. Frans has coached me through the whole writing process in a way that was always frank and sympathetic, while strict and critical when necessary. Frans could always formulate much better what I wanted to say than I could myself. I have greatly appreciated having Frans as a friend, mentor and sparring partner. I would like to thank in particular the following colleagues for commenting on drafts of my chapters in an insightful and constructive manner: Bert van den Brink, Marcus Düwell, André Krom, Lonneke Poort, and Derrick Beyleveld. For helping me out with many of the

Preface

day-to-day problems and letting me air my frustrations, I want to thank Suzanne van Vliet and Judith Zijm.

My empirical research would not have been possible without the interviewees generously donating their time and effort to speak to me: Bert Laeyendecker, Egbert Schroten, Huub Schellekens, Dick Koelega, Tjard de Cock Buning, Henk Verhoog, Marianne Kuil, Koos van der Bruggen, Franck Meijboom, Rosemary Robins, Bidda Jones, Ariel Salleh, Bob Phelps, Ellen Kittson, Glenys Oogjes, Sheena Boughton, and Rob Sward. Thank you very much for your interest and patience. Four native English speakers proof-read chapters of this book and I would like to thank them for their hard work struggling through unfamiliar territory: Justin Rowe, Graham Rowe, Gerald de Jong, and Derek Parker. Thanks is due as well to Graham Smith and Hub Zwart, as well as to two anonymous reviewers from Springer for their helpful comments.

This work took many years to complete and in the process many people supported me both emotionally and logistically. I would like to mention in particular Ilja van der Gaag. Coz, our regular phone conversations about the difficulties of combining motherhood and work, and your help with the children have really pulled me through at times. Dionne van Heteren, Barbara Vriend, Sandra van Heeswijk, and Marvin Bovenkerk were always willing to help out; this has been greatly appreciated. Other friends have given me the opportunity to let off steam and reflect on the writing process: Mandy Bosma, Lien van Eck, Lonneke Poort, Elke Müller, Lynsey Dubbeld, Frederike Kaldewaij, and Tatjana Visak.

The writing process has repercussions not only on the author, but also on her immediate surroundings. I am very grateful for the unfailing support of my parents, Frank and Wiesje. You have always been my source of inspiration and have given me my first lessons in dealing with disagreement. I want to thank Siena and Inuk for giving me a reason to continue. You were both born during the writing of this book and I feel you are inextricably linked with it. Finally, without Justin's support, both financial and emotional, I would not be where I am today.

Contents

1	**Introduction**	1
	1.1 Central Question	1
	1.2 Government Neutrality	3
	1.3 The Liberal vs. the Deliberative Solution	4
	1.4 General Method	5
	1.5 Status of Empirical Results	7
	1.6 Case-Studies Method	8
	1.7 Depoliticisation	10
	1.8 Outline	12
	References	15

Part I   Context of Book

2	**Biotechnology: An Anatomy of the Debate**	19
	2.1 Introduction	19
	2.2 The Debate	21
	2.2.1 Demarcation of Debate Analysis	21
	2.2.2 Historical Context	25
	2.2.3 Rhetoric	27
	2.3 General Positions on Biotechnology	31
	2.3.1 Crossing Species Barriers	32
	2.3.2 Genetic Determinism	33
	2.3.3 Human Hubris	34
	2.4 Animal Biotechnology	35
	2.5 Plant Biotechnology	37
	2.5.1 Justice	38
	2.5.2 Health Concerns and Risk	39
	2.5.3 Environmental Consequences	41
	2.6 Discussion	43
	2.7 Sources of Disagreement	45
	2.8 Unstructured Problems	50
	2.9 Conclusion	54
	References	56

xiii

3 Constraining or Enabling Dialogue? 63
   3.1   Introduction 63
   3.2   The Liberal Neutrality Thesis 65
   3.3   Public Versus Private 66
   3.4   Conversational Restraint 70
   3.5   Morality Versus Ethics 73
   3.6   Broader Views 76
   3.7   Opinion Transformation 78
   3.8   Aggregation and Bargaining 79
   3.9   Deliberation 81
   3.10 Criteria for Dealing with Intractable Disagreement 84
   3.11 Conclusion 85
   References 86

Part II   Theoretical Framework

4 Deliberative Democracy and Its Limits 91
   4.1   Introduction 91
   4.2   Framework 93
        4.2.1   Three Core Questions 93
        4.2.2   Practical Reasons for Public Deliberation 95
        4.2.3   Theoretical Reasons for Public Deliberation 96
        4.2.4   Normative Assumptions 100
        4.2.5   Tensions Between Ideals 102
   4.3   Deliberative Democrats' Answers 103
        4.3.1   Goals of Deliberation 103
        4.3.2   What Types of Argument Are Valid? 106
        4.3.3   Who Should Participate? 109
        4.3.4   Recapitulation 110
   4.4   Critiques of Deliberative Democracy 110
        4.4.1   Goals of Deliberation 111
        4.4.2   What Types of Argument Are Valid? 114
        4.4.3   Who Should Participate? 117
        4.4.4   Recapitulation 120
   4.5   Deliberative Responses 121
        4.5.1   Goals of Deliberation 122
        4.5.2   What Types of Argument Are Valid? 128
        4.5.3   Who Should Participate? 132
        4.5.4   Recapitulation 134
   4.6   Conclusion 134

Intermission: Between Theory and Practice 138
   Differences Between Committees and Consensus Conferences 141
   References 143

Part III Deliberative Fora: Deliberative Democracy Put to the Test

5   Committees: The Politics of Containment 151
- 5.1   Introduction 151
- 5.2   Comparative Analysis Netherlands – Australia 152
  - 5.2.1   Political Cultures 152
  - 5.2.2   Biotechnology in Australia and the Netherlands 154
- 5.3   A Tale of Two Committees 157
  - 5.3.1   Committee for Animal Biotechnology 158
  - 5.3.2   Gene Technology Ethics Committee 161
- 5.4   Goals of Deliberation 164
  - 5.4.1   CAB: Tension Between Different Roles 164
  - 5.4.2   GTEC: A Toothless Tiger 167
- 5.5   What Types of Argument Are Valid? 170
  - 5.5.1   CAB: Narrow Terms of Reference 170
  - 5.5.2   GTEC: Narrow Problem Framing 172
- 5.6   Who Should Participate? 174
  - 5.6.1   CAB: Predominance of Experts 174
  - 5.6.2   GTEC: Tokenism 178
- 5.7   Committees of Containment: Discussion and Conclusion 181
- References 186

6   Consensus Conferences: The Influence of Contexts 189
- 6.1   Introduction 189
- 6.2   Background 190
  - 6.2.1   Deliberative Mini-publics 191
  - 6.2.2   Comparative Analysis 195
- 6.3   Cloning and GM Food: Two Consensus Conferences 199
  - 6.3.1   Australian Consensus Conference 199
  - 6.3.2   Dutch Lay Panel 202
- 6.4   Goals of Deliberation 205
  - 6.4.1   Australia: Focus on Consensus 206
  - 6.4.2   Netherlands: Deepening the Debate 208
- 6.5   Arguments 210
  - 6.5.1   Australia: Trust on Trial 210
  - 6.5.2   Netherlands: A Philosophical Framework 215
- 6.6   Participation 216
  - 6.6.1   Australia: Massive Learning Effect 217
  - 6.6.2   Netherlands: An Elite Affair 221
- 6.7   The Influence of Contexts: Discussion 223
  - 6.7.1   Comparative Analysis 225
  - 6.7.2   Depoliticisation 226
- 6.8   Conclusion 229
- References 230

Part IV Conclusions: Deliberative Democracy Revisited

7 Implications of Empirical Results for Deliberative Theory 235
    7.1   Introduction 235
    7.2   Theory and Practice 236
    7.3   Implications of Comparative Analysis 238
    7.4   Political Culture 242
    7.5   Conclusion 243
    References 244

Appendix A: Animal Biotechnology Debate 245
    Benefits of Animal Transgenesis 246
    Moral Status of Animals 248
    Animal Welfare 249
    Animal Integrity and Dignity 252
    Conclusion 258

Appendix B: Plant Biotechnology Debate 261
    From Green Revolution to Gene Revolution 263
    Justice 267
    Will Biotechnology Feed the Poor? 268
    Golden Rice 274
    Terminator Technology 277
    Patenting 279
    Owning Life 282
    Bioprospecting Versus Biopiracy 283
    Food Safety and Health Risks 287
    Substantial Equivalence 290
    Risks 292
    Precautionary Principle 296
    Labelling 298
    Environmental Consequences 301
    Herbicide and Pesticide Use 302
    Contamination 305
    Loss of Biodiversity 307
    Sustainability 308
    Discussion 310
    Analysis and Conclusion 312

Bibliography to Appendices A and B 317
    Websites 324

Appendix C: Interview Questions 325
    Questions Pertaining to Members of Ethics Committees 325
    Questions Pertaining to Representatives of Interest Groups 327
    Questions Pertaining to Organizers of Public Debates 328

Index 331

List of Abbreviations

AEC	Animal Experimentation Committee
CAB	Committee for Animal Biotechnology (Netherlands)
CBD	Convention on Biological Diversity
COGEM	Committee on Genetic Modification (Netherlands)
FAO	Food and Agriculture Organization
GM	Genetically modified
GMO	Genetically modified organism
GTCCC	Gene Technology Community Consultation Committee (Australia)
GTEC	Gene Technology Ethics Committee (Australia)
GTECCC	Gene Technology Ethics and Community Consultation Committee (Australia)
GTR	Gene Technology Regulator (Australia)
GTTAC	Gene Technology Technical Advisory Committee (Australia)
OGTR	Office of the Gene Technology Regulator (Australia)
TRIPS	Trade-related Aspects of Intellectual Property Rights
WTO	World Trade Organization

List of Tables

Table 2.1	Four problem types of Hisschemöller	51
Table 2.2	Role of committees and consensus conferences in problem types	54
Table 4.1	Conditions and guiding criteria	137
Table 5.1	Assessment model CAB	160
Table 6.1	Passive/active and inclusive/exclusive states	198

Chapter 1
Introduction

Separating morality from politics makes both disciplines incomprehensible.
Jean Jacques Rousseau

One of the most imaginative and penetrating novels of the 20th century, George Orwell's *Animal Farm*, stands to gain a whole new meaning in the 21st century. I imagine a sequel, entitled *Organ Farm*, in which a group of pigs await to become organ donors for people who have suffered a heart attack. The pigs' constitution has been genetically altered to enable their hearts to fit into human bodies. In *Organ Farm*, the pigs will lead an uprising of all donor animals awaiting a similar fate, including baboons, goats and mice, demanding justice and equality for all animals. They consider that every species should take care of its own health problems and claim the right of every species to have a say in its own genetic destiny.

1.1 Central Question

Of course, Orwell's story was an allegory for a different struggle than that of animals against their keepers. However, the story of the organ farm represents a more literal political struggle by humans: the struggle of those who want to 'free' nature and animals from the oppressive forces of the dominant 'class' of humans, who view nonhumans as merely a resource, to dispose of as they please. With the advent of novel technologies, ever more capable of transforming nature and moulding it for human purposes, our relationship with the nonhumans around us has become the subject of critical inquiry. The recent lively debates about issues such as animal experimentation, live animal exports, and cloning illustrate that people disagree about our appropriate treatment of nonhumans. One debate that has been especially contentious and persistent is that about biotechnology, which is a technology that uses living organisms in order to create or change plants, animals, or microorganisms for specific purposes, such as medical research or agricultural production. To be more precise, this debate has focused on *modern* biotechnology, which is carried out in laboratories by molecular biologists at the molecular level, as opposed

B. Bovenkerk, *The Biotechnology Debate*, Library of Ethics and Applied Philosophy 29, DOI 10.1007/978-94-007-2691-8_1, © Springer Science+Business Media B.V. 2012

to traditional forms of biotechnology, which include, for instance, fermenting beer and making yoghurt and cheese.

The biotechnology debate is particularly heated not only because modern biotechnology has a potentially profound influence on people's livelihoods – be it that of the third world farmer, of the organic farmer, the biotechnology company employee, the biomedical scientist, the consumer, or the patient – but also because there are so many different dimensions of disagreement in this debate. As I will show in Chapter 2, where I analyse the debate about animal and plant biotechnology, we should understand this debate as expressing fundamental disagreement. This disagreement has even become intractable for several reasons: the disagreement takes place on many different dimensions (we encounter, for example, scientific, conceptual, value-and worldview related disagreements) and people experience the debate as touching on existential values. Biotechnology ultimately touches on our deeply held beliefs and views about the kind of world we want to live in. Even though this debate is often narrowly construed as one about solely the risks and benefits of biotechnology, the arguments on both sides apply to a wider group of ideas or concerns surrounding, for instance, agriculture, socio-economic relations, nature conservation, and mediatisation. I argue that positions in this debate cannot be seen apart from these broader views about what would constitute a desirable society, about what are proper relationships within society and with nonhuman nature, and even about our own nature. At stake are fundamental views on reality and how to deal with the world and with other humans and non-humans in it and in this context how to deal with scientific uncertainty. In other words, it is a moral and ethical debate about the obligations humans have towards each other, animals, nature, and future generations.

Scientific uncertainty has led to a lack of consensus about the facts involved in this debate, but no consensus exists about the values involved either. In fact, it is becoming ever more clear that in this debate we cannot easily separate facts and values. Normative presuppositions are involved in scientific research, from the stage of problem definition to the choice of methodology, to the interpretation of test results. In the case of biotechnology this is especially true because we are not only dealing with risks (known unknowns), but also with unknown unknowns, which takes us to a whole new level of uncertainty. It appears that the more radical our uncertainty, the more room opens up for normative divergence. We are, therefore, not dealing with a scientific problem that can be solved by scientists and neither with a normative problem that can be dealt with in society; the two are interwoven. According to the useful typology of policy problems by Matthijs Hisschemöller, that I will describe in Chapter 2, treating the scientific and value dimensions of uncertainty separately amounts to treating the wrong policy problem.[1] When we treat a conflict either as primarily based on a lack of scientific consensus or as primarily based on a lack of consensus over values we are treating it as a moderately structured problem,

---

[1] Matthijs Hisschemöller (1993).

whereas in reality the biotechnology conflict is an unstructured problem, where no consensus exists on either facts or values.

I use the debate about the genetic engineering of animals and plants – which for simplicity's sake I will label 'the biotechnology debate' – as a case study because it provides a paradigmatic case of intractable disagreement about novel technologies in pluralistic societies. My underlying assumption is that perhaps no area of contention about novel technologies is as complex as the conflict about biotechnology and if peaceful, democratic ways of dealing with this conflict can be found, these could apply to other, less complex, cases as well.[2] My concern is that intractable disagreement raises the very real prospect of serious conflict and possibly violence and if the state wants to stop the escalation of such conflict it will somehow have to address the disagreement and make a decision about regulation, without estranging the different sides of the debate. This work in political philosophy sets out to explore how governments in democratic societies could rise to this challenge. But more importantly, my concern is not only with the question as to how governments *could* rise to this challenge, but also with how they *should*. In other words, my question is not how we could put an end to intractable disagreements, but rather what would be the fairest way to deal with the inescapable reality of pluralism in this particular context. My central question is, then, considering the fundamental differences of opinion about novel technologies, how should we deal with this politically? As I elaborate below, I assume that governments should, as far as possible, deal with intractable disagreement whilst staying neutral between people's fundamental values. At the same time, I acknowledge that governments will need to make choices that favour some values over others. A subsidiary question of this book is, therefore, how can governments acknowledge divergence of opinions and stay neutral between them for as long as possible while still making regulatory decisions? As I elaborate below, in this context decision-making on the basis of depoliticised debate has been suggested by several theorists and at first sight appears to be helpful. However, disagreement exists about the meaning and potential of depoliticisation. Accordingly, I investigate a further subsidiary question, namely, to what extent should public deliberation be regarded as a form of depoliticisation and to what extent is this successful in practice?

1.2 Government Neutrality

It is my contention that we cannot deal politically with intractable conflict by accepting just one of the competing perspectives involved, because this would do injustice to many people. While it is in the nature of thinking about values that each of us

---

[2] Other complex cases are, for example, radioactive waste, air pollution and the reduction of the ozone layer. According to Pellizzoni, these issues have in common that they have important implications and they are so complex that average citizens do not possess the skills to make adequate judgments about them; furthermore they are characterised by the great levels of uncertainty that exist even among experts. See Luigi Pellizzoni (1999).

takes one position and believes this to be the only true view, governments cannot adopt this attitude; they need to make supra-individual decisions. Governments need to make acceptable regulation for better or for worse. Whether or not we believe in the truth or validity of a specific normative theory, in the political arena no one unifying theory exists, so for political debate regarding values we cannot solve disagreement by declaring one view to be correct on the basis of whatever particular theory. In order to respect the views of their citizens – which is a normative presupposition of any democracy – governments need to participate as a non-party to the conflict for as long as possible. Yet, I cannot envisage a position outside of the debate that governments can take in order to act in a justified way. We lack an external criterion of arbitration between different fundamental values and worldviews, but even if such an external position were to exist, people would still have to convince each other of the correctness of their own views from a position internal to the debate. In a democracy, we cannot simply impose our views on others; we have to provide arguments in order to convince others. Citizens, then, should not expect their government to *simply* take one particular viewpoint or one measurable criterion as definitive in government regulation. Nevertheless, *ultimately* governments need to make choices, particularly when so-called common goods are at stake. While, of course, in day to day policy making governments take active positions, and this is not generally considered illegitimate, in the case of fundamental or even intractable disagreement governments are well advised to stay neutral. How could they do this?

1.3 The Liberal vs. the Deliberative Solution

I will explore two solutions to this problem of neutrality: political liberalism and deliberation. My central argument is that certain deliberative theories and practices, which I develop below, can deal better with the existence of intractable disagreement about novel technologies than general political liberal theories and liberal democratic practices.[3] It is important to note here that liberal democratic theories and deliberative democratic theories operate on two different levels. While political liberalism centers on the political constitutional framework of a just society, deliberative democracy should be understood as primarily a model of political communication or, according to some, political decision-making. The two solutions to the problem of neutrality, therefore, operate on two different levels.

---

[3] As liberal democracy encompasses a range of different models of democracy, it is more appropriate to refer to different analytical models, rather than to 'the theory of liberal democracy'. For our purposes, the following different analytical models can be distinguished. Democracy can be either direct or indirect; in a direct democracy, citizens make most political decisions directly, whereas in an indirect democracy, decisions are made by representatives. Within both types of democracy, decisions can be made by way of aggregation, bargaining, or deliberation. For example, both in a referendum and in elections, preferences are aggregated through a vote, but in the first case citizens vote for policies directly, whereas in the second case they vote for a representative, who will make policy decisions for them.

Deliberative democracy primarily aims to offer an alternative to the methods of political decision-making that are dominant in liberal democracies, namely voting and bargaining. However, as I argue in Chapter 3, the relation between political liberalism and these liberal social choice mechanisms is not a contingent one. The assumptions underlying political liberalism about how governments should remain neutral have led to the liberal focus on voting and bargaining, rather than on deliberation. Furthermore, deliberative democratic theories often operate within a liberal framework, as many of the core principles of liberal democracy are compatible with deliberative ideals, and my critique should, therefore, be regarded as an immanent one. It aims to strengthen rather than replace liberal democracy.

My critique of liberal democracy, then, operates on two levels; firstly, that of the theory of political liberalism and secondly, that of the practice of real world liberal democracies. Similarly, I aim to explore not only the theoretical potential of deliberative democracy to deal with intractable disagreement under ideal circumstances, but also its practical application in real-world contexts. It does so through a comparative analysis of deliberative fora in the biotechnology field in Australia and the Netherlands. As I explain in more detail below, this comparative case study has two purposes; on the one hand it aims to 'test' the theory of deliberative democracy in different settings, while on the other hand it examines to what extent real world deliberative practices conform to the theory of deliberative democracy, and what obstacles we encounter in practice. Because my case-studies offer a double comparison between two countries and two types of deliberative practices, it will also serve to highlight the relevance of broader institutional factors in accounting for the success or failure of such deliberative experiments.

1.4 General Method

The question about the merits of deliberation cannot be answered solely on the basis of theoretical analysis. There is an ongoing, vibrant debate about the theoretical merits of deliberative democracy and many of the important questions surrounding this theory have been studied in depth, but the debate has now reached a point where it is necessary to test the claims of deliberative democrats on the level of real-life practice. As James Bohman argues, 'empirical research is a cure for both a priori scepticism and untested idealism about deliberation... it could enrich and enliven the normative debates about the nature and limits of deliberative problem solving and conflict resolution'.[4] Bohman also notes, however, that surprisingly little empirical case-studies have actually been carried out on the feasibility of deliberative problem solving 'at the appropriate level and scale'.[5] In the empirical part of this book, I want to contribute to filling this gap and I, therefore, use case-studies in the

---

[4] James Bohman (1998), page 422.

[5] Some notable exceptions are offered by Graham Smith and Corinne Wales (2000), Archon Fung (2003) and John S. Dryzek and Simon J. Niemeyer (2003).

field of biotechnology in order to look at the practice of deliberative forums. My aim with this empirical analysis is twofold: first, I want to find out to what extent deliberative fora conform to the ideals of deliberative democracy; and second, I want to examine to what extent these ideals may need to be revised in the light of practice. In this respect, I seek to ground my refined normative theory in a more refined sociological understanding of deliberative practice.

In Part I, I sketch the context of this book and make clear that the biotechnology debate is so intractable because we are dealing with an unstructured problem. In Part II, I examine how political theories could deal with unstructured problems. To this end, I develop a more refined normative theory of deliberative democracy, which flows out of a critique of both political liberalism and deliberative democratic theory. In Part III, I refine my normative theory further by examining its claims – on the basis of my two subsidiary questions – and its feasibility in the light of my comparative case-studies. The practical insights that emerge in Part III are primarily sociological, albeit they are, of course, framed by normative considerations. In other words, I use a mixed method approach, which is on the one hand normative, as it is based on a philosophical critique of and reflection upon political theory, and on the other hand empirical, as it is based on comparative case-studies of two different mini-publics in two different countries. The latter will enable me to explore the influence of different institutional settings and political cultures on both my normative and sociological claims. I use my case-studies in Part III of the book, then, to explore the implications of the normative conditions and criteria that I propose in Part II. This should enable me to both develop a more fine-grained theory that is cognizant of practical realities, and to develop practical insights regarding deliberative mini-publics on the basis of my normative reflection. The latter should help organisers of public debates to attune the design of their debate to its specific aims. My aim is to develop these insights by moving back and forth between theory and practice reflectively. In Part IV, I revisit my deliberative proposal by making explicit what I have learned from this movement between facts and norms.

The empirical part of this book (Part III) consists of a comparative analysis between biotechnology ethics committees (Chapter 5) and consensus conferences (Chapter 6) in the Netherlands and Australia. I use these two particular deliberative 'fora' because they are two of the main ways in which governments and other political actors have responded to conflicts regarding novel technologies and the subsequent call for more public deliberation. Expert committees have been installed to deal with scientific uncertainty while consensus conferences are meant to deal with the lack of consensus about values. As I already mentioned, both treat the policy problem wrongly as a moderately structured problem, while we are in fact dealing with an unstructured problem. My argument is that each type of microcosm deals better with intractable disagreement the more it allows for discussion on the other type of uncertainty. Expert committees should allow for more discussion about values, which entails amongst other things, lay person input. In consensus conferences we should be careful to not let expert discourse dominate the discussion again. In both cases, then, the status of expert knowledge needs to be put into perspective.

1.5 Status of Empirical Results

Before I go on to explain the details of my empirical method, it is necessary to ask whether one can even 'test' deliberative democracy, as I aim to do. As Diana Mutz has argued, those who try to connect deliberative theory to deliberative practice face an almost insurmountable task.[6] On the one hand, deliberative democrats focus on face-to-face discussions between people with opposing viewpoints, which gives the appearance that they build their theory on real-life social contexts. They also make claims about the benefits of deliberation that appear empirical in nature; for example, deliberation would lead to a civic attitude in participants. On the other hand, they define the conditions that are needed for reaching deliberative ideals so strictly that they will probably never be met in real-life circumstances. For instance, power imbalances should not be present and strategic behaviour is not allowed. These conditions cannot be easily replicated in a real-life context and it is, therefore, hard to test the theoretical claims. Researchers can test certain specific aspects of the deliberative process, but then they are open to the accusation that they have not done justice to the larger theory. Nonetheless, empirical research should not be abandoned, because if we want deliberative theory to be more than just an academic exercise we will need to develop insights about its practical application and the claims we make should then be open to empirical scrutiny. In the words of Mutz, 'to the extent that empirical research and political theory fail to speak to one another, both fields are impoverished'.[7]

But what does it mean when we say that empirical research and political theory need to speak to one another? Does it mean that deliberative democrats need to formulate less stringent ideals? In my view ideals are necessary as a guide, to show us what direction it is that we want to move to in the first place. On the other hand, it is not necessarily the case that the closer we move to ideals, the better system we have. Living with only half-fulfilled ideals may be worse than living with no ideals. Still, on a theoretical level, deliberative democrats have internally coherent reasons for proposing their ideals. One of the lessons that will emanate from my comparative analysis is that while deliberative democrats should hold on to their ideals, they also need to appreciate the limits of deliberation; they need to become aware that deliberation is the right form of political interaction in some, but not all, contexts, while in other contexts other forms are necessary. Moreover, they need to accept that types of deliberation should be 'matched' to the right type of problem and that the aims of the particular deliberation has consequences for its specific design, because then they will be in a better position to test aspects of deliberative democracy empirically. In order for theorists of deliberative democracy to make the step to practical implementation they need to differentiate more between the different aims and functions of public deliberation and investigate the different conditions

---

[6] Diana C. Mutz (2006).

[7] Ibid., page 5.

that are necessary for application in specific contexts.[8] In order to examine whether the claims of deliberative democrats are feasible in practice we will, then, necessarily have to focus on partial aspects of the theory and look for contexts where these aspects could be met in practice.

In sum, it should be noted that on the basis of the limited empirical research that I carried out regarding two deliberative fora in two countries one cannot draw sweeping conclusions about deliberative democracy in general. However, the question is whether deliberative democracy as a general political model could ever be empirically analysed. In practice we are faced with hybrid models; in some political cultures there is more room for public debate than in others, some cultures are more elitist than others, and some governments are more willing to relinquish some of their authority than others. The different theories of deliberative democracy cannot all be reduced to one common denominator. This means that we can only test partial aspects of specific deliberative theories in practice in order to understand whether certain claims that deliberative democrats make are feasible. I investigate one such partial aspect, namely the ability to deal with intractable disagreement. I will formulate several conditions and criteria for dealing with intractable disagreement that follow from my theoretical analysis, and in my empirical analysis I test these by asking what the implications of my empirical results are for my theoretical arguments. It is the nature of this research that the range of these implications is limited. Certain obstacles that are encountered in the committee-system and in consensus conferences may not appear in other deliberative settings, while those aspects of my deliberative fora that work well may not be easily replicated in other deliberative settings. My research should, therefore, be regarded as a prompt for further studies with different deliberative fora in different countries. Moreover, it should be noted that I am writing in the context of Western, liberal democracies. In order to carry out public deliberation about novel technologies certain conditions have to be present – citizens must be free to speak their mind, a certain level of equality between citizens must be assumed, there must be a certain level of public education, and the novel technologies should be imaginable in the country where the deliberation is held, for example. These conditions are regrettably not met in many parts of the world and it is, therefore, not feasible to directly apply my conclusions to these countries.

1.6 Case-Studies Method

In order to analyse and compare the Dutch and Australian ethics committees and consensus conferences, I have conducted a literature research of primary and secondary literature. Regarding the committees, the primary literature consists of annual reports and evaluations of the committees in question, including self-evaluations and evaluations by independent parties; the secondary literature deals with the committee-system in general. I have supplemented this review by

---

[8] An example of such research is offered by Fung (2003).

1.6 Case-Studies Method

conducting ten semi-structured qualitative interviews with members of the analysed committees. I have also conducted four interviews with representatives of interest groups that have been directly involved with the functioning of these committees through public consultation procedures, and with government bureaucrats who have dealt with the committee-system. Regarding the consensus conferences, I have consulted official reports and evaluations of both consensus conferences, secondary literature on these exercises, and in the case of the Australian consensus conference, transcripts of the expert presentations, and a recording of a radio programme about the consensus conference by the Australian Broadcasting Corporation (ABC). I have supplemented these with semi-structured qualitative interviews with the facilitator of the Australian consensus conference, an organiser of different public consultations regarding gene technology in the Australian state of Victoria, and with one of the experts who was invited to speak at the consensus conference. In the Netherlands, I have interviewed one of the organisers of the consensus conference, the organiser of another public meeting regarding biotechnology, and an expert who was consulted at the consensus conference. I have chosen not to question members of the lay panels, firstly because they had already taken part in several evaluations and my questions would not generate much added value to these and secondly, because at the time of writing the conferences took place several years ago. For an overview of the interview questions, see Appendix C. The primary aim of the interviews was to test my analysis of the literature review. The interview sample is too small to generate general conclusions, but big enough to detect any significant discrepancy between my interview results and my literature review. No such discrepancies were detected. The purpose of the interviews, in other words, was to provide a check on the information gathered elsewhere, and to supplement it with specific examples.

I have chosen to analyse the Dutch Committee for Animal Biotechnology (CAB) and the Australian Gene Technology Ethics Committee (GTEC) because both are the respective countries' national committees dealing with the ethics of biotechnology. While it will become apparent that both committees have quite a different mandate, this difference in fact serves to highlight problems inherent in the committee-system. I discuss the Dutch committee in more detail than the Australian one, because this committee fulfils more different functions and the subtleties of certain problems become more apparent here. Also, the Dutch committee has been more thoroughly evaluated than the Australian one and I therefore rely more on interviews in the latter case. I have chosen to analyse particularly the consensus conference in the Netherlands about cloning, because this is the most comprehensive one that has been carried out in this country. There has been a public debate about gene technology and food, but this did not take the form of a consensus conference.[9] I analyse particularly the consensus conference in Australia about gene technology

---

[9] See http://www.arbobondgenoten.nl/arbothem/biotech/eindrapport_terlouw.pdf (accessed 30 May, 2007) for the report about the debate about biotechnology and food, called 'Eten en Genen' (Food and Genes).

in the food chain, because it is the first (and at the time of writing only) one that has been carried out in Australia in the field of biotechnology. The Australian consensus conference has been more thoroughly evaluated than the Dutch one, as reports about the latter tend to focus more on the content of the debate than on the debate itself. Therefore, my discussion of the Australian experience will rely more on written documents and the Dutch one more on insights gained from interviews. As will become clear, it is not possible to describe the two conferences in a completely parallel way. Rather, each case highlights a different set of problems, which are largely dependent on contextual factors. I discuss the Australian experience at greater length than the Dutch one, because certain problems are more pronounced in that context.

In Chapters 5 and 6 I defend my selection of particularly the Netherlands and Australia for my comparative analysis in more detail. In short, these countries have certain important similarities, but also enough institutional differences to be able to determine to what extent broader structures and contexts matter. Australia can be characterised as a Westminster style, adversarial political culture coupled with a pluralist interest group system of decision-making while the Netherlands is a consensual type of democracy coupled with a corporatist decision-making model. Presumably, each model places different demands on the way in which decisions are made, such as the need for making compromises and the need for defending one's points of view in public. Because collective decision-making through compromise or consensus, and co-operation between minority groups, are central to consensus democracies, this style of governance has more affinity with deliberative democracy than majoritarian styles of governance. My hypothesis is, therefore, that the Netherlands is better able to deal with intractable disagreement than Australia.

1.7 Depoliticisation

Dealing with intractable disagreement by delegating decisions to committees or by organising consensus conferences could be regarded as an attempt to depoliticise controversial issues. Decision-making on the basis of depoliticised debate may be helpful, especially when we are dealing with complex issues with important social and environmental implications, that involve many vested interests, and that are characterised by great levels of uncertainty, even among experts – all characteristics of the biotechnology debate. When such conditions are involved, depoliticised debate might make more room for the discussion on fundamental values, promise a more considered reflection on generalisable interests, and possibly avoid the promotion of personal interests. Depoliticisation entails a rejection of decision-making based on self-interested motives and because this rejection is central to theories of deliberative democracy as well, depoliticisation at first sight appears to fit well within such theories. James Fishkin and Robert Luskin formulate the depoliticising potential of public deliberation as follows:

> It may even be the mass public that has the greater possibility of real deliberation. Citizens are not bound by constituencies or parties and – in electorates of any size – are not casting

1.7 Depoliticisation

> votes worth surveilling or bargaining over. They have no need to posture or negotiate. Hence they are freer to alter their views, and not just on the merits of concrete legislative proposals but also on more fundamental questions of what is and what should be.[10]

However, while depoliticisation appears conducive to deliberative democracy and several authors have regarded it as a central tenet of deliberative democracy, other authors, in contrast, describe depoliticisation as a move by politicians to contain public debate. Because one's view of the potential of depoliticisation appears to be influential for what one expects from deliberative democracy, as a sub-question in this book I want to investigate to what extent public deliberation should be regarded as a form of depoliticisation and to what extent this is successful in practice.

One of the authors who proposes a move to more depoliticisation is Philip Pettit, who regards depoliticisation as a move away from a situation where 'control is left wholly or mainly to representatives in parliament, or to a government with a parliamentary majority, or to an elected administration'.[11] He argues that even if elected politicians have the good of the whole community at heart, they still have to follow their own (or their party's) interest in being re-elected.[12] Issues that directly pertain to this interest – such as the drawing of electoral boundaries – should therefore be depoliticised, meaning that decisions are taken away from the influence of politicians, by handing them over to an organization that represents 'relevant bodies of expertise and opinion as well as the people as a whole'.[13] Such a body should be accepted by all political parties to ensure that it does not advance sectional interests, but creates policy for the common good of society. It would ultimately remain under parliamentary control, but its decisions would not be tied to one particular politician or party. Pettit argues that the depoliticisation of debate is called for particularly in three areas, because politicians could use decisions they make in these areas to gain electoral power. These areas are firstly, the popular passions, or those issues that the public is highly emotional about and which can easily be misused by politicians – such as heavy sentencing for certain crimes; secondly, aspirational morality, or those issues that involve moralistic arguments – for example about prostitution – which a politician might use to win votes, even if this might go against the common good; and thirdly, those issues that involve sectional interests, and in which lobbying and the self-interest of politicians play a large role.

Even though Pettit does not explicitly refer to novel technologies, his last two areas can be discerned in the biotechnology controversy. In fact, the main reasons why there is a controversy about biotechnology are exactly that normative issues are involved that people feel strongly about – and about which intractable disagreements exist – and that powerful sectional interests are at work lobbying politicians. Depoliticisation of decision-making in this area might be an attractive option for several reasons. Firstly, views on biotechnology do not necessarily follow party

---

[10] James S. Fishkin and Robert C. Luskin (2005), page 286.

[11] Philip Pettit (March 2004), page 52.

[12] Ibid.

[13] Ibid., page 55.

lines, which means that even if politicians could influence decision making, these decisions might go against the views of some of their own party members. Secondly, lobbying might conflate the interests of politicians and the biotechnology industry to a level where opposition is hardly possible. And thirdly, decisions about biotechnology regulation require a flexibility to respond to new developments in research and development. A depoliticised body such as Pettit envisages could monitor such new developments and change policy accordingly, without having to go through a political bargaining process. Even though at first sight depoliticisation might appear to run counter to democracy, because decisions are not made simply on the basis of an equal vote for everyone in the election of representatives, in Pettit's eyes it is democratic in the deliberative sense of the word, because it seeks to move beyond the simple aggregation of preferences (which may be self-interested) to a reasoned agreement based on the force of the better argument and the fulfilment of the common good.

In the next chapter, I will describe the view of Hisschemöller and Hoppe, who argue in contrast that the move to depoliticisation is symptomatic of ill-conceived policy problems; regulators and bureaucrats try to contain public debate by treating unstructured policy problems as moderately structured ones, and one of their strategies is depoliticisation of the problem by delegating them to committees and consensus conferences.[14] Obviously, two different definitions of depoliticisation are at work here. I will also come back to the issue of depoliticisation in Chapter 4, in which I provide a theoretical analysis of deliberative democracy. However, even if it turns out that depoliticisation is a central aspect of deliberative democracies, in order to answer the question whether depoliticisation is in fact feasible, theoretical analysis does not suffice. In other words, even though intractable disagreement may be dealt with best in a depoliticised forum, this is only an advantage if depoliticisation actually stands a chance in practice. In the empirical part I, therefore, also hope to shed light on the potential of depoliticisation to deal with intractable disagreement.

1.8 Outline

I start with an anatomy of the broad debate about the genetic modification of animals in plants in *Chapter 2*, which reveals that we are dealing with disagreement on many different levels and dimensions. I argue that the controversy about biotechnology can be characterised as an unstructured problem. I will argue in *Chapter 3* that the liberal solution of conversational restraint, that we have to privatise conflict about conceptions of the good, is misguided for two main reasons: Firstly, it relies on two dubious distinctions between the public and the private and between morality and ethics. This means that positions are unjustly dismissed as private while they have as much pretension to universality as positions that liberals regard as universal. My critique of the problematic distinction between morality and ethics is supported

---

[14] Matthijs Hisschemöller and Rob Hoppe (Winter 1995).

1.8 Outline 13

by my analysis of the biotechnology debate in Chapter 2. Conversational restraint demands of people that they hold back on points that are most important to them. Secondly, conversational restraint as a neutrality device precludes the possibility of preference or opinion transformation. Particularly in the case of novel technologies where opinion formation is still dynamic the liberal view of preference formation is too static. I argue that these problems could be solved by placing more emphasis on deliberation as a social choice mechanism rather than on the traditionally more dominant mechanisms of voting and bargaining. So while I accept the liberal requirement of state neutrality, I critically reflect on the operation of this concept. For this reason as well my criticism is an immanent one. My aim is to broaden political liberalism in the direction of deliberative democracy, by including a wider group of reasons and people in political debate than political liberals commonly allow. At least three important criteria for dealing with intractable disagreement emanate from my discussion of the liberal versus the deliberative solutions; firstly, no reasonable[15] views should be excluded, secondly, there should be no power differences great enough to distort the decision-making process, and thirdly, the outcomes of the debate should be open-ended.

After arguing that more emphasis should be placed on public deliberation, in *Chapter 4* I explore the political theory, or rather, the group of theories, that center on preference transformation through deliberation, namely deliberative democracy. The core notion of deliberative democracy is that legitimate political decisions are those that are reached through free and uncoerced debate between equals. This theory centers, then, on modes of discourse, which can be defined as John Dryzek does, as 'a shared way of making sense of the world embedded in language'.[16] This means that, ideally, the debate must be free of power imbalances that give some people better chances than others to influence the decision-making process. Following Jürgen Habermas, many deliberative democrats argue that decisions should be made on the basis of the 'forceless force of the better argument'.[17] My exploration of deliberative democracy will revolve around three questions that I argue are relevant for dealing with intractable disagreement: what is the goal of the deliberation, what arguments are valid, and who should participate? I will confront deliberative democracy with critiques that have been mounted against it regarding each of these questions. From this theoretical examination emerge three conditions and a set of criteria that encapsulate what I think public deliberation should encompass. My proposal is based on the three criteria I develop in Chapter 3 (inclusiveness, absence of power imbalances, and open-endedness). In short, I argue that a tension can be discerned in deliberative theory between the goals of consensus, inclusiveness, and quality of debate, and that deliberation serves a different purpose for different types of problems. I conclude that for unstructured problems such as the conflict

---

[15] The term 'reasonable' is of course not a neutral one; what is meant by this will be further clarified in Chapter 3.

[16] John S. Dryzek (April 2005).

[17] Jürgen Habermas (1990), pages 158–159. See also Robyn Eckersley (2000).

over biotechnology it is premature to aim for reaching consensus through deliberation. Both inclusiveness and deepening of understanding are important goals in unstructured problems, amongst other reasons because they contribute to opinion transformation. However, while these two goals can co-exist to a certain extent, we should be aware that including more participants and viewpoints is likely to lead to a loss of quality of argument. Therefore, sometimes we need to limit the group of participants, but by the same token deliberation should be regarded as but one avenue for political participation, and other forms, such as protest and activism should be accorded an important function as well.

In *Chapters 5 and 6*, I will compare my normative claims against real life deliberative settings. From this empirical analysis it becomes clear that committees tend to contain rather than stimulate public debate. Instead of making disagreements more tractable, then, committees exclude certain viewpoints and thereby create a false sense of making disagreements more tractable. Compared to the committee system, consensus conferences appear to create more room to discuss issues related to fundamental values or worldviews. However, in these fora forces are at work also to exclude groups with extreme viewpoints and to gain control over the exercise, for example by a strict demarcation of topics. Amongst other things, my examination reveals that these exercises in public deliberation do not actually lead to depoliticisation. Especially in the antagonistic political climate characteristic of Australia, deliberative fora will themselves become politicised; in such a culture committee or lay panel reports will either not have any influence and be toothless or become a strategic tool of politicians who interpret findings to suit their own agenda. In the Netherlands no clear line between politicised and depoliticised fora, such as envisaged by Pettit, can be discerned. In fact, I follow Hisschemöller in arguing that depoliticisation is not the best way to deal with unstructured policy problems. As I will argue in Chapters 4 and 6, these exercises need to be *re*politicised rather than *de*politicised. Repoliticisation needs to take place in two directions, within both the scientific and the normative discussions.

In *Chapter 7*, I revisit deliberative democracy by seeking the implications of my case-studies for both theory and practice, and draw my final conclusions. In general, my empirical analysis suggests that political culture is very influential for the way public deliberation is regarded and for its success. We cannot expect a grand theory of deliberative democracy that is equally applicable to each country. The question as to whether the conditions and criteria of deliberative democracy that I formulated are warranted is dependent, then, on the political culture of the country in which it is applied. Dryzek argues that discursive democracy calls for a drastic restructuring of society. He argues against liberalism, as 'the most effective vacuum cleaner in the history of political thought', meaning that liberalism incorporates or co-opts all theories that appear to criticise it, such as feminism or environmentalism, and thereby weakens their capacity to question existing political institutions.[18] However, in my view it is not the liberal framework that needs to be abandoned, but rather that,

---

[18] John S. Dryzek (2000), page 27.

firstly, more emphasis needs to be put on its deliberative elements, and secondly, that successful deliberation calls for careful attention to the political culture of countries. Particular challenges exist for countries that can be characterised as antagonistic, but where consensus democracies have elitist tendencies, these need to be dealt with as well. The aims, form, and scope of the deliberation should, then, be chosen carefully, dependent on the specific context. Tailoring deliberative democracy to specific contexts may be the next big challenge for deliberative democrats.

References

Dryzek, John S. (2000), *Deliberative Democracy and Beyond: Liberals, Critics, Contestations* (New York: Oxford University Press).

Dryzek, John S. (2005), 'Deliberative Democracy in Divided Societies. Alternatives to Agonism and Analgesia', *Political Theory*, 33 (2), 218–242.

Dryzek, John S. and Niemeyer, Simon J. (2003), 'Pluralism and Consensus in Political Deliberation', paper given at Annual Meeting of the American Political Science Association, August 28–31.

Eckersley, Robyn (2000), 'Deliberative Democracy, Ecological Representation and Risk: Towards a Democracy of the Affected', in Michael Saward (ed.), *Democratic Innovation: Deliberation, Representation and Association* (New York: Routledge), 117–132.

Fishkin, James S. and Luskin, Robert C. (2005), 'Experimenting with a Democratic Ideal: Deliberative Polling and Public Opinion', *Acta Politica*, 40, 284–298.

Fung, Archon (2003), 'Survey Article: Recipes for Public Spheres: Eight Institutional Design Choices and Their Consequences', *The Journal of Political Philosophy*, 11 (3), 338–367.

Habermas, Jürgen (1990), *Moral Consciousness and Communicative Action* (Cambridge, MA: MIT Press).

Hisschemöller, Matthijs (1993), 'De Democratie van Problemen. De relatie tussen de inhoud van beleidsproblemen en methoden van politieke besluitvorming (The Democracy of Problems. The Relationship Between the Content of Policy Problems and Methods of Political Decision-Making)', PhD Thesis (University of Amsterdam).

Hisschemöller, Matthijs and Hoppe, Rob (1995), 'Coping with Intractable Controversies: The Case for Problem Structuring in Policy Design and Analysis', *The International Journal of Knowledge Transfer and Utilization*, 8 (4), 40–60.

James, Bohman (1998), 'The Coming of Age of Deliberative Democracy', *The Journal of Political Philosophy*, 6 (4), 400–425.

Mutz, Diana C. (2006), *Hearing the Other Side. Deliberative Versus Participatory Democracy* (Cambridge: Cambridge University Press).

Pellizzoni, Luigi (1999), 'Reflexive Modernization and Beyond. Knowledge and Value in the Politics of Environment and Technology', *Theory, Culture & Society*, 16 (4), 99–125.

Pettit, Philip (2004), 'Depoliticizing Democracy', *Ratio Juris*, 17 (1), 52–65.

Smith, Graham and Wales, Corinne (2000), 'Citizens' Juries and Deliberative Democracy', *Political Studies*, 48, 51–65.

Part I
Context of Book

Chapter 2
Biotechnology: An Anatomy of the Debate

Genetic engineering represents our fondest hopes and aspirations as well as our darkest fears and misgivings. That's why most discussions of the new technology are likely to be so heated.
Jeremy Rifkin, The Biotech Century

Once a new technology rolls over you, if you're not part of the steamroller, you're part of the road.
Stewart Brand, Author and Futurist

2.1 Introduction

Apart perhaps from the biblical snake that seduced Adam and Eve into tasting the forbidden fruit – and thereby starting humankind's ambiguous relationship with nature – no animal has given rise to as much moral debate as Dolly, the cloned sheep. Dolly's birth unleashed a turbulent debate in the media, parliaments, schools, social clubs, and even local green grocers all over the world. Proponents argue that cloning would provide many potential benefits: it could boost livestock production, would in the long run lead to medical innovations, solve human infertility problems and lead to a decline in the use of laboratory animals. Opponents reason that cloning is harmful for animals, demonstrates an inappropriate, instrumental, attitude towards them, will lead to a loss of genetic diversity, and will disrupt the social hierarchy within a herd.[1] The debate surrounding Dolly's birth and her untimely death,[2] sheds light on a larger debate about developments in a relatively new technology field, namely that of biotechnology.

In this chapter, I will analyse 'the biotechnology debate': the debate between opponents and proponents of the use of biotechnological procedures. The prefix

---

[1] For discussions on cloning, see Martha C. Nussbaum and Cass R. Sunstein (1998), Henriëtte Bout (1998), and Gina Kolata (1997).

[2] Sadly, Dolly had to be put down in 2003, when she was 6, due to her suffering the results of a progressive lung disease. Source: *The Economist*, 22 February, 2003.

'bio' is derived from the Greek word bios, meaning 'life'. Biotechnology can be defined as 'any technique that uses living organisms or parts of organisms to make or modify products to improve plants or animals or develop micro-organisms for specific uses'.[3] Even though Roger Straughan, from whom this definition was derived, claims this is a value-neutral definition, not everyone would agree that biotechnology necessarily involves 'improving' plants or animals, or that plants and animals need to be improved in the first place. It should, therefore, be noted at the outset that the way in which one defines the term 'biotechnology' can serve a rhetorical function. Moreover, the above-mentioned definition is so general that it also includes traditional and uncontroversial techniques such as brewing beer and making yoghurt, and therefore 'traditional' biotechnology is usually distinguished from 'modern' biotechnology. A more detailed, and neutral, definition of modern biotechnology is given by the Cartagena Protocol on Biosafety, as 'the application of: a) in vitro nucleic acid techniques, including recombinant deoxyribonucleic acid (DNA) and direct injection of nucleic acid into cells or organelles, or b) fusion of cells beyond the taxonomic family, that overcome natural physiological reproductive or recombination barriers and that are not techniques used in traditional breeding and selection'.[4]

When people refer to modern biotechnology they often imply genetic engineering, which is the horizontal transfer of genes between different organisms, sometimes belonging to different species, and which is made possible by recombinant DNA technology, a 'technique for joining DNA molecules in vitro and introducing them into living cells where they replicate'.[5] However, biotechnology also involves techniques such as xenotransplantation (the transplantation of cells, tissues or organs from one species into another, including animal-to-human transplantation) and cloning (the making of a near-exact replica of an organism or group of cells).[6] 'Biotechnology' is a broad term that points to many different practices; it is used, among other things, in the fields of medicine, agriculture, food enhancement, pest management, and fundamental scientific research. All of these fields come with their own advantages, disadvantages, difficulties, promises, interests, and moral or ethical concerns.

In this chapter, I provide an analysis of the biotechnology debate in order to show that it involves many different dimensions of disagreement, often touching on fundamental values and worldviews, and certain biases. Furthermore, I argue that biotechnology should be regarded as an example of a so-called 'unstructured problem'. As the biotechnology debate is too broad and multi-faceted to do justice in one chapter, for reasons of space and balance between the different focal points of this study, I can only give a very general overview of all the different positions and arguments here. Nevertheless, as one of my main goals is to demonstrate the great

---

[3] Roger Straughan (1992), page 2.

[4] Secretariat of the Convention on Biological Diversity (2000).

[5] Robert C. King and William D. Stansfield (1997), page 291.

[6] Xenotransplantation Working Party (2003), page 1.

range of issues that are part of the biotechnology controversy, I have included a more detailed analysis in the appendix. Appendix A focuses on animal biotechnology and Appendix B concerns plant biotechnology. These appendices can also be used as a referral when the reader wishes to become more informed about one of the various specific discussions that I can only touch on here.

2.2 The Debate

Before I turn to the specific arguments that are used in the debates about animal and plant biotechnology, I first need to explain what it is exactly that I am analysing and I need to sketch the historical context of this debate. As a background to the analysis of the biotechnology debate, I also point out the rhetoric used on both sides.

2.2.1 Demarcation of Debate Analysis

Because the background of my research is an interest in the question as to what should be the proper relationship between human beings and animals, and more broadly, the natural world, I will limit myself to an analysis of nonhuman animal and plant biotechnology. But even this limitation would provide me with enough material to fill several books, and therefore, I will further limit my discussion to the most contentious issues: genetic engineering of animals and plants, in particular for medical research and agriculture. Even though I started this discussion with the case of Dolly, who for many exemplifies the development of modern biotechnology, I will, therefore, not further deal with animal cloning, nor will I elaborate on the increasingly popular field of genomics, which uses information from the science of genetics in order to help breeders select more effectively and efficiently for favourable traits, without using recombinant DNA-techniques.[7] In order not to overburden this text with terminological detail, I will at times refer to 'animal and plant biotechnology', or simply 'biotechnology' when in fact I mean the genetic modification of plants and animals.

I interchange the terms 'genetic modification, engineering, and alteration'. In the past, I have always preferred to use the term 'genetic modification' because it appeared the most neutral one. However, as I learned more about the history of genetic modification, I realised that none of these terms can be regarded as neutral. The original terms 'genetic engineering' and 'genetic manipulation' were abandoned by the biotechnology industry because of their negative connotations and they were replaced by 'genetic modification' or even 'genetic improvement' in order to counter public suspicion.[8] As I am trying to keep an open mind in my depiction of

---

[7] See Godfred Frempong, (Spring–Summer 2006) and Pim Lindhout and Daniel Danial (Spring–Summer 2006).

[8] Les Levidow (2001), page 47.

the biotechnology debate, I will not use the more polemic pro-biotechnology term 'genetic improvement' or anti-biotechnology term 'genetic manipulation'. Also, as Michael Reiss and Roger Straughan note, the terms 'genetic engineering' and 'modification' (and, in my view, also 'alteration') seem to capture best what is actually being done: the genetic make-up of an organism is changed (modified or altered), or an organism with a different genetic structure is created (or 'engineered').[9]

When I refer to 'the' biotechnology debate, I of course realise that there is no one such all-encompassing debate. Rather, we encounter countless smaller debates about specific issues involving biotechnology. What particular debates will I analyse and how will I analyse them? One of the main aims of this chapter is to provide a background to the analysis of particular debates as carried out in biotechnology ethics committees (Chapter 5) and consensus conferences (Chapter 6) in Australia and the Netherlands. One cannot properly understand these debates without some prior knowledge of the global context in which these debates take place. This chapter will, so to speak, 'map' the main issues that play a role in 'the global biotechnology debate' and which are relevant to the Dutch and Australian debates.

In my overview of the main issues that are in dispute regarding the genetic modification of animals and plants, I will present the arguments that have been given pro and contra each position. In this analysis, I will distinguish arguments that are directed at the consequences of biotechnology from more fundamental arguments concerning the technology itself. A similar distinction has been made by Straughan, who terms the first extrinsic and the last intrinsic concerns. Straughan addresses only objections *against* biotechnology as either 'intrinsically wrong in itself or extrinsically wrong because of its consequences'.[10] In my view, however, the same distinction between intrinsic and extrinsic concerns can be made for arguments *in favour of* biotechnology.[11]

It should be mentioned that the extrinsic/intrinsic distinction here employed is not completely unproblematic. Les Levidow and Susan Carr criticise the distinction made by Straughan, because it perpetuates the controversial distinction between fact and value, and its corollaries science and ethics, thereby portraying science as pure fact-finding and denying that values play a role in science. The consequence of this distinction, in Levidow's eyes, has been that value-related arguments have been labelled as belonging to the realm of subjective feelings, while technical issues

---

[9] Michael J. Reiss and Roger Straughan (1996), page 2.

[10] Straughan (1992), page 9. The distinction between intrinsic and extrinsic is similar to, but not precisely the same as, the distinction between consequentialist and deontological arguments. One can use deontological arguments when discussing extrinsic objections (for example, when one argues against certain applications of biotechnology on the basis of justice-arguments, one is using deontological arguments, but the objections are not directed at biotechnology in itself).

[11] Even though arguments in favour of biotechnology are usually stated in terms of its benefits, and hence its consequences, defences of biotechnology can also be of an intrinsic nature, stating that the technology is inherently good, for example.

2.2 The Debate

have been deemed objective, and therefore have been accorded a higher status.[12] However, I believe that the distinction can serve a useful heuristic purpose and does not necessarily lead to the division that Levidow is concerned about. As will become clear in the remainder of this book, I do not subscribe to the viewpoint that science is value-neutral, nor that all value-related concerns are mere expressions of personal feelings. By making the intrinsic/extrinsic distinction, I do not mean to suggest that technical and ethical concerns should be addressed separately, let alone that value-related concerns would somehow be less important. On the contrary, I think that organising and analysing the different topics in this way will make clear that many extrinsic arguments are intertwined with intrinsic ones, and that there is a bias in favour of extrinsic ones. In order to show this, it is useful to separate the two categories at first.

My focus will be on the *arguments* advanced pro or contra biotechnology and I will, therefore, not discuss the particulars of specific 'scandals' involving, for instance, the deliberate or accidental release of GMOs into the environment, or the activities of specific agents such as Monsanto or Greenpeace, except when these are illustrative of the arguments analysed. Rather than tracing the specific 'players' in the debate, I instead focus on the content of the arguments, because in this book I examine the way in which deliberative democracy could deal with disagreement and in this context the emphasis of the theory is primarily on arguments. Moreover, while I am aware of the great influence of power relations and hold that deliberative democracy should not be blind to them, I will not focus on them independently, because this topic could be the subject of a whole separate book.[13] The material I draw on consists in the first place of scientific and academic publications, and secondly on government reports. I have not made a study of propaganda materials of biotechnology companies or ardent critics of biotechnology, but I will refer to scientific studies that have been made in this vein. Neither have I dissected reports in newspapers or other media, but I do refer to other studies that have. Moreover, my focus is not on the question of how many people are in favour of or against certain applications of biotechnology, but on the content of the actual arguments that have been advanced, and I therefore do not study results of opinion polls.

While I do not focus on public attitudes towards biotechnology, it should be mentioned that all parties in the debate at various points in time claim to have public opinion on their side. These claims are problematic, because they carry the pretence that they are able to give an objective account of what public opinion is and also assume that there is one uniform public, whereas publics can in reality be different, depending on, for example, whether one looks at people in their role of consumer or of citizen. While the most common method for discovering 'public opinion' is the use of opinion polls, one can wonder whether this is always the most appropriate instrument for analysing attitudes and opinions regarding such a complex issue as biotechnology. The fact that different surveys have led to opposing results suggests

---

[12] Les Levidow and Susan Carr (March 1997).

[13] As indeed it is. See Richard Hindmarsh (1994).

that we should approach opinion poll results with caution.[14] Aidan Davison et al. have studied opinion polls and surveys regarding biotechnology and have come to the conclusion that 'far from being objective tools for the mapping of public opinion, opinion polls reflect ideologically grounded conceptions of 'public opinion''.[15] They found that the way in which polls were being carried out actively contributed to constructing public opinion rather than merely describing it in a neutral fashion.

Heather Dietrich and Renato Schibeci go even further by arguing that surveys 'play an important role in the shaping of public discussion of biotechnology applications, by helping to legitimate new biotechnology products and displace "interest group" opposition to biotechnology'.[16] Surveys make many implicit assumptions, such as the view that science and technology are neutral tools, and they leave out issues that many people find important, such as views about the control of multinational companies over gene technology.[17] Moreover, not distinguishing between different 'publics' and merely focusing on 'opinion poll publics' does not do justice to public debate. It discounts the fact that there are active, interested public groups that are more knowledgeable of the problematic issues in the debate and that 'despite the more adversarial attitudes of such groups, they do play a crucial role in shaping public opinion'.[18] Another problem with surveys is that their results can be interpreted very differently, depending on who reads them.[19] It should also be noted that it is disputed what the effects are of increasing the public's knowledge about biotechnology. Proponents often assume that if the public is better informed about the 'realities' of genetic engineering the public will embrace it,[20] whereas opponents believe that more knowledge will lead to a rejection, while the reality is far from clear-cut.[21]

Finally, I should mention that it is the disagreement that this technology gives rise to, and the ways in which we can deal with this disagreement, that I am ultimately interested in, rather than in influencing the reader's point of view. Therefore,

---

[14] Janet Norton (1998).

[15] Aidan Davison, Ian Barns, and Renato Schibeci (Summer 1997), page 328.

[16] Heather Dietrich and Renato Schibeci (2003), page 386.

[17] Ibid.

[18] Davison, Barns, and Schibeci (1997), page 332.

[19] Dietrich and Schibeci (2003), page 386.

[20] For example, Jim Peacock argued in 1994 that education would solve the public's 'irrational' anxiety about GM foods (Jim Peacock 1994).

[21] Jeffrey Burkhardt (2001), Reiss and Straughan (1996) Claire Marris argues that 'both sides think that direct benefits to the consumer are a central determinant of public acceptance; thus the 'pros' seek to communicate the benefits, whilst the 'antis' try to demonstrate that these benefits will not be realised or that they will benefit commercial corporations rather than ordinary citizens'. Claire Marris (2001). However, Marris' study found that the concerns of the public with GMOs were not based on a lack of technical knowledge and could therefore not be solved by providing more scientific education. Nevertheless, research of public attitudes regarding biotechnology, such as the Eurobarometer, does suggest that the more citizens learn about GMOs the more critical they become of it. See George Gaskell, Martin Bauer, and John Durant (1998), Sue Mayer (2002), Alison Shaw (2002).

2.2 The Debate

it is unnecessary for me to present my own position on biotechnology. Still, I am aware that one cannot completely sever one's analysis of a topic from one's opinion about it; one of my points will be that normative positions play a role in empirical research, and that one cannot remain completely value-neutral. Therefore, it is in order to state my own position, which – as may become apparent – is critical of and concerned about many applications of biotechnology, but does not constitute a wholesale rejection. If there are any knockdown arguments for or against biotechnology that I fail to mention, this is because I focus on the main arguments that are actually encountered in the debate, and particularly on those that are relevant to the more particular debates in the Netherlands and Australia.

2.2.2 Historical Context

At the outset of the biotechnology era, in the 1970s, both proponents and opponents of biotechnology were primarily scientists. Of course, before the 1970s pioneering research had already been done, starting with Georg Mendel's pea plant crossing experiments in the 1860s and greatly advanced by Watson and Crick's discovery of the double helix structure of DNA in 1953. However, the actual technique of recombining DNA from two different organisms was not carried out until 1972, by Stanley Cohen and Herbert Boyer using bacteria, and by Paul Berg using viruses.[22] Berg himself was one of the first scientists to raise concerns over the safety of the new technique. He wrote a letter to the journal *Science* expressing his concerns and calling for a temporary voluntary moratorium on tests involving recombinant DNA-technology. This led to the famous Asilomar Conference, where Berg's call for a moratorium was accepted by a worldwide network of scientists.[23] The main fears were that the bacteria used as vectors for genetic modification would escape from laboratories and cause epidemics and that genetic modification would be used for producing biological weapons.[24] In this period, the main rift in the debate was between scientists who wanted to err on the side of caution and those who felt that a moratorium would interfere with their scientific freedom. It should be noted that at this conference, the problem field was restricted to technical issues and value-related concerns were explicitly left out of the debate.[25]

---

[22] Bastiaan C.J. Zoeteman, Miranda Berendsen, and Pepijn Kuyper (2005).

[23] See Eugene Russo (3 April 2000), Marcia Barinaga (March 3 2000), John Coulter (March 31 2000), and Alexander M. Capron and Renie Schapiro (Spring 2001).

[24] Zoeteman, Berendsen, and Kuyper (2005). It should be noted that there was already enough knowledge to make biological weapons without having to resort to genetic modification.

[25] Hindmarsh argues that at the conference 'the genetic engineering "problem" was framed in terms of finding a technical response to a technical problem'. This gave scientists 'legitimacy as sole arbiters of evidence and makers of policy'. He concludes that 'the conflict had been managed before it started'. Hindmarsh (1994).

In the 1980s, field testing of genetically modified plants was initiated and the fear of new diseases was replaced by a fear of environmental destruction. Because genetic modification was now taken out of the scientists' laboratories, and brought into people's environments, a larger public became involved in the biotechnology debate. Zoeteman et al., authors of *Biotechnology and the Dialogue of the Deaf*, divide the participants in the biotechnology debate from the 1980s into two groups (with many positions in between): progress-optimists and nature conservationists.[26]

The 1990s saw an increase in the public indignation with genetic modification. At first, possible environmental risks were the main concern, but at the end of the decade, when genetically modified foods became a reality, risks to public health were the focal point.[27] According to Richard Hindmarsh, in the 1990s, the focus was not on the question *whether* biotechnology should be adopted, but *how*. In Hindmarsh's eyes, this is 'simplistic, because it would imply that the issues raised in the earlier phases have actually been resolved'.[28] In the mid-1990s, Greenpeace became involved in the debate, and this organization continues to play an important role in calling criticism of genetic engineering to the attention of the wider public. Critics of biotechnology represent a wide range of groups, such as environmentalists, some (particularly organic) farmers, consumer groups, animal-rights groups, critics of globalisation, aid and development organizations, some scientists (such as the Union of Concerned Scientists), feminists, religious groups, and indigenous peoples. At the end of the 1980s, companies involved in the commercialisation of genetically modified (GM) products also became major players in the biotechnology debate; they formed associations that took part in the debate in order to protect the industry's interests.[29]

We could say that biotechnology industry representatives are on the progress-optimist end of the scale, and Greenpeace and other nature and animal-interest organizations on the nature conservationist-end of the scale; whereas scientists, politicians, and the larger public populate different places on this spectrum.[30] For example, many scientists are principally pro-biotechnology, but do warn that we should act with caution regarding some applications of biotechnology.[31] Also, some see potential benefits, but are critical of the socio-economic implications of biotechnology.[32] Critics of biotechnology can be divided into different categories, from wholesale rejection to mild criticism. Guido Ruivenkamp and Joost Jongerden propose to distinguish between three different ways of objecting to biotechnology:

---

[26] Zoeteman, Berendsen, and Kuyper (2005).

[27] Ibid.

[28] Hindmarsh (1994).

[29] Examples are the NIABA in the Netherlands. Zoeteman, Berendsen, and Kuyper (2005), and AusBiotech in Australia. http://www.ausbiotech.org/ (accessed several times between 2004 and 2007).

[30] Maggie Scott and Susan Carr (2003).

[31] See Geeta Bharathan et al. (2002), page 175.

[32] For example, Steve Hughes and John Bryant (2002) and Gordon Conway (2000).

(1) the act of outright rejection, in which one refuses to accept a certain technology completely; (2) the act of resistance, which refers to 'opposing technologies that are disapproved of or disagreed with'; and (3) the act of redesign, or 'developing new and different kinds of technologies', such as the so-called tailor-made biotechnologies.[33] In reality, of course, there are many subtle variations of these ideal types. The debate about biotechnology is more often than not presented as one between just two extreme parties, while in reality many different positions in between are held. Nevertheless, the many subtle intermediary positions do not prevent this debate from being generally quite polarised; the more extreme opinions tend to characterise the debate.

Whilst keeping in mind the range of possible positions between absolute support and outright rejection, for reasons of clarity, in my account of views on biotechnology I will present 'proponents' and 'opponents' as though they were a unified group. I may seem to give a disproportionate amount of attention to positions critical of biotechnology. This is explained by the fact that opponents of biotechnology adopt a more active attitude in the debate than proponents and therefore advance more – and more complex – arguments. This is a logical consequence of the fact that opponents are fighting a battle against a technology that is being strongly pursued by powerful players such as (multinational) companies and governments, while proponents seem to have more leeway.

2.2.3 Rhetoric

Biotechnology advocates (and in, fact, many opponents as well) claim that biotechnology is part of an inevitable process that cannot be stopped. It is often argued that 'you cannot turn back the clock', suggesting that those who oppose biotechnology want us all to live again like humans did in the Stone Age, or the pre-industrial era. While opponents of biotechnology usually are not interested in turning back the clock at all, one could say that they harbour an initial suspicion towards this new technology, leading some to characterise critics of biotechnology as anti-science; even former British Prime Minister Tony Blair has proclaimed that the anti-science climate in the UK was growing, thereby threatening scientific and economic progress.[34] This corresponds to the oft-heard accusation that opponents are (neo-) Luddites who think that all technology is wrong in itself. The next step is usually to point out that while attacks on biotechnology entail a general distrust of technology, these attacks are written on computers, and communicated over the internet.[35] Despite the fact that Luddism is meant to be a derogatory term, those that are the subject of this accusation tend to embrace it, and argue that, like the original Luddites, they do not oppose all technology as such, but only those

---

[33] Guido Ruivenkamp and Joost Jongerden (Spring–Summer 2006).

[34] Mayer (2002, page 141).

[35] Gary Comstock (2000, pages 185–6).

new technologies that potentially have negative consequences for society or the environment.[36] Luddites reject the idea that technology is autonomous and outside of society's control.[37]

It appears, then, that the accusation of Luddism mainly serves a rhetorical function. Daniel Kleinman and Jack Kloppenburg, who have analysed publicity material of biotechnology company Monsanto, encountered numerous references to the term Luddism, and conclude that they use this term in a rhetorical way in order to gain acceptance for biotechnology through discursive means.[38] My goal of providing a broad reconstruction of the arguments that play a role in the biotechnology debate is complicated by the rhetoric that is encountered on both sides of the divide. Even though Michael Bruner and Max Oelschlaeger try to revive the use of rhetoric by casting it in terms of 'effective discourse', and convincingly argue that it can serve a useful purpose – in particular for pressing the environmental cause – in the context of biotechnology the excessive amount of rhetoric used primarily appears to polarise the debate.[39] It is important to be aware of the rhetoric that is used, as it can be quite successful at either gaining public support or creating public outrage. In this debate the stakes are high and rhetoric could play a decisive role. As the UN Food and Agricultural Organization (FAO) proclaims, 'the ongoing "war of rhetoric" about agricultural biotechnology may pose a greater threat than the technology itself does'.[40] On the other hand, as Bruner and Oelschlaeger contend, rhetoric can be employed critically in order to counter the negative consequences of the use of rhetoric itself. 'Critical rhetoric' can expose the 'discourse of power' in a particular debate.[41]

Biotechnology critics could respond to biotechnology advocate rhetoric by using critical rhetoric, exposing the rhetorical moves of their opponents, and by offering alternative images and paradigms.[42] However, one problem of the way rhetoric is employed in this particular debate is that it is directed at the whole spectrum of biotechnological procedures at once and therefore discounts, firstly, the more subtle positions that are held in between sheer advocacy or complete rejection, and secondly, the fact that some applications of biotechnology might be considered

---

[36] For example Mae-Wan Ho (1999), Richard Hindmarsh (2001), Daniel Lee Kleinman and Jack Kloppenburg, Jr. (September 1991). Kirkpatrick Sale argues that Luddites did not oppose all machinery, but only the machinery that they could not control. See Kirkpatrick Sale (Spring–Summer 2006), page 74.

[37] Many ex-Luddites later became politicians aiming at social reform. Joost Jongerden (Spring–Summer 2006), pages 64–67.

[38] Kleinman and Kloppenburg (1991).

[39] See Michael Bruner and Max Oelschlaeger (Winter 1994).

[40] As quoted in Scientific American (August 2004). This statement might actually be considered rhetorical itself, as it suggests that biotechnology is unproblematic.

[41] Bruner and Oelschlaeger (1994, page 388).

[42] Bruner and Oelschlaeger (Ibid.) drive this point home when they state, 'our point is simple: whoever defines the terms of the public debate determines its outcomes. If environmental issues are conceptualised, for example, in terms of 'owls versus people', then the owls (and the habitat that sustains them) do not have much of a future'.

2.2 The Debate

acceptable and others unacceptable. Jon Turney writes in this context that calling GM food 'Frankenfoods', 'invites an all-or-nothing response to a whole complex of developments, when we should be insisting on our right to choose some, and block others'.[43]

One form of rhetoric that is drawn on in the biotechnology debate is calling the viewpoint of one's opponent a myth. For example, McAfee accuses biotechnologists of clinging to the 'myth of genetic-engineering "precision"', whereas in her eyes it is not a precise science at all.[44] Even though claims like McAfee's are based on scientific evidence, biotechnology proponents often accuse their opponents of basing their arguments on 'bad science'. Fedoroff and Brown, for example, state that there are a handful of scientists that oppose GM food, but those that do are 'rarely those who know this new science well'.[45] Moreover, proponents often assume that when people oppose biotechnology, it is for non-scientific or emotional reasons.[46] In this line, biotechnology advocates have accused critics, such as Greenpeace, of leading a 'misinformation campaign' and stimulating public fears, thereby assuming that only the pro biotechnology industry and scientists have adequate knowledge.[47] Indra Vasil, for example, holds that 'anti-biotechnology activists' do not offer 'any credible scientific evidence to support their *allegations*'. He speaks of 'unsubstantiated claims of dangers ... rather than any *real* scientific concerns' and argues that 'regulatory decisions should .... be based on science rather than emotions and *perceived* risks'.[48]

The kind of language and metaphors used in order to denote genetic alteration can serve a rhetorical function as well. The metaphor of DNA as the 'book of life' or genetics as a computer code is often drawn. Implicitly this suggests that life is a book or a code that can be rewritten according to our wishes.[49] The prefix 'bio' is put in front of many words, such as 'bio-utopia', 'bio-colonisation', or 'bio-piracy', and this suggests some sort of campaign or conspiracy of biotechnologists to create a world in which everything is determined by biology.[50] GM foods are dubbed 'Frankenfoods', in order to conjure up images of the scientific project of creating life going horribly wrong and its products getting out of control.[51] Another example is calling rice to which a human protein has been added 'cannibal rice'.[52]

---

[43] Cited in Nina Fedoroff and Nancy Marie Brown (2004), page 8.

[44] Kathleen Mcafee (2003), pages 204–205.

[45] Fedoroff and Brown (2004, page xii).

[46] Henry I. Miller (November 1999), page 1042.

[47] For example, Jorge E. Mayer (September 2005), page 726.

[48] Indra K. Vasil (2003), page 850–851. My italics.

[49] Stephen Crook (2001).

[50] See for example Hindmarsh (2001), in Peter Wheale and Ruth Mcnally (1990), Vandana Shiva (2000).

[51] Phil Cohen (31 October 1998), Gordon Graham (2002), Robyn Rowland (2001).

[52] True Food Network, 28 May 2007, http://truefoodnow.org (accessed several times between 2004 and 2007).

Both sides of the debate also use rhetoric when they refer to notions of 'the natural'. Biotechnology advocates disqualify the argument explained in the next section, that crossing the species barrier would be unnatural, as rhetorical, as merely an emotional reaction based on a fear of everything that is new; they even term it the 'yuk factor' of genetic engineering.[53] Paul Thompson argues, however, that many people have an aversion to eating dogs and cats as well and this could be attributed to the yuk factor, but still no-one would force them to eat their companion animals on the basis that it is just an emotion.[54] Moreover, objections to the use of the unnaturality argument are often couched in rhetoric terms themselves. The Nuffield Council on Bioethics, for example, states that we should dismiss the argument that genetic modification is unnatural, because 'the "natural/unnatural" distinction is one of which few practicing scientists can make much sense'.[55] The weakness of this line of argument has been exposed by Deckers, who argues firstly, that the Council does not take into account that perhaps the alteration of plants through non-GM means is not accepted by everyone; and secondly, that assuming it is accepted this does not automatically make it acceptable. I would add that the fact that scientists cannot make sense of something does not mean it is not a valid viewpoint.

Moreover, advocates also resort to naturality arguments; either directly, by claiming that genetic mutations occur naturally, or indirectly, by suggesting that genetic engineering simply 'helps nature along a little'. Monsanto, for example, points out in a brochure, that 'BST has always occurred naturally in milk' and some of its advertisements contain pictures of landscapes and other natural phenomena, suggesting a link between its products and nature.[56] An ad from company Novo Nordisk claims to be 'fighting for a better world, naturally'.[57] One of the conclusions of Kleinman and Kloppenburg's study of Monsanto's publicity materials revealed that the following line of reasoning was advanced: biotechnology is a natural science; the fact that it is natural makes it safe; in fact, biotechnology uses nature 's means more efficiently than nature itself.[58] Vocal biotechnology critic Vandana Shiva points out that the biotechnology industry presents biotechnology as either natural or unnatural according to what suits their interests: 'when biological organisms are required to be owned they are treated as not natural; when the responsibility for consequences of releasing GMOs is to be owned they are treated as natural'.[59] After this more general depiction of the biotechnology debate, we will now turn to specific arguments concerning genetic engineering of, firstly, animals and, secondly, plants.

---

[53] For example Hughes and Bryant (2002), page 126.

[54] Paul B. Thompson (July–August 1997a).

[55] Nuffield Council on Bioethics (1999), page 15.

[56] Kleinman and Kloppenburg (1991).

[57] Levidow and Carr (1997), page 33.

[58] Kleinman and Kloppenburg (1991), pages 434 and 439. See also Levidow and Carr (1997), page 33.

[59] Vandana Shiva (1995) as cited in John Barry (2001), page 16.

2.3 General Positions on Biotechnology

Scientists often argue that they should have the freedom to experiment with new techniques; behind this appeal to scientific freedom lies the view that science is value-neutral and that the increase of knowledge is an intrinsic good. Scientists, in this view, are not responsible for the potential abuse of the knowledge they generate. After criticism to this view, however, it appears that scientists nowadays generally agree that they have a certain scientific responsibility and that they have to justify their choices of research to the public. Still, scientists often point out the risk of lagging behind international developments if biotechnology is too heavily regulated, and they usually prefer self-regulation through 'gentlemen's agreements'.[60] Critics argue that even if knowledge increase would have intrinsic value, scientific freedom still has limits; it cannot be allowed at any cost – for example to human or animal well-being. They also argue that scientists are too involved with commercial ventures for an unbiased self-regulation to be viable. Much of the biotechnology research at universities is sponsored by private industry. This may actually run counter to scientific freedom, as business interests and confidentiality requirements sometimes stand in the way of a free and open exchange of knowledge and collaboration between scientists.[61] Moreover, the patenting system is regarded by some as a stifling factor for scientific progress.[62]

According to many scientists, genetic modification is simply an extension of conventional plant breeding; we have been 'tinkering' with nature for our own advantage for thousands of years and genetic modification is merely the next step in this process.[63] Opponents, on the other hand, regard it as a radical break with conventional breeding methods.[64] Reiss and Straughan explain that there are three ways in which modern biotechnology differs from 'traditional biotechnology'[65]: Firstly, the species that were crossed by traditional biotechnological methods were closely related to one another, whereas using modern biotechnology, geneticists can, for example, put a human gene into that of a pig and bacterial genes into plants. Secondly, the time-frame within which changes were made in traditional biotechnology was much longer than in modern biotechnology; in other words, modern biotechnology has greatly accelerated the speed with which changes to organisms are made. Thirdly, the scale on which changes to organisms are made is much larger in modern biotechnology; modern biotechnology is applied to many different fields.[66] Nevertheless, proponents of biotechnology think that too much emphasis

---

[60] Zoeteman, Berendsen, and Kuyper (2005), page 30.

[61] See José E. Trias (1996), page 152.

[62] See Rebecca Eisenberg (1996).

[63] Fedoroff and Brown (2004).

[64] Mae-Wan Ho (1998); Jeremy Rifkin (1998).

[65] Reiss and Straughan (1996), page 2.

[66] Ibid., page 5.

is put in the debate on inter-species gene transfer, whereas most transgenesis carried out on crops involves transferring genes between different varieties of the same crop, or switching off genes.[67]

2.3.1 Crossing Species Barriers

One source of the disagreement over whether or not genetic modification constitutes a qualitatively new technique is a deeper disagreement over the question whether it is unnatural to cross species barriers. In order to understand what this means, we should ask three questions. Firstly, what exactly is meant by crossing a species barrier; secondly, does the crossing of species-barriers occur in nature; thirdly, what is the moral relevance of crossing species-barriers? First of all, no unequivocal definition of 'species' can be given; biologists differ over how species should be demarcated. Essentialistic notions of species are based on the idea that certain characteristics demarcate species. However, we cannot point out one specific characteristic that each individual member of a species possesses. Another way of defining a species is as groups of populations that form a reproductive community. A problem of this definition is that on occasion individuals of different species (such as horses and donkeys) interbreed. Both definitions disregard that species are located on a temporal and spatial continuum and, therefore, it appears impossible to demarcate the boundaries between two species unambiguously.[68] If we look at the genetic level, there do not seem to be species barriers. Different species share the same 'building blocks' of genes and it is the specific order and amount of genes that create phenotypical differences between species. From the perspective of geneticists, then, the boundaries between species are fluid.[69] Others argue, however, that in practice it is usually very well possible to determine that a species boundary has been transgressed; it is clear to everyone that a jellyfish and a rabbit are not part of the same species, for example.[70]

Advocates of genetic engineering point to the possibility of gene transfer between different species in nature, for example between mice and rats, in order to argue that crossing the species barrier is not unnatural. Opponents argue that this phenomenon takes place very rarely in nature and should be considered an 'error of nature'. Moreover, some interspecies transfers would never happen in nature. However, it does not automatically follow that transgenesis in a laboratory is morally unacceptable. What is unnatural is not necessarily morally unacceptable, no more than what is natural is automatically morally acceptable. What takes place in nature should not be taken as a criterion for human behaviour. Those who use this type of reasoning

---

[67] Hughes and Bryant (2002).

[68] G.D. Van Staveren (1991).

[69] Comstock (2000).

[70] Rob De Vries (2006), page 486.

2.3 General Positions on Biotechnology 33

from an 'is-' to an 'ought-statement', in which what 'is' in nature is necessarily good and therefore 'ought' to be, commit the naturalistic fallacy.

Despite the naturalistic fallacy, the unnaturality argument is so prominent in debates about the (un)desirability of genetic modification that apparently it refers to some kind of basic intuition.[71] As Theo van Willigenburg pointedly remarks, 'not that which is natural determines what is morally permissible and what is not, but our moral convictions about what is and is not permissible, and about what is good and bad, determine what we consider 'natural' and 'unnatural'.[72] In my view, the objection to the 'unnaturalness' of genetic modification is not in the first place about our acting *in accordance with* nature, but about our actions *towards* nature. Still, we cannot simply say that something is bad because it interferes with nature; an extra moral premise or argument needs to be given. The 'unnaturalness' argument is in fact often employed to express a range of different, yet related, concerns with genetic engineering, such as concerns over the process and the uncertain outcome of genetic modification, and the overconfident attitudes of scientists.[73]

The discussion about the relevance of crossing species barriers is complicated by the fact that proponents and opponents argue on two different levels. Geneticists do not see a problem because on a genetic level there is great similarity between all organisms. Critics, such as Henk Verhoog, on the other hand, think this is irrelevant: What is important in order to reach a moral judgment is what effect interventions on the genetic level have on the phenotypical level. If one reduces an organism to a mere sum of its genes, then indeed it appears unproblematic if only one of its, say, 10,000 genes is altered, but this small alteration can have a large effect on the organism as a whole.[74]

2.3.2 Genetic Determinism

Like Verhoog, many critics of biotechnology express dissatisfaction with the way in which molecular biology conceptualises the natural world; they qualify biotechnology as a dangerous 'abstraction from material reality'.[75] They think biotechnology operates from a flawed view of science: 'People start to forget the organism altogether by thinking of it as a mere collection of genes'.[76] Critics label this flawed view of science 'genetic determinism', which entails the belief that genes can be studied and manipulated in isolation from other genes, and that one gene, or a limited set of genes, encodes for one specific characteristic.[77] Reality is much more

---

[71] Frans W.A. Brom (1997), page 305.

[72] Theo Van Willigenburg (1997), page 170.

[73] Jan Deckers (2005)

[74] Henk Verhoog (1991).

[75] Marcello Buiatti (Winter 2005), page 14.

[76] Ho (1999), page 74.

[77] P.R. Wills (2002).

complex: genes interact with each other within a network of genes, genes are influenced by environmental factors, and these influences can become hereditary. Also, some genetically engineered organisms turn out to be able to 'repair' themselves and only less than two percent of illnesses are caused by a defect in a single gene.[78] According to critics of genetic determinism, this paradigm is overly reductionistic; it reduces organisms to a mere collection of genes and pictures the struggle for survival as one that is fought out on the level of genes.[79]

Underlying this debate appears to be a conflict between the fields of molecular biology, with its emphasis on the level of molecules, and ecology, with its emphasis on relations and re-appreciation of holistic worldviews. Even though most geneticists today do not underestimate the complexity of genomes, and many in fact support the alternative theory of the 'Fluid Genome', critics think they still operate within a reductionistic framework. Molecular biologists argue in response that if one wants to solve a scientific problem, one will have to use a reductionistic methodology at some point, only to later place this in the broader context of the organism and its environment.[80] It is argued that the simplistic portrayal of gene technology takes place when the science is translated into policy decisions and when scientific breakthroughs are reported in the media.[81] Moreover, it should be mentioned that many members of the public who oppose genetic engineering actually share the simplistic view of genetics.

One important reason why critics oppose the genetic deterministic framework is that they fear that interfering with natural processes while a lot about these processes is still unknown may lead to human health problems and ecological catastrophes. This fear seems to be tied in with a general pessimistic outlook on the controllability of technological developments. These fears express technological determinism – the idea that science and technology are autonomous systems.[82] Ironically, the idea of technological determinism is put forward by many biotechnology advocates as well; either it is stated that biotechnology is an inevitable development and it is, therefore, naïve to reject it, and/or it is stated that it is an inevitable development and that this is a positive thing, because it will bring us prosperity.

2.3.3 Human Hubris

An intrinsic argument against biotechnology is that it reflects a 'dominator-attitude' towards nature.[83] This claim expresses the unease that some people feel with humanity's quest for ever more control over nature. On the one hand, this unease

---

[78] Ho (1999); Wills (2002); McAfee (2003).

[79] This latter view is expressed, for example, by Richard Dawkins (1976).

[80] Interview with a committee member on 5/4/2003.

[81] Rosaleen Love (2001).

[82] Wouter Achterberg (1994).

[83] For an elaboration of different 'basic attitudes' towards nature in the context of biotechnology, see H. Jochemsen (2000), pages 55–64.

seems to be based on the view that nature is not in essence controllable; on the other hand, it is based on the view that humankind is not showing proper humility in the face of nature's grandeur. The aim to achieve as much control as possible over nature is seen as misguided, because it assumes that there is something wrong with nature and that it is humans' place to improve this situation. This in turn reflects human arrogance, or *hubris*, which could be explained as 'an unwillingness to recognize humanity's limits'.[84]

A related intrinsic argument is that genetic modification is like 'playing God'. This objection is usually interpreted as a religious argument: when we are genetically engineering plants and animals we do not understand our role as beings subordinate to God and we are acting as if we are the Creator.[85] Many find this argument uncompelling, because a plurality of views about God exists, not all religious groups reject genetic engineering, and it is unclear why genetic engineering should be considered playing God, but not other acts of creation, such as artistic ones.[86] However, Frans Brom argues that, despite the way in which they are usually portrayed, those that put forward the 'playing God-argument' need not have a religious background. He thinks this argument is used in order to express the intuition that there are boundaries that humans should not cross.[87] Its purpose is often not to reject biotechnology on religious grounds, but to express the idea that humans should not pretend to be able to direct developments in nature. This objection expresses the fear that biotechnological knowledge will remain in the hands of a 'technologically educated elite' and that this 'knowledge will be controlled by economic market forces of supply and demand'.[88] Moreover, this objection seems to refer to a broader underlying question, namely: 'what kind of world do we want to live in?'

2.4 Animal Biotechnology

The application of genetic modification to animals is primarily carried out for medical purposes and animal husbandry. Proponents of animal biotechnology in the medical field point out that genetically modified animals are better and more precise 'models' for human and animal diseases. As a benefit to animals in agriculture, it is argued that animals can be created that are more robust and can therefore deal better with the farm environment. Two notions play an important role in the debate about animal biotechnology, namely: animal welfare and animal integrity.

What exactly is meant by animal welfare is disputed. According to Brom it refers to, firstly, the absence of negative experiences; secondly, the presence of positive experiences; and thirdly, the extent to which an animal is capable of functioning

---

[84] Thompson (1997a), page 36.

[85] Comstock (2000), page 184.

[86] Ibid.

[87] Brom (1997), page 152.

[88] Ibid., page 156.

'normally'.[89] In the debate about biotechnology, the second and third senses of animal welfare become more pressing, as it is (at least theoretically) possible to create transgenic animals that experience very little pain and are well suited to intensive farming practices, but in the process have been stripped of their capacity to experience positive experiences and of their species-specific behavioural traits.[90]

Proponents argue that animal biotechnology could lead to less suffering, because transgenic animals could be made that experience less pain and are more resistant to diseases, and because fewer animals will be used in the long run.[91] Animal rights groups and many members of the public, on the contrary, believe genetic modification will bring about more animal suffering, and will actually lead to more animal use, because it opens up many new possibilities for investigation. Paul Thompson and Bernard Rollin argue, however, that 'it is commercial production that poses the most serious threats to animal welfare'.[92] Proponents often point out that animal biotechnology for agricultural purposes does not cause more harm than already existing intensive farming practices. However, a lot of the criticism of animal biotechnology should be viewed in the light of broader concerns surrounding animal use and it is, therefore, misleading to use conventional farming (or research) methods as the standard by which to judge new biotechnological techniques.

One of the main issues in the animal biotechnology debate centres on the question of whether or not it is wrong to adjust an animal's make-up to its (farm or laboratory) environment instead of the other way around. A new notion that has been introduced in this context is 'animal integrity'. Bart Rutgers describes integrity as 'the wholeness and intactness of the animal and its species specific balance, as well as the capacity to sustain itself in an environment suitable to the species'.[93] This notion has also been adopted by the Dutch Committee for Animal Biotechnology, which I will focus on in Chapter 5. As I have argued elsewhere, integrity is never invoked on its own, but always in relation to a *violation* of integrity and this presupposes a link to human action and refers to intentions behind actions.[94] Also, a violation of an animal's integrity should be understood as a notion that permits gradations, in order for this moral category to be helpful in a balancing context. Integrity, finally, seems to refer to an intuition that we should not change the 'species being' of an animal for our benefit: we should not tamper with the characteristics that make a chicken a chicken or a pig a pig. Opponents of the notion of animal integrity often reject its basis in emotions. It is argued, in reply, that emotional responses to issues like animal biotechnology are important intuitions that signal that something might be wrong, which needs to be more precisely articulated. Others argue that instead of

---

[89] Ibid., page 118.

[90] Ibid.

[91] Jeffrey Burkhardt (2003).

[92] Paul B. Thompson (1997b), page 6.

[93] F.J. Grommers, L.J.E. Rutgers, and J.M. Wijsmuller (1995); L.J.E. Rutgers and F.R. Heeger (1999).

[94] Bernice Bovenkerk, Frans W.A. Brom, and Babs J. Van Den Bergh (2002).

integrity, the concepts of 'animal dignity' or *telos* give a better basis for the moral intuition in question.[95]

The appeal to integrity, dignity, and *telos* all reflect the intuition that an animal is not just a thing, or an instrument, and should not be treated as such. This latter point is also encountered in the debate by reference to another group of terms: the 'instrumentalisation', 'objectification', 'commodification', or 'mechanisation' of animals, meaning that animals are used *as if* they are things or are even changed to the extent that they *become* things.[96] This concern connects to a set of broader concerns. There are fears that introducing transgenic animals in agriculture will intensify the tendency to make agricultural practices high-tech and this might lead to social injustice.[97] The discussions I have merely been able to touch on here are analysed in more detail in Appendix A.

2.5 Plant Biotechnology

Some of the goals for which plants are genetically modified, or are expected to be in the future, are enhancement of flavour, colour, or nutrition of fruits, improved shelf-life, creating herbicide-resistant crops, creating crops that are resistant to pests, creating crops that can deal better with environmental stresses such as drought, heat, cold, or high salinity in soils, production of drugs and vaccines in plants ('biopharming'), creating plants that can clean up environmental pollution, or that contribute to sustainability, and even manufacturing plastics from genetically modified plants and micro-organisms instead of petroleum.[98] Transgenic plants have so far primarily been employed for agricultural purposes, mainly for herbicide tolerance of crops and insect resistance. At the moment, most applications of agricultural biotechnology do not seem to confer a direct benefit to consumers, but only to biotechnology companies and certain groups of farmers.[99] This could provide one explanation for consumer rejection of transgenic foods, which is particularly strong in Europe. Besides questioning the potential benefits of the application of genetic engineering, critics also perceive many potential problems in the areas of interhuman relationships, and human and environmental health. Due to limitations of space, here I will only briefly describe the topics about which controversy exists; for a more detailed discussion on plant biotechnology, see Appendix B.

---

[95] Phillipp Balzer, Klaus Peter Rippe, and Peter Schaber (2000); Sarah Elizabeth Gavrell Ortiz (2004); Thompson (1997b).

[96] Brom (1997).

[97] Wolfgang Goldhorn (1990), page 85.

[98] Hughes and Bryant (2002). Huges and Bryant give quite an extensive list of (possible) applications of plant genetic modification. See also: Nuffield Council on Bioethics (1999); Straughan (1992); Rifkin (1998); Hughes and Bryant (2002); Vasil (2003); Lino Paula and Frans Birrer (2006).

[99] Lindhout and Danial (2006).

2.5.1 Justice

The application of genetic engineering to crops is widely regarded as a continuation of the Green Revolution, which refers to developments in agriculture from the 1950s in which scientific inputs, that enabled the creation of 'high yielding varieties' of crops, mechanisation, widespread use of chemical herbicides and pesticides, and more intensive irrigation were employed in order to increase yields for a growing world population. Despite its success in increasing crop yields, the Green Revolution has led to environmental degradation and a concentration of land in the hands of a few large farms producing crops for export.[100] Currently the rate of yield increase is in decline, leading some to call for a second revolution based on biotechnology – a Gene Revolution – that avoids environmental problems but has even higher productivity. Critics, however, argue that we should not use a 'technical fix' for problems cause by modern technology in the first place.

Proponents of biotechnology list many economic benefits of growing GM crops, resulting from higher yields, longer shelf-life, tolerance to early maturing of plants, and tolerance to environmental stress. Critics argue that these applications primarily benefit the biotechnology industry and rich farmers in the West and that they stand to increase the gap between the rich and the poor, and therefore lead to injustice.[101] This criticism is rejected by biotechnology advocates, who argue that one of the most important aims of biotechnology is to reduce poverty.[102] They list 'Golden Rice', which is meant to counter the Vitamin A deficiency that plagues many people in poor countries, as one of their success stories. On several aspects of Golden Rice, scientific controversy exists: about its nutritional benefits, the amount of Golden rice that needs to be consumed to reach the recommended daily amount of vitamins, and on the conditions for vitamin A uptake. Another contested topic is the development of so-called 'terminator technology', which consists of inserting genes into crops that make them infertile, so that germination of the next generation of seeds does not occur and farmers will no longer be able to save seeds for next year's sowing.[103] Finally, many commentators, both of the pro-GM and the anti-GM camp lament the concentration of power in the biotechnology field.

A hot topic in the biotechnology debate is the question whether patenting of gene constructs should be allowed. An often heard intrinsic objection to patenting is that it amounts to owning life, which is deemed to be morally objectionable, because it denies the sacredness of life or because it turns living organisms and parts of nature into saleable commodities.[104] Many also argue that living creatures should not be patentable, because they can only be discovered, but not invented. Proponents of the patent-system argue that patents stimulate inventions which ultimately benefit the

---

[100] Comstock (2000), page 157; Jack Ralph Kloppenburg, Jr. (1988), page 6.

[101] Robert Vint (2002).

[102] A. Trewavas (8 August 2002), page 669.

[103] Vint (2002).

[104] Hindmarsh (1994); Ho (1999).

general public, because they protect the intellectual property rights without which companies would have no motivation to invest in biotechnology research and development. Critics, on the other hand, argue that patents only benefit large multinational companies.[105] They also lament the fact that farmers can be held liable for their unintended use of patented GM seeds when their crop has accidentally cross-bred with a nearby GM-crop.

Critics of biotechnology tend to call the collecting of genetic resources from developing countries by developed world biotechnology companies and public sector institutions 'biopiracy', whereas advocates term it 'bioprospecting'. Spokespersons of the biotechnology industry claim that genetic resources only get market value after they have been engineered into useful products or applications and that, therefore, the people or the countries who supply the genetic material need not be compensated.[106] Moreover, they consider genetic resources 'the common heritage of humankind'. Critics argue that local communities should be rewarded for their plant breeding efforts, instead of penalized for breaking a patent when they use resources they have used and improved for centuries.[107]

2.5.2 Health Concerns and Risk

Proponents of biotechnology claim that it is safe to eat GM foods; consumers in the United States have eaten GM food for decades, without experiencing any health problems.[108] Critics argue that it is a fallacy to think that the fact that no cases of illness after eating GM foods have been documented must mean it is safe. It would be near impossible to determine whether a particular GM food was causing illness in the community; in the case of GM foods there is no surveillance system, because we do not even know what disease to look for.[109] Furthermore, according to critics, many of the safety assessments of GM foods were of dubious scientific quality and have not been carried out by independent scientists, but by industry scientists who are likely to be biased.[110]

Proponents of biotechnology argue that more extensive safety testing is not necessary for GM foods, because there have not been significant changes to GM crops and there is no 'conceptual distinction' between organisms that have been traditionally bred and those that have been genetically altered.[111] The term that has been used to describe this view is 'substantial equivalence'. Critics argue that this term misleadingly gives the impression of being scientific and that products

---

[105] Gitte Meyer (2001), page 25; Frans W.A. Brom (2003), page 124.

[106] Rifkin (1998), page 37.

[107] Sigrid Sterckx (2004).

[108] Florence Wambugu (1999).

[109] Judy Carman (2004).

[110] Ibid.

[111] Miller (1999).

highly equivalent to their non-GM counterparts can still be dangerous.[112] In order to determine whether the consumption of GM foods and the introduction of GMOs into the environment are safe, scientists make a risk analysis. However, the conventional approach to risk assessment, which treats risks as an 'objective probability of harm' has come under increasing scrutiny for neglecting the normative, social, and cultural dimensions of risks.[113] The questions of what are unwanted consequences or harms and what is the acceptability of certain risks are based on a value judgment, including, amongst other things, the perceived benefits involved.[114] Risk communication has proven to be ineffective and even to have contrary effects if social or cultural dimensions are overlooked. The conventional 'one-way sender/receiver' model of communication does not lead to public trust. Especially when the public perceives these institutions as functioning along lines of social dominance and control – as is the case with biotechnology – in order to be effective, risk communication needs to take a more interactive shape in which information flows in both directions.[115]

What makes risk analysis in gene technology especially problematic is that it is characterised by low levels of certainty – we are dealing not only with known uncertainties, but also with 'unknown unknowns'[116] – and low levels of 'consensus with respect to the parameters of the scientific issues to be addressed'.[117] Research shows that lay citizens tend to take into account the existence of unknown unknowns while scientists and policy makers tend to ignore them.[118] Differences between risk perceptions of experts and lay people have traditionally been seen as stemming from lay persons' irrational fears, whereas they can also be regarded as the result of a different interpretation of the social implications of technology and the different framing of policy issues.[119] Ulrich Beck, who coined the concept of the 'risk society', argues for more democracy in science through open expert controversy and a role of scientists in public debate as consultants rather than as authoritative decision-makers. Brian Wynne goes even further and argues that we need to question the priority given to the scientific model that is used by experts. The values that are inherent in risk analysis have to be made more explicit. Also, lay people can provide experiential knowledge that forms an indispensable contribution to the risk analysis of experts. Grove-White and Szerszynski in this context argue that 'many social conflicts, overtly about the technical determination of environmental risks, can be more usefully seen as conflicts between commitments to certain models of how society

---

[112] Erik Millstone, Eric Brunner, and Sue Mayer (1999); Carman (2004).

[113] Gillian Turner and Brian Wynne (1992), page 111.

[114] Lawrence Bush et al. (2004), pages 5 and 115.

[115] Turner and Wynne (1992); Frans W.A. Brom (August 2004a).

[116] Brian Wynne (2001).

[117] Bush et al. (2004), page 5.

[118] Wynne (2001).

[119] Sheila Jasanoff (2001).

is – or should be – ordered'.[120] According to Levidow and Carr, the biotechnology industry and many government agencies have ignored the normative aspects of risk and have actively sought to separate risk from ethics and to reduce both to matters that could be dealt with by experts.[121]

Because there are so many scientific uncertainties involved in gene technology and the possibility of unknown consequences is perceived to be high, many propose to use the so-called 'precautionary principle' when making decisions in this field. In the context of biotechnology the principle has been defined as follows: 'Where there are threats of serious or irreversible damage, lack of full scientific certainty shall not be used as a reason for postponing cost-effective measures to prevent environmental degradation or harms to human health'.[122] However, this definition is open to widely varying interpretations, depending on what we mean by 'serious' or by 'irreversible', 'damage' or 'cost-effective'. How we interpret these terms is, again, based on value judgments.

The question of whether and how GM foods should be labelled has been one of the most discussed and most contentious topics in the biotechnology debate. Advocates of the 'product approach' argue that analysing the product is the only way to measure whether there are GMOs in food and that health risks can only be based on the components of the food consumed and not on the process by which the food was made.[123] It is also pointed out that a precondition of labelling is that GM and non-GM grains can be segregated effectively, which can be problematic and expensive.[124] Critics argue that lax labelling laws violate consumer sovereignty and religious liberty by preventing people from making informed food choices on the basis of their own moral or religious convictions.[125] Biotechnology advocates claim that rather than aiming to protect consumer autonomy, countries use mandatory labelling in order to erect trade barriers.[126]

2.5.3 Environmental Consequences

While biotechnology advocates argue that the use of agro-chemicals can be reduced by using genetically engineered variants, critics point out that the fact that crops are tolerant to herbicides means that farmers can spray as much as they like.[127] The scientific evidence is not clear-cut; results can vary depending on what particular

---

[120] Robin Grove-White and Bronislaw Szerszynski (1992), page 292.

[121] Levidow and Carr (1997), page 40.

[122] United Nations (1992), principle 15.

[123] Mayer (2002).

[124] Alan Mchughen (2000); Martin Brookes (1998).

[125] Thompson (1997a).

[126] See, for example, Australian Financial Review (2002).

[127] Martin Brookes and Andy Coghlan (1998).

aspect of herbicide or pesticide use is examined, the amount of active ingredients, and the number of pounds per acre sprayed.[128]

Particularly much debated is the use of the soil bacterium *Bacillus thuringiensis* (*Bt*) as an inbuilt insecticide in crops. The environmentally benign *Bt* was originally only used sparingly by organic farmers in the form of a spray. The main drawback of so-called 'biopesticides' is that pests can more easily become resistant; insect populations will then be harder to control, calling for ever more aggressive methods of pest control. This disadvantages organic farmers, because it renders useless one of the few pesticides they are allowed to use.[129] Generally, it is very difficult to predict the environmental results of different pest management regimes, as there are many unintended side-effects.

Another possibly harmful effect of transgenic crops on the environment is caused by hybridisation – rhetorically termed 'contamination' or 'genetic pollution' by critics – which can take place because pollen travel by wind and are carried by insects, seeds are spilled around fields during harvest or transport, and bacteria and viruses exchange genetic material and transfer this to plants.[130] Some fear that this latter type of transfer will lead to transgenesis of non-target species and the unintended creation of novel genetic combinations that are potentially dangerous.[131] They are afraid that the 'escape' of transgenic plant genes could lead to 'superweeds', a decline of native species populations, and the extinction of rare plant species.

Many proponents as well as opponents of genetic engineering acknowledge that GM crops are associated with a loss of biodiversity, but many think this is due to the wider context of intensive agriculture in which genetic modification is applied and not to the technology per se.[132] Spokespersons from the biotechnology industry claim that sustainable agriculture is only possible with the use of biotechnology, but opponents argue that the biotechnology industry frames problems in terms of the symptoms rather than the underlying problems. One of the underlying causes of the different views on the sustainable promise of gene technology is that different interpretations of sustainability are used. For some it refers to the capability to provide food security for future generations, or even to economic sustainability, while others also look at the sustainability of a healthy environment. Some proponents argue that conventional farming is actually more sustainable and better able to conserve wildlife than organic farming, because it uses land more efficiently. Critics find this approach highly positivistic, demanding strong scientific proof and eschewing anecdotal evidence, whereas proponents of organic farming 'rely on personal experiences and beliefs that make them more receptive to the idea that there is a difference between organic and conventionally produced food'.[133]

---

[128] Charles Benbrook (2001).

[129] Richard Hindmarsh (1991).

[130] Brookes (1998).

[131] McAfee (2003).

[132] Hughes and Bryant (2002).

[133] Annette Mørkeberg and John R. Porter (2001), page 677.

2.6 Discussion

It should have become clear that we are dealing with a complex, multi-faceted and heated debate in which many different positions are defended, ranging from outright rejection to moderate criticism, and from a qualified support of some applications to an optimistic embrace of all forms of gene technology. In short, we are dealing with a very diverse debate which involves many disagreements on many different levels and dimensions. The advent of biotechnology has not only led to a heated debate, but also to name calling, vocal protests, and acts of civil disobedience, or (in the eyes of some) 'eco-sabotage'. Field tests of GM crops have been destroyed by action groups, farmers who have discovered secret field trials of GM crops on the borders of their land are outraged, citizens' concerns have been responded to by calling the public irrational, and scientists critical of genetic modification have been ostracised.[134]

Many of the arguments in the biotechnology debate appear to turn on issues of power and control. Both biotechnology advocates and its critics claim that self-interested or political motives lie behind their opponents' arguments; some of their accounts almost read like conspiracy theories. For example, the European moratorium on the import of transgenic crops is regarded as a 'protectionist manoeuvre' and a bowing to certain political parties.[135] Also, critics describe methods of the industry to push biotechnological developments, including 'co-optation' of environmental groups onto gene technology committees – as these groups were always in the minority, their views could be said to be represented, but they would not have an effective voice in discussion outcomes – blocking influential critics (such as Peter Singer) from committees, blocking public education or awareness campaigns, or exactly the opposite, stimulating public 'indoctrination' under the guise of education.[136]

It is my contention that biotechnology has given rise to such a heated controversy because a lot is at stake: not only economic interests and the health of people and their environment, but ultimately our deeply-held beliefs and views about the kind of world we want to live in. Even though the debate is often constructed narrowly as being about the merits of a novel technology, the arguments on both sides apply to a wider group of ideas or concerns surrounding agriculture, socio-economic relations, nature conservation, medicalisation, and so on. Positions in this debate cannot be seen apart from broader views about what would constitute a desirable society, about what are proper relationships within society and with nonhuman nature, and even about our own nature. For example, many intrinsic objections, such as the objection to 'playing God' and the objection against the instrumentalisation of animals, refer to fundamental questions about the boundaries of scientific endeavour and our proper relationship to nonhumans. Also, one's estimation of the potential

---

[134] See Geraldine Chin (2000).

[135] Vasil (2003).

[136] Hindmarsh (1994).

and dangers of biotechnology seem to be influenced by one's optimism or pessimism about technological developments and their perceived inevitability. Furthermore, how one estimates the potentials of, and even the need for, the application of gene technology to agriculture is strongly influenced by how one perceives the development of modern agriculture in general. This evaluation is supported by a Canadian study into public attitudes towards genetic engineering, which concluded that 'attitudes to genetic engineering were affected by a respondent's "core beliefs", that is, their knowledge of science and technology, attitude to nature and attitude to God'.[137] This finding is also pointedly summarised by the Food Ethics Council: 'A key concern in the debate about the use of GM technology is the fundamental question of the kind of society that we wish to live in'.[138]

Several studies have concluded that very little genuine dialogue between opponents and proponents of biotechnology is possible and this contributes to the persistence of the disagreements.[139] Part of the problem seems to be that these groups argue on different levels. Geneticists for example, look at organisms on a genotypical level, while ecologists and many critics of biotechnology look at them on the phenotypical level. Similarly, Levidow and Carr argue that a lot of scientific disagreement over biotechnology arises from 'divergent cognitive frameworks': ecologists tend to view the environment as a 'fragile 'ecological balance'', whereas geneticists see it as 'resilient, capable of stabilizing itself'.[140] Some of the basic assumptions of the participants in the debate, such as assumptions about the precision of rDNA techniques, are contrary to each other, the participants tend to focus on different levels – for example, on an atomistic as opposed to a holistic level – and they draw the boundary between facts and values differently. What this reconstruction has also aimed to make clear is that many arguments that are ostensibly limited to scientific or empirical questions in fact have normative aspects. Examples are the debate about the supposed unnaturality of crossing the species barrier, and the discussion about patenting, which is based on diverging normative assumptions about distributive justice. But perhaps this point is best illustrated in the debate about risks; the framing of questions regarding risks tends to determine the outcomes of risk assessments and this framing carries – often implicit – value judgments.[141] Framing determines amongst other things what types of consequences are considered, how much risk is deemed acceptable, and what benefits justify what level of risk. Fundamental views about the status of scientific research and the level of precision of gene technology influence empirical assessments of the safety of GMOs, and these assessments are, therefore, not as straightforward as they are often presented.

---

[137] Janet Norton, Geoffrey Lawrence, and Graham Wood (1998). Norton et al. cite Decima Research (1993).

[138] Peter Lund and Ben Mepham (1998), np.

[139] For example, Norton (1998); Zoeteman, Berendsen, and Kuyper (2005).

[140] Levidow and Carr (1997).

[141] See Soemini Kasanmoentalib (1996).

2.7 Sources of Disagreement

Even though many of the specific issues that I have discussed are unique to biotechnology, the diversity of opinions encountered in the biotechnology debate is a reflection of the diversity of opinions found in society in general. Especially in modern Western societies, we are faced with religious, cultural, and normative diversity. The terms 'cultural pluralism' or 'social pluralism' are often used to refer to the plurality of diverging values and worldviews stemming from this diversity.[142] The existence of this social pluralism can have several different sources. People can disagree about something because they lack all the relevant facts bearing on the issue, because they base their viewpoint on different sources of evidence, or because they interpret the facts differently. We could label this 'factual disagreement'. In the biotechnology debate factual disagreements exist, for example, about whether or not GM crops lead to more herbicide and pesticide use in the long run and whether or not contamination between GM and non-GM crops occurs. At first sight this type of disagreement appears to be easily solved by pointing to scientific evidence. However, disagreement also exists on the level of scientific theory. Biologists inspired by an ecological worldview approach genetics from a different, more holistic, perspective than molecular biologists, for example. This could be termed 'scientific disagreement'. Other examples of this kind of disagreement include the questions whether risk assessment is objective or value-laden and whether or not the current methods used for testing the safety of GMOs are adequate.

People do not only disagree about factual knowledge and the methods of knowledge generation. Their disagreement may already begin at the level of definitions of the core terms used in the debate. Our choice of definitions determines how the debate is framed. For example, core concepts such as 'substantial equivalence' and the 'precautionary principle' are open to several different interpretations, and there is disagreement about whether to define GMOs as inventions or discoveries. All of these could be called 'definitional disagreements'. How one decides to define certain core notions in a debate and one's initial position on the topic are typically inter-related: one can define terms to suit one's own interests or values. On a deeper level, such definitional disagreements are not only about how we describe certain core terms, but also about power. The dominant discourse in society tends to be that of powerful groups; when these groups choose to employ certain definitions they are determining the terms of discussion. Defining core concepts in a debate is therefore at the same time about agenda setting. The biotechnology debate illustrates this point well. Proponents of biotechnology define the technique of biotechnology as 'the use of biological processes to produce useful organisms or products'.[143] Opponents would not agree that this technology necessarily creates useful products, however, and therefore the disagreement already starts on the basic level of definition. In a similar vein, Kleinman and Kloppenburg show that

---

[142] Matthew Lawrence (1996), page 337.

[143] See http://www.agresearch.co.nz (accessed on 1 May 2005).

by dominating the discourse, biotechnology company Monsanto 'has attempted to define biotechnology, and thereby shape political debate and stifle opposition. In its effort to define biotechnology, Monsanto has sponsored science museum exhibitions and lectures by company scientists'.[144] Kleinman and Kloppenburg analyse Monsanto's publicity materials in order to expose the ways in which the company is seeking to gain acceptance for biotechnology through discursive means. In these materials Monsanto, for instance, defines biotechnology as a natural process, while at the same time emphasising its scientific 'added value': biotechnology 'does what nature does – genetic selection – but it does so more efficiently than nature'.[145]

Because of the bias that is often inherent in definitional disagreements, they can sometimes be redefined in terms of 'interest-based disagreements', which are disagreements that are based on some sort of interest, or at least on the claim that material or economic interests are advanced, whether or not this is the case. Both parties in this debate accuse each other of self-interested motives, even though they appear to advance arguments in the name of the greater good. An example is the claim that countries that enforce labelling rules for GM products are not doing so for the sake of the consumer's free choice, but in order to throw up trade barriers. Another example is the claim that biotechnologists use the argument that they want to feed the world, when in reality they primarily want to feed themselves.

Of course, not all arguments that are based on interests are illegitimate, in the sense that they are merely self-serving. Arguments that appeal to the value of justice, for instance, can very well be focused on the interests of a certain (usually disadvantaged) group. For example, when Third World farmers reject terminator technology because they do not want to purchase seeds annually, they are trying to secure their own interests. However, in the name of equality in the context of the global division of wealth, this particular appeal to interests is justified, because it refers to the more generalisable values of justice and equality. Admittedly, it is sometimes difficult to distinguish between merely self-serving interest-based arguments and legitimate interest-based arguments. Arguments that are motivated by some sort of sectarian interest are sometimes submitted to a process of self-censorship and disingenuously expressed publicly as in the public interest. This suggests that when the actual motivation behind an argument is exposed, the argument may appear less justified.

Of course, it could be maintained that many of the disagreements we encounter in the biotechnology debate are simply due to conflicts of interest – for example, between the interests of farmers to improved harvests and of consumers to safe food. However, many of the conflicts of interest in this debate can also be traced to opposing values or worldviews; they are interrelated in the sense that one's values and worldviews are influenced by one's more material interests, even though, of course, one's values can also clash with one's material interests. For example, some farmers embrace the technology of genetic engineering, because they consider it to

---

[144] Kleinman and Kloppenburg (1991), page 429.

[145] Ibid., page 436.

be a more efficient way to carry out their jobs and they consequently see it as a way to reduce poverty. Their own material interests and the value of equality therefore reinforce each other. On the other hand, some farmers, while embracing the financial incentives that GM crops could offer them, lament the loss of biodiversity that they believe GM crops bring about, and their own interests therefore clash with their other values, in this case with the value they attach to preserving biodiversity.

Arguments in the biotechnology debate that appeal to values include those that resist the commodification of nature or the mistreatment of animals, on the basis that they possess intrinsic value. We could term disagreements based on this type of argument 'value disagreements', because they can be traced to conflicts between different, often fundamental, values. For example, disagreements about the question of whether animal testing should be allowed are often based on conflicting views about the proper moral status accorded to animals. As we have seen, there is a certain overlap between the category of interest-based and value disagreements. Similarly, factual, definitional, and scientific disagreements have value aspects, in the sense that evaluative judgments are involved, such as in the experimental design of scientific research or in the interpretation of test results. While the different types of disagreement classified here often merge together in real-world conflicts, they nonetheless remain analytically separable and are, therefore, of considerable heuristic value in narrowing down the more fundamental sources of disagreement. It can then become clear that, for instance, a conflict about the risk of introducing a specific GM crop into the environment is not based only on scientific disagreement, but also on an underlying value disagreement about the evaluative steps in the type of risk analysis employed. A similar point is made by Ann Clark and Hugh Lehman about the assessment of GM crops:

> Differing values lead people who accept the same scientific opinions as true to reach contradictory conclusions regarding the acceptability of use of the same product. However, differing values also make it difficult to achieve agreement as to what are the facts.[146]

However, even when people share the same values, they can still disagree about the question of what they ought to do: How they ought to behave in a certain situation, what should and what should not be allowed. In other words, they can disagree about the question as to what duties we have towards others. What do we owe to each other and who are these others? In the biotechnology debate disagreement exists, for example, about what bioprospectors owe to indigenous communities who have bred certain plants through the generations; are these communities owed compensation for their work and for passing on their knowledge? Another example can be found in the animal biotechnology debate; even though most agree that animals have moral status, people differ as to what commitments this gives rise to on the part of humans; does this mean we should not use animals in research at all, does it mean that we should not change their telos?, *etcetera*. This type of disagreement could be termed 'moral disagreements'. Moral disagreements do not simply flow from value disagreements. While we can have a moral disagreement even if we have consensus

---

[146] Ann E. Clark and Hugh Lehman (2001), page 15.

on values, we can also have value disagreement and still agree on what we ought to do. One person might argue for a moratorium on commercially grown GM-crops because she values biodiversity, while another person might support a moratorium because she values consumer choice. Even when people's values are diametrically opposed they can still reach moral agreement; for example, one person may value every living organism as something sacred while another person may see organisms as simply a set of genes; both may object to patenting genetically engineered organisms – the first because it disrespects the sacredness of life and the second because it stifles innovation.

Both value and moral disagreements can be based on conflicts between different worldviews. For example, proponents of biotechnology generally have an optimistic view about human progress and they picture science as a controllable instrument to further human development. Opponents tend to paint a more pessimistic picture of human progress through science and technology, based on past technological failures and the view that science and technology are not always controllable.[147] Also, those who are in favour of genetically engineering animals and plants usually adopt an anthropocentric worldview, in which human beings are the central locus of value,[148] whereas many of their adversaries have non-anthropocentric worldviews, in which not only human beings but also other entities or even groups of entities and the relations between them – such as ecosystems – have independent moral standing. As a corollary to these different worldviews, people include a varying range of entities in the moral community: Anthropocentrists argue that we have direct duties only to human beings, atomistic non-anthropocentrists either include only humans and other sentient animals,[149] or only individual living entities,[150] whereas holistic non-anthropocentrists – also called ecocentrists – include not only individual organisms, but also ecosystems and sometimes even inanimate parts of nature, such as rocks.[151] Whereas atomists give precedence in our moral deliberations to individuals, for holists 'the whole is more than the sum of its parts' and the interests of the whole should determine the outcome of moral deliberations. Often – albeit not necessarily – ethical holism has its basis in metaphysical holism, which asserts that wholes exist independently from and are 'as real as their parts'.[152]

---

[147] These critics think that technology is out of control especially in conjunction with a capitalistic market, in which these 'forces' acquire their own autonomous dynamics. See Achterberg (1994).

[148] Note that by anthropocentrism is not simply meant that only human beings are capable of ascribing value, for example to animals or other living entities – for this is what non-anthropocentric worldviews also have to acknowledge – but that they are also the primary recipients of value, and thus that human interests always ought to take precedence in our moral deliberations. See Robyn Eckersley (1992).

[149] For example, Peter Singer (1975).

[150] Examples of this biocentric theory are those of Paul Taylor (1986) and Robin Attfield (1998).

[151] For example, J. Baird Callicott (1989) and Lawrence Johnson (1992).

[152] Joseph R. Des Jardins (1993), page 181.

2.7 Sources of Disagreement

In the biotechnology debate, then, we can discern not only diverse values and worldviews, but even distinct metaphysical views, which can be described as 'pictures' of the fundamental building blocks of the world and the relationships between these building blocks. A worldview refers to the way we think societies and nature are constituted and function, and the roles of humans and other entities in society and nature. For example: 'are we living in a 'dog-eat-dog world' or do we have a more cooperative picture of our society (and of the relation between different societies)?', or 'what do we mean by 'progress'?'. Worldviews are related to our conceptions of the good life. Metaphysical views influence our worldviews, but they operate on a more abstract level. Metaphysical views are answers to questions such as 'is everything predetermined or not?' and 'are rivers and mountains living entities in their own right or only the living entities within them?'. Similarly, our opinions of the merits of biotechnology are ultimately embedded in our views of the proper place of human beings in the world; of course, for many this does not only relate to our relationship to non-human entities, but also to God. The objection that biotechnology amounts to 'playing God' can, therefore, be traced back to their metaphysical worldview (in this case, their religion). However, as has become clear from the foregoing, the objection to 'playing God' can be endorsed not only by the religious, but also by non-anthropocentrists, although the latter usually prefer to describe it as an objection to the display of human *hubris* that biotechnology represents in their eyes. Another objection that can be traced back to diverging metaphysical views is that 'interfering with nature's building blocks' is unnatural. Disagreements based on worldviews and metaphysical views could be termed 'metaphysical disagreements'. Our metaphysical views influence our normative viewpoints and vice versa.[153] Metaphysical views will not determine our normative viewpoints, however, as two people with the same metaphysical views can base different value rankings on these. Nonetheless, metaphysical views inform and influence our normative viewpoints by structuring the field of possibilities.

The different sources of disagreement that I have distinguished are related to each other in complex ways. While we could say that often definitional disagreements are in reality based on interest-based disagreements, or scientific disagreements on value disagreements, for example, one does not always follow from the other. What does become clear from my analysis is that many disagreements that are ostensibly about opposing definitions, facts, or scientific methods turn out to 'hide' arguments about values, morals, or worldviews. For instance, an ostensibly simple debate about the hybridisation of GM and non-GM crops may be based on an underlying disagreement about what we consider to be natural or what we think is the proper role of farms in our society; do they only play a role in food provision, or also in the maintenance of wildlife, for example? While value, moral and metaphysical disagreements are not by definition more fundamental than the other categories of

---

[153] In some cases our moral viewpoints could influence our metaphysical views. An example given to me by Betsy Postow † is that a certain view of moral responsibility could imply a certain 'metaphysical' conception of the self as a continuing being (rather than a more Parfit-influenced conception of the self, for example).

disagreement, then, they do often appear to underlie debates about factual, definitional or scientific issues, and often remain implicit in the background. Particularly in the context of regulatory decisions, the biotechnology debate tends to revolve around pragmatic issues, such as labelling and risk analysis, without the different parties acknowledging that even in these cases value, moral or worldview disagreements play a role. The likely reason for this is that when disagreements are based on these underlying categories it becomes harder to generalise the different viewpoints, as they appeal to deep-rooted values that are perceived to be subjective. This is not to say that these values cannot be defended by rational argument, but merely that there will be more disagreement already about the premises of such arguments. In general, one would expect, therefore, that the more a disagreement is based on differences in values, moral views, worldviews and metaphysical views, the more likely this disagreement is to become intractable.

The focus on pragmatic issues does not do justice to the wider reasons why people take a stance for or against biotechnology. A person's position, even on such a practical issue as labelling of GMOs, cannot be seen apart from her broader set of assumptions about biotechnology, about our place in the world, about the moral status of nature, and even about the proper role of food in our lives. The debate then does not always reflect the actual sources of the disagreements and obscures the real reasons behind arguments. Limiting the debate to more pragmatic issues, and to those issues that can be dealt with by way of making a cost/benefit analysis, tends to create a bias in favour of proponents of biotechnology. Proponents' arguments tend to be of an extrinsic nature – they focus on the extrinsic benefits of biotechnology – while opponents' arguments are both of an extrinsic and an intrinsic nature. Especially intrinsic arguments do not lend themselves for the calculation of a cost/benefit analysis, because they cannot so easily be quantified. Someone who does not find biotechnology problematic in principle is likely to have fewer qualms about casting the debate in pragmatic terms than someone who opposes it on intrinsic grounds. After all, the question in such a debate is no longer whether biotechnology is desirable at all, but rather how we can regulate it in order to lower associated risks and enhance people's autonomy. Intrinsic arguments, on the other hand, tend to address the desirability of biotechnology in itself. Framing the biotechnology debate solely in pragmatic terms, in other words, is inherently biased, and therefore unfair, because it tends to favour pro-biotechnology standpoints.

2.8 Unstructured Problems

I described the debate about the merits of biotechnology as having reached a state of intractability, because various different sources and dimensions of disagreement, from disagreement over definitions to disagreement over worldviews, are involved. We could say that definitional, factual, and scientific disagreements primarily belong in the realm of science, while interest-based, value, moral, and worldview disagreements are primarily normative in character and belong in the realm of politics. In the biotechnology debate we encountered both scientific and normative disagreements

2.8 Unstructured Problems

and uncertainties and it became clear that the two are intertwined. This means that we are not only dealing with a factual or a scientific problem that can be solved by scientists, and neither with a purely normative problem that could be solved in the political realm. Matthijs Hisschemöller makes a useful distinction between four different types of policy problems based on these two types of uncertainties.[154] He distinguishes policy problems by the extent to which they are 'structured'. The four problem types are schematically displayed in Table 2.1 below. It should be mentioned that these four problem types are ideal types, because in reality facts and values are always interwoven.

In a structured problem, broad agreement exists about the facts and values involved – about the problem and the solution – and solving the problem essentially comes down to applying standard techniques and procedures to the problem; policy decisions can then largely be left to bureaucrats and experts. Examples are the maintenance of roads or the application of rules regarding the distribution of housing.

In unstructured problems, no such agreement about facts and values is present, and even the question as to how the policy problem should be defined in the first place is disputed. In other words, an unstructured problem is based on both normative and scientific uncertainties, and it is unclear where one category of uncertainty ends and the other begins. According to Hisschemöller, unstructured problems, which have also been termed 'wicked' or 'complex decision problems', are characterised by 'factual uncertainty, interdependency, complexity, ambiguity, and political, organisational, and institutional differences of judgement'.[155] They involve conflicts about the goals of policy, the procedures that should be followed, and the instruments that should be used, and they involve a large number of political actors.

Besides unstructured and structured problems, Hisschemöller, together with Rob Hoppe, distinguishes two types of moderately structured problems.[156] In the first, consensus exists about the values and goals that are involved and dissensus about

Table 2.1 Four problem types of Hisschemöller

	Normative certainty (normative consensus)	Normative uncertainty (no normative consensus)
Scientific certainty (consensus about facts)	*Structured problem* Can be solved by experts or bureaucrats	*Moderately structured problem (means)*
Scientific uncertainty (no consensus about facts)	*Moderately structured problem (ends)*	*Unstructured problem* Learning strategy/public debate

---

[154] Hisschemöller (1993).

[155] Ibid., page 28.

[156] See also Hisschemöller and Hoppe (1995).

causal aspects, or about the means to achieve the goals. In other words, agreement exists about the problem, but not about the solution. In the second, dissensus exists about the values involved and consensus about the facts involved. In other words, disagreement exists about the problem, but agreement exists about the solution. An example of the latter, seemingly logically impossible, type of problem is the debate about abortion that has been 'solved' in the Netherlands by the creation of procedures in which all parties can formulate their own goals.

Hisschemöller and Hoppe argue that policy makers have the tendency to treat unstructured problems as more structured ones in order to gain more control over the policy process and outcome. In their diagnosis, 'intractable controversies occur when policymakers stubbornly continue to address the "wrong" policy problem'.[157] While they focus on the siting of hazardous waste facilities, biotechnology is another clear example of an unstructured problem; it should have become clear from my analysis that disagreement exists about facts and values, about definitions, problem framing, and possible solutions, and that many political actors are involved. According to Hisschemöller and Hoppe, policy makers often employ two strategies (or as they term it, 'biases') in order to treat an unstructured or moderately structured problem as structured and effectively depoliticise the conflict: Firstly, a 'division of labour' is made between experts and lay people, and policy elites and experts dominate the policy process, while the lay public is excluded. However, the authors point out that the distinction between experts and lay people in unstructured problems is an artificial one; no consensus even exists about who should be considered experts in the matter. When those labelled lay people reject this division of labour this strategy backfires resulting in even more conflict than policy makers were trying to avoid. Secondly, the policy problem is narrowly defined, for example as a simple problem of weighing costs and benefits, while broader, often value-related concerns are excluded. While this second strategy allows for broader public participation, 'discussants are bound by consensus about the policy goal orientation which prescribes what topics and interests are at stake'.[158] According to the authors, both strategies should be regarded as efforts to contain the policy problem, but 'when such "containment" efforts fail, problems become unstructured and controversies flare up'.[159] The exclusion of value-related concerns makes that citizens do not feel their viewpoints are taken seriously. This in turn leads to distrust of authorities and non-compliance or protest and this in turn leads to intractability.

The authors think that the solution can only be found in a public debate about the values involved. They argue that unstructured problems call for a so-called 'learning strategy', which involves broad public debate at an early stage, in which experts

---

[157] Ibid., page 42. Paula misleadingly terms moderately structured/means problems 'badly structured problems'. Lino Paula (2008). In Hisschemöller's terminology a problem turns into a badly structured problem when policymakers do not acknowledge that they are dealing with an unstructured policy problem, giving rise to type III mistakes, or solving the 'wrong' problem. See Hisschemöller (1993), page 27. I would like to thank Lonneke Poort for pointing this out to me.

[158] Hisschemöller and Hoppe (1995), page 46.

[159] Ibid., page 42.

2.8 Unstructured Problems

have a similar status as lay people, and which eventually feeds into political decision making.[160] They believe that the benefit of this approach is that misunderstandings about different parties' motives and beliefs are cleared up, and that this means that even if consensus is often unlikely, at least the policy problem is defined more realistically and will be perceived as more legitimate. In such a scenario, it is also acknowledged that lay citizens possess a special kind of experiential expertise that can supplement the knowledge of official experts.

When we apply Hisschemöller's distinction of problem types and look at the way governments deal with the biotechnology controversy at the moment, it appears that they treat it either as a moderately structured/ends or a moderately structured/means problem and they use both biases or strategies. However, while Hisschemöller and Hoppe match the 'division of labour-strategy' to moderately structured/means problems and the 'narrowing definition-strategy' to moderately structured/ends problems, it appears that in the biotechnology controversy at least governments have used the biases the other way around. In the moderately structured/ends case, it is assumed that the uncertainties are primarily scientific in character. For this reason, expert committees are installed so that the problem can be looked at and discussed by a variety of different experts with different scientific backgrounds. This could be regarded as an attempt to depoliticise the problem. In the moderately structured/means case, it is assumed that no consensus exists about values, but a solution can still be found using a procedural solution; it is assumed that not many scientific uncertainties exist. Often this is achieved by focussing on a single aspect of the problem or demarcating the problem narrowly. In fact, as will become clear in the empirical part of this book, in the case of consensus conferences both biases can be discerned, because while lay persons are involved, in the end the division of labour between experts and lay persons remains.

While Hisschemöller and Hoppe, then, define depoliticisation as the *exclusion* of lay perspectives, as we saw in the introduction, Pettit regards the *inclusion* of lay views as depoliticisation. Even though he does not specify any examples of depoliticised bodies, it appears to follow from his views that consensus conferences could also be regarded as such. Organising consensus conferences is in fact one way in which governments have sought to acknowledge dissensus about values. In Chapters 5 and 6, I will analyse each of these methods of dealing with the biotechnology controversy and I will ask to what extent they could be regarded as instances of depoliticisation and to what extent this is successful. When we supplement Hisschemöller's problem types with my positioning of committees and consensus conferences, the table looks as follows:

It is my contention that both methods underestimate the degree of intractability involved, because they fail to treat biotechnology as an unstructured problem. As will become clearer in Chapters 5 and 6, when the problem is delegated to expert committees, governments fail to acknowledge that different interpretations of the problem itself exist and that different stakeholders' values and worldviews are at

---

[160] Ibid.

Table 2.2 Role of committees and consensus conferences in problem types

	Normative certainty (normative consensus)	Normative uncertainty (no normative consensus)
Scientific certainty (consensus about facts)	*Structured problem* Can be solved by experts or bureaucrats	*Moderately structured problem (means)* **Consensus conferences** Narrowing definition-bias
Scientific uncertainty (no consensus about facts)	*Moderately structured problem (ends)* **Expert committees** Division of labour-bias	*Unstructured problem* Learning strategy/public debate

odds with each other; they fail to acknowledge, in other words, that the problem cannot only be solved by formal experts alone. When the problem is discussed in a consensus conference, even though lay people are involved, the strict distinction between lay people and experts is maintained and the status of expert knowledge is left unchallenged. Both liberal democrats and deliberative democrats also appear to treat biotechnology as a moderately structured/means problem, because they fail to acknowledge that no scientific consensus exists and to challenge the status of expert knowledge. If we take a look again at Table 2.2, it is my hypothesis that committees that move more to the right (accepting normative uncertainty and allowing for public input) and consensus conferences that move down (accepting scientific uncertainty) function better, because they treat the problem more appropriately. So expert committees try to solve scientific uncertainty and consensus conferences try to solve normative uncertainty, but each would function better the more they engage with the other type of uncertainty as well.

2.9 Conclusion

In this chapter, I have analysed the debate about the merits of the genetic modification of animals and plants. It has become clear that this debate is multi-dimensional and that different sources of disagreement underpin the pluralism in our society regarding novel technologies such as biotechnology. I have distinguished seven of these sources of disagreement and argued that they are related to each other in complex ways. Often disagreements that are ostensibly about pragmatic issues are in reality based on underlying interest-based, value, moral, or metaphysical disagreement. This is one reason why the conflict about biotechnology has become intractable. I have argued that in the biotechnology debate, particularly in the regulatory context, the emphasis is on pragmatic arguments, while the underlying reasons often remain unaddressed. This has led to a bias in favour of biotechnology advocates, who tend to use extrinsic arguments and tend to focus on only the factual or scientific aspects of biotechnology. I have, furthermore, argued that the first three sources of disagreement, namely definitional, factual, and scientific disagreement primarily belong to the realm of facts and science, while interest-based, value, moral, and metaphysical disagreement primarily belong to the realm of values and politics. However, it has also become clear that these two categories cannot be so neatly separated; in reality, fact and value are interwoven. For example, defining a problem in a certain way can serve a political function. The interplay between facts and values becomes especially clear when we look at the normative aspects of scientific research.

Science is not objective and value-free; in all stages of scientific research value judgments are made. When a decision is made about what phenomenon will be researched, values are involved already; why is this phenomenon researched and not something else? This choice can be influenced by interests of those that finance the research. The definition of the problem that is researched and the core concepts that are used in the research are also based on value judgments. In the biotechnology debate, we saw, for example, that core concepts such as sustainability, the precautionary principle, and animal welfare are all open for different interpretations. If we want to test the effects of genetic modification on animal welfare and we only consider animal welfare in the first sense of the word – as the absence of suffering – our research parameters will be different than if we also take welfare in the other senses into consideration. Next, the choice of what scientific discipline is appropriate to study a phenomenon includes a value judgment. For example, if we want to find out whether 'contamination' of GM-free by GM-crops occurs, it makes a difference whether we carry out ecological or molecular biological research. Furthermore, the specific research set-up is not value-free; are we going to test the environmental effects of GM-crops in a laboratory, on a small plot of land, or on a large plot? Also, if we want to determine the risks of consuming GM-food, do we measure this in terms of morbidity, mortality, or nuisance? The interpretation of test results is also based on value judgments. For example, a study that showed that the larvae of Monarch butterflies were killed after the consumption of GM-crops was interpreted in opposite ways, as either showing that the crops were dangerous or showing that they were not. Also, if epidemiological studies show no adverse affects of human consumption of GM-foods, should this be interpreted as proving that GM-foods are safe to eat or that the effects might not show up until decades later? The question as to how scientific knowledge should be applied is also based on the values and worldviews of the actors involved. Moral disagreement exists as to whether we *should* do everything that we *can* do. Finally, the way in which test results are made public is influenced by 'contextual values'; for example, a researcher may not be free to publish her test results if she is bound by patents or trade secrets. Also, if a researcher produces test results that could have serious social implications, does she have a duty to make these public?

In general, then, it could be said that our values and worldviews can influence our interpretation of the 'facts' and our judgment about what we ought to do. The biotechnology debate brings together many different fundamental questions and conflicts and is ultimately about the kind of world we want to live in. In this context, it is acknowledged more and more that expert knowledge is not purely objective, that it is not infallible, and that lay persons can contribute certain experiential knowledge that is often overlooked by scientific experts. For this reason, Hisschemöller and Hoppe argued that in the case of problems that they qualify as unstructured – of which biotechnology, in my view, is an example – we need to adopt a learning strategy; we need broad public debate at an early stage involving both lay persons and experts, who have a similar status. The debate needs to focus not only on solutions, but also on problem definition. These authors show that intractability can occur on a policy level when governments address the wrong policy problem. This happens when they treat unstructured problems as structured or moderately structured, by employing the strategies of depoliticisation by relegating the problem to the domain of experts, or of narrow problem demarcation. These strategies lead citizens to feel that their viewpoints are not taken seriously, which leads to distrust of authorities, non-compliance or protest, and ultimately to intractability. In this book, my question is not how we could put an end to intractable disagreements, but rather what would be the fairest way to deal with the inescapable reality of pluralism. My central question is, therefore, considering the fundamental differences of opinion about novel technologies, how should governments deal politically with the intractable disagreement following from this? My hypothesis is that this works best when the right policy problem is addressed; structured problems should be treated as structured ones and unstructured ones as unstructured. This means that expert committees will need to allow for more discussion on values and for more public input, while in consensus conferences the existence of scientific dissensus is acknowledged and the status of formal expert knowledge can be questioned. Let us now have take a closer look at how political theories have proposed to deal with pluralism.

References

Achterberg, Wouter (1994), *Samenleving, Natuur en Duurzaamheid (Society, Nature and Sustainability)* (Assen: Van Gorcum).

Attfield, Robin (1998), 'Intrinsic Value and Transgenic Animals', in Alan Holland and Andrew Johnson (eds.), *Animal Biotechnology and Ethics* (London: Chapman & Hall), 172–189.

Australian Financial Review (2002), 'Sticky Labels: The New Barrier to Trade and Aid', *Australian Financial Review*, November 2.

Balzer, Phillipp, Rippe, Klaus Peter, and Schaber, Peter (2000), 'Two Concepts of Dignity for Humans and Non-Human Organisms in the Context of Genetic Engineering', *Journal of Agricultural and Environmental Ethics*, 13, 7–27.

Barinaga, Marcia (2000), 'Asilomar Revisited: Lessons for Today', *Science*, 287 (5458), 1584.

Barry, John (2001), 'GM Food, Biotechnology, Risk and Democracy in the UK: A Sceptical Green Perspective', paper given at ECPR (European Consortium for Political Research), Joint Sessions, Grenoble.

Benbrook, Charles (2001), 'Do GM Crops Mean Less Pesticide Use?' *Pesticide Outlook*, October, 204–207.

Bharathan, Geeta et al. (2002), 'Crop Biotechnology and Developing Countries', in John Bryant, Linda Baggott la Velle, and John Searle (eds.), *Bioethics for Scientists* (Chichester: Wiley), 171–198.

Bout, Henriëtte (ed.) (1998), *Allemaal Klonen. Feiten, meningen en vragen over kloneren (All Clones. Facts, opinions, and questions about cloning)* (Amsterdam/Den Haag: Boom/Rathenau Instituut).

References 57

Bovenkerk, Bernice, Brom, Frans W.A., and Van den Bergh, Babs J. (2002), 'Brave New Birds. The Use of 'Animal Integrity' in Animal Ethics', *The Hastings Center Report*, 32 (1), 16–22.

Brom, Frans W. A. (1997), *Onherstelbaar verbeterd. Biotechnologie bij dieren als een moreel probleem (Irrepairibly Improved. Animal Biotechnology as a Moral Problem)*. (Utrecht: University of Utrecht).

Brom, Frans W. A. (2003), 'The Expressive-Communicative Function of Bio-Patent Legislation: The Need for Further Public Debate', in Christoph Baumgartner and Dietmar (Hrsg.) Mieth (eds.), *Patente am Leben? Ethische, rechtliche und politische Aspekte der Biopatentierung* (Paderborn, Germany: Mentis), 117–126.

Brom, Frans W.A. (2004a), 'WTO, Public Reason and Food. Public Reasoning in the 'Trade Conflict' on GM-Food', *Ethical Theory and Moral Practice*, 7 (4), 417–431.

Brookes, Martin (1998), 'Running Wild', *New Scientist*, 160 (2158), 38–41.

Brookes, Martin and Coghlan, Andy (1998), 'Live and Let Live', *New Scientist*, 160 (2158), 46–49.

Bruner, Michael and Oelschlaeger, Max (1994), 'Rhetoric, Environmentalism, and Environmental Ethics', *Environmental Ethics*, 16, 377–396.

Buiatti, Marcello (2005), 'Biologies, Agricultures, Biotechnologies', *Tailoring Biotechnologies*, 1 (2), 9–30.

Burkhardt, Jeffrey (2001), 'Agricultural Biotechnology and the Future Benefits Argument', *Journal of Agricultural and Environmental Ethics*, 14, 135–145.

Burkhardt, Jeffrey (2003), 'The Inevitability of Animal Biotechnology? Ethics and the Scientific Attitude', in Susan J. Armstrong and Richard D. Botzler (eds.), *The Animal Ethics Reader* (London/New York: Routledge), 332–341.

Busch, Lawrence, Grove-White, Robin, Jasanoff, Sheila, Winickoff, David, and Wynne, Brian (30 April 2004), 'Amicus Curiae Brief Submitted to the Dispute Settlement Panel of the World Trade Organization in the Case of EC: Measures Affecting the Approval and Marketing of Biotech Products'. http://www.lancs.ac.uk/fss/ieppp/wtoamicus/amicus.brief.wto.pdf

Callicott, J. Baird (1989), *In Defense of the Land Ethic: Essays in Environmental Philosophy* (Albany, NY: State University of New York Press).

Capron, Alexander M. and Schapiro, Renie (2001), 'Remember Asilomar? Reexamining Science's Ethical and Social Responsibility', *Perspectives in Biology and Medicine*, 44 (2), 162.

Carman, Judy (2004), 'Is GM Food Safe to Eat?' in Richard Hindmarsh and Geoffrey Lawrence (eds.), *Recoding Nature: Critical Perspectives on Genetic Engineering* (Sydney: UNSW Press), 82–93.

Chin, Geraldine (2000), 'The Role of Public Participation in the Genetically Modified Organisms Debate', *Environmental and Planning Law Journal*, 17 (6), 519.

Clark, Ann E. and Lehman, Hugh (2001), 'Assessment of GM Crops in Commercial Agriculture', *Journal of Agricultural and Environmental Ethics*, 14, 3–28.

Cohen, Phil (1998), 'Strange Fruit', *New Scientist*, 160 (2158), 42–45.

Comstock, Gary (2000), *Vexing Nature? On the Ethical Case Against Agricultural Biotechnology* (Boston/Dordrecht: Kluwer).

Conway, Gordon (2000), 'Crop Biotechnology: Benefits, Risks and Ownership', Speech delivered March 28, 2000 in Edinburgh, Scotland at: *GM Food Safety: Facts, Uncertainties, And Assessment*; The Organization for Economic Co-Operation and Development (OECD) Edinburgh Conference on the Scientific and Health Aspects of Genetically Modified Foods. http://www.agbioworld.org/biotech-info/articles/biotech-art/conwayspeech.html

Coulter, John (2000), 'Asilomar Revisited', *Science*, 287 (5462), 2421.

Crook, Stephen (2001), 'Risks, Regulations & Rhetorics', in Richard Hindmarsh and Geoffrey Lawrence (eds.), *Altered Genes II: The Future?* (Melbourne: Scribe Publications), 126–142.

Davison, Aidan, Barns, Ian, and Schibeci, Renato (1997), 'Problematic Publics: A Critical Review of Surveys of Public Attitudes to Biotechnology', *Science, Technology, & Human Values*, 22 (3), 317–348.

Dawkins, Richard (1976), *The Selfish Gene* (Oxford: Oxford University Press).

De Vries, Rob (2006), 'Genetic Engineering and the Integrity of Animals', *Journal of Agricultural and Environmental Ethics*, 19, 469–493.

Decima Research (1993), 'Final Report to the Canadian Institute of Biotechnology on Public Attitudes towards Biotechnology' (Ottawa: Canadian Institution of Biotechnology).

Deckers, Jan (2005), 'Are Scientists Right and Non-Scientists Wrong? Reflections on Discussions of GM', *Journal of Agricultural and Environmental Ethics*, 18, 451–478.

Des Jardins, Joseph R. (1993), *Environmental Ethics. An Introduction to Environmental Philosophy* (Belmont, CA: Wadsworth Publishing Company).

Dietrich, Heather and Schibeci, Renato (2003), 'Beyond Public Perceptions of Gene Technology: Community participation in public policy in Australia', *Public Understanding of Science*, 12, 381–401.

Eckersley, Robyn (1992), *Environmentalism and Political Theory. Toward an Ecocentric Approach* (New York: UCL Press).

Eisenberg, Rebecca (1996), 'Patents: Help or Hindrance to Technology Transfer?' in Frederick B. Rudolph and Larry V. McIntire (eds.), *Biotechnology. Science, Engineering, and Ethical Challenges for the 21st Century* (Washington, DC: Joseph Henry Press), 161–172.

Fedoroff, Nina and Brown, Nancy Marie (2004), *Mendel in the Kitchen. A Scientist's View of Genetically Modified Foods* (Washington, DC: Joseph Henry Press).

Frempong, Godfred (2006), 'Tailoring Biotechnology in Ghana: Implications for Genomics Development', *Tailoring Biotechnologies*, 2 (1), 51–62.

Gaskell, George, Bauer, Martin, and Durant, John (1998), 'Public Perceptions of Biotechnology in 1996: Eurobarometer 46.1', in John Durant, Martin Bauer, and George Gaskell (eds.), *Biotechnology in the Public Sphere: A European Sourcebook* (London: Science Museum), 189–214.

Goldhorn, Wolfgang (1990), 'The Welfare Implications of BST', in Peter Wheale and Ruth McNally (eds.), *The Bio-Revolution. Cornucopia or Pandora's Box?* (London: Pluto Press), 82–86.

Graham, Gordon (2002), *Genes. A Philosophical Enquiry* (London: Routledge).

Grommers, F.J., Rutgers, L.J.E., and Wijsmuller, J.M. (1995), 'Welzijn – Intrinsieke Waarde – Integriteit: Ontwikkeling in de Herwaardering van het Gedomesticeerde Dier' (Welfare – Intrinsic Value – Integrity: Developments in the revaluation of the domesticated animal). *Tijdschrift voor Diergeneeskunde,* 120, 490–494.

Grove-White, Robin and Szerszynski, Bronislaw (1992), 'Getting Behind Environmental Ethics', *Environmental Values*, 1 (4), 285–296.

Hindmarsh, Richard (1991), 'The Flawed "Sustainable" Promise of Genetic Engineering', *The Ecologist*, 21 (5), 196–205.

Hindmarsh, Richard (1994), 'Power Relations, Social-Ecocentrism, and Genetic Engineering: Agro-Biotechnology in the Australian Context', Doctoral Thesis (Griffith University).

Hindmarsh, Richard (2001), 'Constructing Bio-utopia: Laying Foundations Amidst Dissent', in Richard Hindmarsh and Geoffrey Lawrence (eds.), *Altered Genes II: The Future?* (Melbourne: Scribe Publications), 36–53.

Hisschemöller, Matthijs (1993), 'De Democratie van Problemen. De relatie tussen de inhoud van beleidsproblemen en methoden van politieke besluitvorming (The Democracy of Problems. The relationship between the content of policy problems and methods of political decision-making)', University of Amsterdam.

Hisschemöller, Matthijs and Hoppe, Rob (1995), 'Coping with Intractable Controversies: The case for problem structuring in policy design and analysis', *The International Journal of Knowledge Transfer and Utilization,* 8 (4), 40–60.

Ho, Mae-Wan (1998), *Genetic Engineering Dream or Nightmare? The Brave New World of Bad Science and Big Business* (Bath: Gateway Books).

Ho, Mae-Wan (1999), *Genetic Engineering: Dream or Nightmare?* (Second edn.; Dublin: Gateway), 385.

References

Hughes, Steve and Bryant, John (2002), 'GM Crops and Food: A Scientific Perspective', in John Bryant, Linda Baggott la Velle, and John Searle (eds.), *Bioethics for Scientists* (Chichester: Wiley), 115–140.

Jasanoff, Sheila (2001), 'Citizens at Risk: Reflections on the US and EU', in Matias Pasquali (ed.), *Third Congress of the European Society for Agricultural and Food Ethics. Food Safety, Food Quality and Food Ethics* (Florence, Italy: University of Milan), 43–49.

Jochemsen, H. (ed.) (2000), *Toetsen en Begrenzen. Een ethische en politieke beoordeling van de moderne biotechnologie (Testing and Limiting. An Ethical and Political Assessment of Modern Biotechnology)* (Amsterdam: ChristenUnie).

Johnson, Lawrence (1992), 'Toward the Moral Considerability of Species and Ecosystems', *Environmental Ethics*, 14 (2), 145–157.

Jongerden, Joost (2006), 'Luddites, or the Politics in Technology. An Introduction', *Tailoring Biotechnologies*, 2 (1), 63–68.

Kasanmoentalib, Soemini (1996), 'Science and Values in Risk Assessment: The Case of Deliberate Release of Genetically Engineered Organisms', *Journal of Agricultural and Environmental Ethics*, 9 (1), 42–60.

King, Robert C. and Stansfield, William D. (1997), *A Dictionary of Genetics* (Fifth edn.; New York/Oxford: Oxford University Press).

Kleinman, Daniel Lee and Kloppenburg, Jack, Jr. (1991), 'Aiming for the Discursive High Ground: Monsanto and the Biotechnology Controversy', *Sociological Forum*, 6 (3), 427–447.

Kloppenburg, Jack Ralph, Jr. (1988), *First the Seed. The Political Economy of Plant Biotechnology 1492–2000* (Cambridge: Cambridge University Press).

Kolata, Gina (1997), *Clone. The Road to Dolly and the Path Ahead* (London: Penguin Press).

Lawrence, Matthew (1996), 'Pluralism, Liberalism, and the Role of Overriding Values', *Pacific Philosophical Quarterly*, 77, 335–350.

Levidow, Les (2001), 'The GM Crops Debate: Utilitarian Bioethics?' *A Journal of Socialist Ecology*, 12, 44–55.

Levidow, Les and Carr, Susan (1997), 'How Biotechnology Regulation Sets a Risk/Ethics Boundary', *Agriculture and Human Values*, 14 (1), 29–43.

Lindhout, Pim and Danial, Daniel (2006), 'Participatory Genomics in Quinoa', *Tailoring Biotechnologies*, 2 (1), 31–50.

Love, Rosaleen (2001), 'Knowing Your Genes: Who Will Have the Last Laugh?' in Richard Hindmarsh and Geoffrey Lawrence (eds.), *Altered Genes II: The Future?* (Carlton North, VIC: Scribe Publications), 112–125.

Lund, Peter and Mepham, Ben (1998), 'Comments of the Food Ethics Council Submitted to the Nuffield Council on Bioethics Enquiry into Genetically Modified Crops: Social and Ethical Issues'.

Marris, Claire (2001), 'Public Views on GMOs: Deconstructing the Myths. Stakeholders in the GMO Debate Often Describe Public Opinion as Irrational. But Do They Really Understand the Public?' *EMBO Reports*, 2 (7), 545–548.

Mayer, Jorge E. (2005), 'The Golden Rice Controversy: Useless Science or Unfounded Criticism?' *BioScience*, 55 (9), 726–727.

Mayer, Sue (2002), 'Questioning GM Foods', in John Bryant, Linda Baggott la Velle, and John Searle (eds.), *Bioethics for Scientists* (Chichester: Wiley), 141–152.

McAfee, Kathleen (2003), 'Neoliberalism on the Molecular Scale. Economic and Genetic Reductionism in Biotechnology Battles', *Geoforum*, 34, 203–219.

McHughen, Alan (2000), *A Consumer's Guide to GM Food: From Green Genes to Red Herrings*, (Oxford: Oxford University Press).

Meyer, Gitte (2001), '*Fighting Poverty with Biotechnology? Report from a Copenhagen Workshop on Biotechnology and the Third World*' (Copenhagen: University of Copenhagen. Centre for Bioethics and Risk Assessment).

Miller, Henry I. (1999), 'Substantial Equivalence: Its Uses and Abuses', *Nature Biotechnology*, 17 (11), 1042–1043.

Millstone, Erik, Brunner, Eric, and Mayer, Sue (1999), 'Beyond 'Substantial Equivalence'', *Nature*, 401, 525–526.

Mørkeberg, Annette and Porter, John R. (2001), 'Organic Movement Reveals a Shift in the Social Position of Science', *Nature*, 412, 677.

Norton, Janet (1998), 'Throwing Up Concerns About Novel Foods', in Richard Hindmarsh, Geoffrey Lawrence, and Janet Norton (eds.), *Altered Genes. Reconstructing Nature: The Debate* (St. Leonards, NSW: Allen & Unwin), 173–185.

Norton, Janet, Lawrence, Geoffrey, and Wood, Graham (1998), 'Australian Public's Perception of Genetically-Engineered Foods', *Australasian Biotechnology*, 8 (3), 172–181.

Nuffield Council on Bioethics (1999), '*Genetically Modified Crops: The Ethical and Social Issues*' (London: Nuffield Council on Bioethics).

Nussbaum, Martha C. and Sunstein, Cass R. (eds.) (1998), *Clones and Clones. Facts and Fantasies About Human Cloning* (New York/London: W. W. Norton and Company).

Ortiz, Sarah Elizabeth Gavrell (2004), 'Beyond Welfare. Animal Integrity, Animal Dignity, and Genetic Engineering', *Ethics and the Environment*, 9 (1), 94–120.

Paula, Lino (2008), *Ethics Committees, Public Debate and Regulation: An Evaluation of Policy Instruments in Bioethics Governance* (Amsterdam: Free University).

Paula, Lino and Birrer, Frans (2006), 'Including Public Perspectives in Industrial Biotechnology and the Biobased Economy', *Journal of Agricultural and Environmental Ethics*, 19, 253–267.

Peacock, Jim (1994), 'Genetic Engineering of Crop Plants Will Enhance the Quality and Diversity of Foods', *Food Australia*, 46 (8), 379–381.

Reiss, Michael J. and Straughan, Roger (1996), *Improving Nature? The Science and Ethics of Genetic Engineering* (Cambridge: Cambridge University Press).

Rifkin, Jeremy (1998), *The Biotech Century. Harnessing the Gene and Remaking the World* (London: Victor Gollancz).

Rowland, Robyn (2001), 'The Quality-Control of Human Life: Masculine Science & Genetic Engineering', in Richard Hindmarsh and Geoffrey Lawrence (eds.), *Altered Genes II: The Future?* (Melbourne: Scribe Publications), 85–98.

Ruivenkamp, Guido and Jongerden, Joost (2006), 'Editorial: From Rejection and Resistance Towards Redesign', *Tailoring Biotechnologies*, 2 (1), 5–8.

Russo, Eugene (2000), 'Reconsidering Asilomar. Scientists See a Much More Complex Modern-Day Environment', *The Scientist*, 14 (7), 14.

Rutgers, L.J.E. and Heeger, F.R. (1999), 'Inherent Worth and Respect for Animal Integrity', in Marcel Dol et al. (eds.), *Recognizing the Intrinsic Value of Animals: Beyond Animal Welfare* (Assen: Van Gorcum), 41–51.

Sale, Kirkpatrick (2006), 'The Achievements of 'General Ludd'. A Brief History of the Luddites', *Tailoring Biotechnologies*, 2 (1), 69–78.

Scientific American, Editors (2004), 'The Green Gene Revolution', *Scientific American*, 291 (2), 8.

Scott, Maggie and Carr, Susan (2003), 'Cultural Theory and Plural Rationalities: Perspectives on GM among UK Scientists', *Innovation*, 16 (4), 349–368.

Secretariat of the Convention on Biological Diversity (2000), *Cartegena Protocol on Biosafety*, (Montreal, Canada), Article 3 (I).

Shaw, Alison (2002), 'It Just Goes Against the Grain. Public Understandings of Genetically Modified (GM) Food in the UK', *Public Understanding of Science*, 11, 273–291.

Shiva, Vandana (2000), *Stolen Harvest. The Hijacking of the Global Food Supply* (Cambridge, MA: South End Press), 127.

Singer, Peter (1975), *Animal Liberation: A New Ethics for Our Treatment of Animals* (New York: New York Review).

Sterckx, Sigrid (2004), 'Ethical Aspects of the Legal Regulation of Biodiversity'. Plenary Lecture at the Fifth Congress of the European Society for Agricultural and Food Ethics: Science, Ethics & Society (Leuven, Belgium).

Straughan, Roger (1992), 'Ethics, Morality and Crop Biotechnology' (UK: Reading University).

References 61

Taylor, Paul (1986), *Respect for Nature: A Theory of Environmental Ethics* (Princeton, NJ: Princeton University Press).

Thompson, Paul B. (1997a), 'Food Biotechnology's Challenge to Cultural Integrity and Individual Consent', *Hastings Center Report*, 27 (4), 34–38.

Thompson, Paul B. (1997b), 'Ethics and the Genetic Engineering of Food Animals', *The Journal of Agricultural and Environmental Ethics*, 10, 1–23.

Trewavas, A. (2002), 'Malthus Foiled Again and Again', *Nature*, 418, 668–670.

Trias, José E. (1996), 'Conflict of Interest in Basic Biomedical Research', in Frederick B. Rudolph and Larry V. McIntire (eds.), *Biotechnology. Science, Engineering, and Ethical Challenges for the 21st Century* (Washington, DC: Joseph Henry Press), 152–160.

Turner, Gillian and Wynne, Brian (1992), 'Risk Communication: A Literature Review and Some Implications for Biotechnology', in John Durant (ed.), *Biotechnology in Public: A Review of Recent Research* (London: Science Museum), 109–141.

United Nations (1992), 'Rio Declaration on Environment and Development', paper given at Conference on Environment and Development (UNCED) (Rio de Janeiro, Brazil, June 3–14 1992).

Van Staveren, G.D. (1991), *Het overschrijden van soortgrenzen. De rol van het soortbegrip binnen de discussie rond genetische manipulatie (Crossing of Species Barriers. The Role of the Notion of Species in the Discussion About Genetic Manipulation)*.

Van Willigenburg, Theo. (1997), 'Natuurberoep in de Bio-ethiek en Verantwoordelijkheid voor Handelen (The Appeal to Nature in Bio-ethics and the Responsibility for Action)' in Jozef Keulartz and Michiel Korthals (eds.), *Museum Aarde. Natuur: criterium of contructie?* (Meppel: Boom), 166–180.

Vasil, Indra K. (2003), 'The Science and Politics of Plant Biotechnology – A Personal Perspective', *Nature Biotechnology*, 21, 849–851.

Verhoog, Henk (1991), 'Genetische Manipulatie van Dieren (Genetic Manipulation of Animals)', *Filosofie in Praktijk*, 12 (2), 87–106.

Vint, Robert (2002), 'Force-Feeding the World. America's 'GM or Death' Ultimatum to Africa Reveals the Depravity of Its GM Marketing Policy', *Genetic Food Alert*, 1. http://www.ukabc.org/forcefeeding.pdf.

Wambugu, Florence (1999), 'Why Africa Needs Agricultural Biotech', *Nature*, 400, 15–16.

Wheale, Peter and McNally, Ruth (eds.) (1990), *The Bio Revolution. Cornucopia or Pandora's Box?* (London: Pluto Press).

Wills, Peter (2002), 'Biological Complexity and Genetic Engineering', *Environment, Community and Culture* (Brisbane).

Wynne, Brian (2001), 'Public lack of confidence in science? Have We Understood Its Causes Correctly?' in Matias Pasquali (ed.), *Third Congress of the European Society for Agricultural and Food Ethics. Food Safety, Food Safety, Food Quality and Food Ethics* (Florence, Italy: University of Milan), 103–106.

Xenotransplantation Working Party (2003), *Animal-to-Human Transplantation: A Guide for the Community. Public Consultation on Xenotransplantation 2003–2004* (Canberra: National Health and Medical Research Council).

Zoeteman, Bastiaan C.J., Berendsen, Miranda, and Kuyper, Pepijn (2005), *Biotechnologie en de Dialoog der Doven. Dertig jaar genetische modificatie in Nederland (Biotechnology and the Dialogue of the Deaf. Thirty years of genetic modification in the Netherlands)* (Bilthoven: Commissie Genetische Modificatie), 120.

Chapter 3
Constraining or Enabling Dialogue?

> *In the very act of sustaining diversity, liberal unity*
> *circumscribes diversity. It could not be otherwise. No form of*
> *social life can be perfectly or equally hospitable to every human*
> *orientation.*
> William A. Galston (1991), *Liberal Purposes*

3.1 Introduction

As the foregoing chapter has made clear, in the discussion about the merits of biotechnology we are faced with a diversity of opinions and attitudes, giving rise to persistent disagreements. Central to this debate are diverging fundamental ways of looking at the world and fundamental opinions about obligations between people and our obligations towards non-human entities. Many of the disagreements in this debate are based on differences in fundamental opinions about what constitutes 'the good life'. It should have become clear that the debate about biotechnology is a complex one in which multidimensional disagreements exist. The multidimensionality of these disagreements is exactly what makes them intractable. I acknowledge that my description of the biotechnology debate is not theory-neutral, but my aim is to show that this reading of it is a good way to qualify the debate as a debate involving fundamental differences in values and worldviews. My aim in this book is not to argue for the superiority of one particular value, worldview, or conception of the good life; I want to leave the question open whether there is in fact one such superior view. My concern here is that given the multidimensionality of views about novel technologies, governments are faced with intractable disagreement between their citizens and they have to make decisions whilst at the same time respecting their citizens.

The best answer to the question of how governments should deal politically with the existence of pluralism is widely thought to be liberalism. Indeed, liberalism is a political theory that is concerned with decision-making under conditions of pluralism. Liberals recognize that citizens have different value-hierarchies and that it is possible that several possible 'reasonable' answers can be given to one important question. Liberals hold that rather than imposing one particular value-ranking – for

---

B. Bovenkerk, *The Biotechnology Debate*, Library of Ethics and Applied Philosophy
29, DOI 10.1007/978-94-007-2691-8_3, © Springer Science+Business Media B.V. 2012

example that of the leading elite, or that of the major political party – on citizens, governments should remain neutral on matters of morality, unless they bear upon fundamental liberal rights. As will become clear in this chapter, according to some political liberals – in particular Bruce Ackerman, but also, for example, John Stuart Mill – the way to remain neutral is by constraining dialogue about those values. Another possible political solution to pluralism that has been gaining support in recent years in fact proposes the opposite strategy; rather than constraining dialogue about value conflicts, governments should enable it. This latter, deliberative, solution is the one I support and in this chapter I will argue why this should be so.

My line of reasoning follows two sets of arguments. First, I show that the two essential related distinctions that the liberal neutrality thesis is based on – namely the distinction between the public and the private spheres and between morality and ethics – are problematic. The liberal choice to limit politics to morality and the public sphere is misguided, because the requirement of conversational restraint that these distinctions support demands of people that they bracket precisely those views that are most important to them. While a basic premise in modern democracies is that governments should remain a certain level of neutrality, and I wholeheartedly support this premise, I argue that it should be understood differently than political liberals have; my criticism is, therefore, an immanent one. Second, I argue that liberals employ too static a view of opinion formation; they assume that in the public sphere citizens have fixed preferences, opinions, and interests that merely need to be weighed against each other rationally. This denies the dynamics of opinion formation that is at work in unstructured problems like the biotechnology controversy. The static view of opinion formation is reflected in the emphasis of actually existing liberal democracies on the social choice mechanisms of aggregation and bargaining. The deliberative emphasis on preference or opinion transformation is not only procedurally superior to aggregation and bargaining, but also offers more hope of reaching mutual understanding.

As I will argue in the next chapter, most theories of deliberative democracy have liberal underpinnings and deliberative democracy should therefore not be regarded as an alternative to liberal democracy. My main theoretical target here is political liberalism, which is a theory of the liberal democratic state. Deliberative democracy, in contrast, can be understood as a model of political communication and decision-making. Political liberalism makes certain normative assumptions that I challenge here and that are reflected in the models of political decision-making that most actually existing liberal democracies favour. My criticism on a more practical level, then, is directed towards the social choice mechanisms of aggregation and bargaining. While in a liberal democracy deliberation is also used as a social choice mechanism, including sometimes in Parliament, in a deliberative democracy there would be more emphasis on deliberation and it would substitute voting and bargaining in certain contexts. In other words, while my criticism is not directed at the liberal democratic state as such, it is directed at some of the normative assumptions of political liberalism, that in my opinion have led to an emphasis on voting and bargaining at the expense of opinion transformation through deliberation. If we would want to strengthen liberal democracy by making it more deliberative, this would

3.2 The Liberal Neutrality Thesis

require a rethinking of government neutrality through the public/private distinction and an alternative view of opinion (trans)formation.

Finally, following from my two sets of arguments, I propose three criteria for dealing with intractable disagreement, that will inform my investigation in the next chapter.

3.2 The Liberal Neutrality Thesis

From my description of the biotechnology debate it became clear that people disagree on many different dimensions, that they cannot always convince each other of their points of view, and that they use arguments that they themselves characterise as moral or ethical. Of course, the plurality of viewpoints that we encounter in the debate does not automatically entail that all positions are equally correct or that not one true or correct point of view exists. However, while each participant in the debate should strive to acquire the truth, or the best position, and should therefore assume that there is one right answer to the dilemmas in the debate, the acknowledgement that there are other viewpoints than one's own should at least entail that people are open towards the possibility that they could be mistaken. There is a difference between an opinion that is lived from the inside, so to speak, and one that is viewed from the outside; on the inside we make claims to having the right or true opinion, while at the outside we acknowledge that other people have different opinions and that these people may have good reasons for them. Moreover, we should make a distinction between truth questions in the existential sense and truth questions on the meta-level. People tend to be deeply convinced of the truth of their own worldview that answers fundamental existential questions about life, but different people have different worldviews and on the meta-level we do not have an external criterion to answer such questions.[1] It is not an exaggeration to assume that we may never be able to settle such fundamental truth questions once and for all and this is due to circumstances such as what Egbert Schroten terms the '*condition humaine*' and John Rawls 'the burdens of judgment'.[2] Humans have limited knowledge, different judgments, and various experiences and many problems are so complex that deliberation will never yield one unequivocal solution.

The lack of an external criterion of arbitration between different fundamental values or worldviews entails that there is no position external to the debate that governments could take in order to act in a justified manner. In a situation of intractable disagreement regarding an unstructured problem, even if a government would be convinced of the truth of a certain position, it should not simply press this view on its citizens, but should acknowledge that on the meta-level a diversity of opinions should be accepted. This is for several reasons. Firstly, it is out of respect for persons that we do not simply dismiss or overrule their opinions. Respect for persons

---

[1] Frans W.A. Brom (2004). Brom bases his account on an article by Egbert Schroten (1984).

[2] Schroten (1984) and Stephen Mulhall and Adam Swift (1996).

is a normative premise that I take as a given, because as has been argued elsewhere, each political theory in a democracy should adopt it.[3] Secondly, even if there were a unifying moral theory that could solve dilemmas involving values and the good life, no such theory exists in politics; for ethical-political debates such disagreements cannot be solved simply by declaring one moral theory as the truth. Under existing pluralism even those who are wholly convinced of their own viewpoints have to be willing to give arguments in a discussion with others. In this book I am looking for the political circumstances under which such a discussion could be held adequately in relation to questions involving intractable disagreement. I think that governments should not take one specific political view as superior in a discussion about greatly contested issues. As I will argue in what follows, political liberalism and the related liberal understanding of universalisability are not the only vantage points that should be accepted. I will argue that in liberalism many citizens' viewpoints are unjustly delegated to the private sphere, while they have universality pretensions, just like liberal views.

3.3 Public Versus Private

Given the differences of opinion regarding values, worldviews and the good life, how should we deal with this politically when we still want to live together in one society? The liberal solution to the problem of pluralism is – crudely stated – to argue that the state should remain neutral on those issues on which we cannot agree. According to John Rawls, in debates about political institutions and the basic rules of co-existence we should only allow participants to draw on those values that all free and equal citizens could reasonably endorse, and those values that they could not all endorse are off limits.[4] We could distinguish three levels on which political debate can take place: the level of specific policy questions, of decision-making procedures themselves, and on 'constitutional essentials' or 'basic questions of justice'.[5] While Rawls is only concerned with the last two, other liberals, in particular Ackerman, argue that neutrality requires that citizens also draw only on universalisable values when discussing specific policy questions (such as biotechnology regulation). As will become clear later, my criticism is primarily pointed at Ackerman and his requirement of conversational restraint. Nevertheless, the liberal solution that Ackerman takes to its most extreme depends on two related distinctions that are central to liberalism more generally; the distinction between morality and ethics and between the public and the private. In this section and the next, I will argue that these distinctions are problematic.

---

[3] See Will Kymlicka (2002) and Wibren Van der Burg (1991).

[4] This position is defended by liberal philosophers such as John Rawls, Bruce Ackerman, Ronald Dworkin and Robert Audi.

[5] The latter are Rawls' terms. See John Rawls (Spring 1987).

3.3 Public Versus Private

Since John Stuart Mill's case for liberty, liberals have been concerned to protect citizens' personal beliefs from state force. Based as it is on the idea of religious toleration, the thesis of state neutrality entails that the state should not interfere with its citizens' comprehensive doctrines, or in other words with their conceptions of the good. People's pursuit of the good life can only justifiably be limited by the state when it interferes with other people's ability to pursue their good life. This is derived from John Stuart Mill's harm principle.[6] As long as actions carried out by a person in the private realm do not harm anyone, the state should not interfere with these actions, even if other citizens find this person's actions morally reprehensible. The two paradigmatic cases for this position are religion and homosexuality. The state should not interfere in people's religious beliefs, because one person's privately held beliefs are of no consequences for another person's ability to hold her personal beliefs. And as long as homosexual sex takes place between consenting adults, and nobody gets hurt, it is not the state's business to interfere, even if the majority of a society were to oppose homosexuality on moral grounds.

The way in which state neutrality can be achieved, according to most liberals, is by only allowing reasons that can be publicly justified to play a role in political decisions – although as noted above, liberals diverge over whether this applies to constitutional essentials only or also to more general public policy issues. Only those reasons that all citizens could reasonably agree on are publicly justified. Citizens are not reasonable when they want to force their own comprehensive doctrines, such as their own religion, upon other citizens. The underlying reason for liberals to exclude comprehensive views, then, is that if state power is justified on the basis of comprehensive views that are not shared by all citizens, force is used towards a group of people that do not accept that state's legitimacy. The liberal concern with protecting minorities against arbitrary majoritarian rule becomes apparent here. State neutrality, as we have seen, according to many liberals, relies on allowing only arguments that do not refer to comprehensive doctrines to be put forward within the political realm, and it therefore relies on making a distinction between the public and the private spheres, or as Rawls prefers to term it, between public and non-public reasons.[7]

This distinction has come under increasing criticism, for example from communitarians and feminist thinkers. In fact, a critique of the public/private distinction (or in their case the public/domestic distinction) is a central tenet of feminist theory, symbolised by the slogan 'the personal is the political'. Feminists challenge this distinction for being exclusionary and for favouring the dominant class of human beings, men: 'liberalism constructs an opposition between public and private that has naturalized women's subordination and refused them the ostensibly

---

[6] David Bromwich and George Kateb (eds.) (2003). The Harm Principle states that the state is only justified in restricting a person's autonomy in order to prevent harm to others caused by that person.

[7] John Rawls (1999).

universal status of free and equal individual'.[8] This is because the public sphere has traditionally been associated with men, rationality, equality, and justice, whereas the private sphere has been associated with women, emotivism, subordination, and care. The separation of these domains meant that the principles that ruled in the public domain were withheld from the private domain, which, amongst other things, enabled the perpetuation of domestic violence. Feminists, therefore, argue that the state should interfere in the private sphere sometimes, for example to stop the abuse of wives and children. Also, they argue that what happens in the private sphere will influence someone's prospects in the public sphere and vice versa, and therefore they are not two distinct spheres at all.[9] In the words of Joan Landes: 'every action has a public and private consequence'.[10]

As Raymond Guess shows, the public/private dichotomy is not a pre-existing, objective distinction that determines when the state should or should not intervene in somebody's life.[11] Rather, what belongs to the public and what to the private sphere depends on what is valued and what is deemed in need of regulation or protection. We can, therefore, not simply argue that the state should not interfere with an action because it is private, because this will inevitably lead to the further question of why this action should be regarded as private in the first place. It appears, therefore, that it is not the public/private dichotomy per se that feminists object to, but rather the answers that have been given to the question of why certain things should be regarded as private and not others.

The public/private dichotomy as understood by feminists is in fact a different one than that championed by many liberals. In Rawls' eyes, in a pluralist society, there are three main areas of disagreement or conflict. The first is disagreement over comprehensive views, the second is a conflict based on unequal positions, such as class, race or gender positions, and the third derives from the burdens of judgment.[12] As touched on before, the burdens of judgment refer to limited human knowledge and sources of disagreement, such as the facts that evidence in a particular case can be weighed or interpreted in different ways or that the central concepts bearing on this case are vague.[13] Rawls' theory of political liberalism is primarily a response to disagreements of the first kind; he believes that if these can be dealt with – by distinguishing between the public and the private – conflicts of the second kind can

---

[8] Kate Nash (1998), page 33. These claims were originally made by Carole Pateman. See, for example, Carole Pateman (1988).

[9] In short, feminists have pointed out the connection between women's subordinate positions in the workforce and in the family, which reinforce each other. Power dynamics are influential in both domains. Also, the family is the place where early socialization takes place, so what happens within the family will have repercussions on people's behavior and attitudes in public life. See Susan Moller Okin (1989) and Kymlicka (2002).

[10] Joan B. Landes (1996), page 307.

[11] Raymond Guess (2001).

[12] Rawls (1999), page 612.

[13] Mulhall and Swift (1996).

3.3 Public Versus Private

also be resolved.[14] The feminist attack of the public/private distinction, on the other hand, focuses on the second area of conflict.

Because of Rawls' focus on the conflict between comprehensive views, he primarily distinguishes between public and private *reason*. According to Rawls, the way to deal with disagreement between different comprehensive doctrines is by getting citizens to find shared public reasons of a political conception of justice. His answer to the question as to how it is possible for people with irreconcilable, yet 'reasonable', views to live together in a just society is that 'political power is fully proper only when it is exercised in accordance with a constitution the essentials of which all citizens as free and equal may reasonably be expected to endorse in the light of principles and ideals acceptable to common human reason'.[15] Rawls, therefore, presumes that universal principles and ideals exist that should be shared by all reasonable people and that these are sufficiently substantive to ground a constitution. In his view, those reasons that reasonable people could not come to agree on in an 'overlapping consensus' are non-public or private reasons and should be left out of political debate about the basic structure of society.[16] If citizens were to rely on their comprehensive doctrines this would jeopardise the social and political stability of a society.[17] Rawls argues that state neutrality only applies to 'constitutional essentials and matters of basic justice'.[18] However, at other points he states that the idea that reasons that can be used in debate should be public, only applies to debate in the 'public political forum', which is broader than merely constitutional matters and matters of basic justice. This forum includes debate by judges, especially Supreme Court judges, debate between 'government officials' and between 'candidates for public office'.[19] Discourse in the background culture of civil society is not bound by the criterion of public reason. Moreover, Rawls allows non-public

---

[14] In light of feminist critiques emphasising the pervasiveness of gender inequality, one can wonder whether this belief of Rawls does not testify of a certain amount of naïveté.

[15] John Rawls (1996), page 137.

[16] This is explained in more detail in Rawls (1987). The idea behind the overlapping consensus is that citizens who disagree on comprehensive doctrines do not simply find a modus vivendi, but actually each support the institutional arrangements of their state on the basis of their own comprehensive doctrines. Rawls does not say that people should be coercively silenced when they want to bring in their comprehensive doctrines. However, in effect they are silenced when their views can never be reflected in the basic institutions of society.

[17] Fred D'Agostino and Gerald F. Gaus (1998).

[18] Rawls (1999), page 580.

[19] Rawls takes the Supreme Court to be the pre-eminent example of a sphere where only universally endorsable reasons should be used. This seems difficult to square with the fact that the position of Supreme Court judge is appointed by whichever president is in power at the moment of transition and that it is usually common knowledge what attitude the new judge holds towards contentious issues such as abortion.

reasons in the public political forum in some cases as well, provided that in due time they will be stated in terms of public reason.[20] Not all liberals make such allowances for non-public reasons.

3.4 Conversational Restraint

A liberal who, unlike Rawls, does want to constrain debate on matters that fall outside of the scope of constitutional essentials is Bruce Ackerman.[21] In his early work, particularly in 'Why Dialogue?', Ackerman argues that the fact that we live in a pluralistic society means that any decision that is made by the state that draws on ethical views (or as Rawls would say: on comprehensive doctrines) will oppress some members of society.[22] We need dialogue to find out what rules everyone can agree to live by, and not in order to convince each other of specific ethical views. Dialogue therefore only serves a pragmatic function: 'If you and I disagree about the moral truth, the only way we stand half a chance of solving our problems in coexistence in a way both of us find reasonable is by talking to one another about them'.[23] However, the only way in which we can successfully talk about our problems in coexistence is if we exercise 'conversational restraint', which means that we should

> 'simply say *nothing at all* about [disagreement about the moral truth] and put the moral ideals that divide us off the conversational agenda of the liberal state. In restraining ourselves in this way, we need not lose the chance to talk to one another about our deepest moral disagreements in countless other, more private contexts'.[24]

Ackerman, then, argues that the only way citizens can co-exist in a pluralistic society without oppressing any citizen's moral convictions is by ensuring state neutrality, and this is achieved by moving ideals that cannot be agreed on by everyone off the political agenda, into the private sphere. As Ackerman himself concedes, we can wonder whether the conditions of conversational restraint leave us very much to talk about in politics. Leaving values that people disagree about out of political debates could lead to an under-determination of political decisions.

Moreover, as I already pointed out, the question of what counts as a public and private issue is more problematic than it would seem at first sight. The distinction

---

[20] An example Rawls uses of a legitimate appeal to this 'proviso' is when civil rights leaders such as Martin Luther King use religious reasons to argue against discrimination, because they can use public reasons to support their arguments as well. See Rawls (1996). The reason why Rawls allows non-public reasons in this and similar cases, is that their use 'strengthens the ideal of public reason itself'. See Philip L. Quinn (1995), page 45.

[21] Another liberal that constrains all political issues in this way is Robert Audi. Robert Audi (2000).

[22] Bruce Ackerman (1989).

[23] Ibid., 16.

[24] Ibid., 16.

between public and private functions quite well in the domains for which it was first proposed, namely sexual conduct between consenting adults and religious freedom. In these domains people can hold conflicting values without compromising each other's capacity to hold their own values, or in other words, to lead their good lives. In these cases the state does not need to take action or make decisions that are mutually exclusive. Decriminalizing homosexuality has not led to a limitation on heterosexuals' opportunity to enjoy heterosexual sex, for example. There are many cases, however, where people's conceptions of a good life are incompatible and where one person's ability to live according to her conception of the good life interferes with that of another.

The public/private distinction becomes especially problematic for a government when state action is necessary beyond the granting of negative freedom, or refraining from interference. This is the case where public goods, such as natural or environmental goods, are involved. People attribute different, often mutually exclusive, values to nature, but decisions still have to be made on environmental policy. If one person holds that a certain nature area needs to be put aside as a wilderness reserve because the good life consists in finding a balance between humans and nature, while another person holds that this area should be 'developed' because the good life consists in meeting our basic needs for housing, a political decision will have to be made that is not neutral towards at least one of their conceptions of the good life. Decisions have to be made about how we deal with public goods, and these decisions cannot be left up to individuals. Any position that is taken on environmental policy necessarily includes value judgments. In summary, many issues over which people are divided can simply not be moved to the private sphere, because they involve public goods. The same holds for disagreement about biotechnology. What is ultimately at issue in the biotechnology debate is the organization of our surrounding biological reality, which is at the same time constitutive of our own identity. This is hardly a private issue.

Furthermore, even when the state merely refrains from interference, some citizens' values are given priority over those of others. Relegating an issue to the private sphere can mean that government has in fact already made a normative choice. Not taking a stand can effectively mean choosing one value over another. For example, if the government decides not to interfere with the biotechnology industry's decisions to plant genetically modified crops, it in fact penalizes farmers who try to keep their crops GM-free. In this context, several philosophers argue that bracketing values, like political liberals propose, can lead to a non-neutral 'liberal bias' in decision-making. As Amy Gutmann and Dennis Thompson point out, state inaction is not as neutral as it seems: 'the failure of the state to act can subject citizens to as much coercion and violation of their rights as a decision to act'.[25] This is illustrated well by the example of abortion: When a government leaves the decision whether or not to accept abortion up to individual women, it is effectively giving priority to women's right to choose and not to the fetus' 'right' to life. This refers to the

---

[25] Amy Gutmann and Dennis Thompson (1990), page 68.

familiar critique that liberalism cannot be as neutral as it purports to be. Thomas Nagel summarizes this critique well by stating that

> liberals ask of everyone a certain restraint in calling for the use of state power to further specific, controversial moral or religious conceptions – but the results of that restraint appear with suspicious frequency to favour precisely the controversial moral conceptions that liberals usually hold.[26]

Similarly, Bert van den Brink argues that in liberal democratic theory the problem of abortion is necessarily decided in favour of the pro choice stance, because liberalism has an inherent bias towards individual autonomy – and hence choice – and it is therefore not as neutral as it purports to be on questions of the good life.[27] Or, as communitarian Michael Sandel, points out, giving women freedom of choice in the matter of abortion in practice favours the idea that abortion is morally permissible over the idea that it is not: 'for defenders of abortion, little comparable is at stake [as for opponents]; there is little difference between believing that abortion is morally permissible and agreeing that, as a political matter, women should be free to decide the moral question for themselves'.[28] Against this critique it can be argued that even though liberal viewpoints and the condition of restraint often do lead to the same result, the justifications are different. In other words, liberal positions just happen to be favoured more often, because they tend to conform more to the idea of neutrality. Nevertheless, as I mentioned before, it is doubtful whether the question as to what issues should be 'bracketed' in political dialogue in order for a state to remain neutral, can be decided in a neutral way.[29] In the examples of religious freedom and homosexual conduct it is usually clear that the state can and should remain neutral by delegating decisions to the private sphere – although even here the boundary between the public and the private can be open for discussion, as is shown for religion in the case of the Israel-Palestine conflict and for homosexuality regarding the question whether or not homosexual couples should be allowed to adopt children. But the boundary between the private and the public becomes even more problematic in the examples of the nature lover and abortion. Whatever decision the state makes in these matters – even just the decision to do nothing – one conception of the good life will be favoured over another.

In sum, both Ackerman and Rawls argue that the state should not interfere in people's beliefs, or in their capacity to lead their lives according to their own comprehensive moral doctrines – as long as this does not harm others. Ackerman takes this idea much further than Rawls. For him, it means that the state's decisions should only be based on reasons that all could reasonably endorse and this entails that no

---

[26] Thomas Nagel (Summer 1987), page 216.

[27] Bert Van den Brink (2000).

[28] M.J. Sandel (1998a), page 20. The same holds for other issues that liberals do not want the state to inference in, such as pornography, homosexuality, or contraception. See Nagel (1987).

[29] This was argued by Guess (2001), and has also been defended by Michael J. Sandel (1998b).

appeals to comprehensive doctrines should be made in the public political arena. While my criticism of conversational restraint applies to a lesser degree to Rawls, because he argues that public reason should only be used for constitutional essentials and the 'public political forum', Rawls does rely on the distinction between the public and the private that I have called into doubt.[30] Ultimately, when decisions favouring certain policies are made, even if comprehensive notions of the good are not expressly excluded from political debate, within a liberal structure certain views will be favoured over others. This means that while participants in a debate regarding, for example, abortion may be free to bring up religious arguments, they will know beforehand that a 'neutral' liberal government will never adopt their viewpoints. The same could be argued for biotechnology. Extrinsic objections are often more in line with liberal views, because they are perceived to rely less on comprehensive notions of the good than intrinsic objections. While participants in a debate may be allowed to use intrinsic arguments, they will experience a liberal bias for extrinsic arguments when the government takes policy decisions, because liberal governments will seek to remain neutral between different comprehensive doctrines. However, in my eyes the state can remain neutral while still considering reasons drawn from comprehensive doctrines. Why this is so, will become clearer when we take a closer look at the second, related, distinction liberals draw on, that between ethics and morality.

3.5 Morality Versus Ethics

There are two main ways to distinguish ethics from morality. In the first, more traditional approach, morality refers to a system of moral norms, which tell us how we should act in order to bring about or express a value, which can be described as a state of affairs that is an end in itself – such as happiness or life. Ethics is the systematic or scholarly study of morality. However, in recent decades, in moral and political theory a second distinction between ethics and morality has been employed, which, stated simply, equates ethics with the particular and morality with the universal. Robert Piercey defines morality as,

> a peculiarly modern type of practical reasoning, one in which rights, universal duties, and categorical obligations are central. Ethics, by contrast, is an older and fuzzier-edged kind of practical thinking. It reflects on the good life more broadly, and it is intimately bound up with the values and self-understandings of concrete historical communities.[31]

---

[30] However, what remains unclear in Rawls' work is why comprehensive doctrines should not be relied on when it comes to constitutional essentials, but they should when it comes to particular policy decisions. In *The Idea of an Overlapping Consensus,* he argues that it would be unreasonable for people who draft the constitution to want to rely on their comprehensive doctrines because then they would use state power to correct those who disagree with them. However, why would these people not be unreasonable if they wanted to force their comprehensive views on others when specific public policy decisions are involved? Rawls (1987).

[31] Robert Piercey (2001), page 53.

According to this modernist conception of morality, there are universal goods: goods that every person can recognize as goods. As ethics, on the other hand, is not universal, but rather context-dependent, ethical values constitute people's private morality. One scholar who proposes to use this distinction is Jürgen Habermas, who reserves the term 'ethics' for 'notions of the good life' or 'substantive ideas of personal excellence'.[32] Every person gives shape to her own notion of the good life, and comprehensive notions are therefore not generalisable. Rather, they could be regarded as 'personal preferences'.[33] Morality, on the other hand, is comprised of reasoning based on moral categories that every rational person should accept.

This distinction between ethics and morality plays a role in the debate about the extent to which governments can, and should, remain neutral regarding normative matters. Liberals emphasise the universal character of morality, which forms the basis of, for example, universal human rights, while at the same time recognizing the plurality of conceptions of the good life that exist in society. These different conceptions of the good life are a matter of people's private morality that states should not interfere with. Communitarians, on the other hand, ascribe a more perfectionistic role to the state; the state should favour certain goods over others, and it can, and should, therefore, not remain neutral regarding conceptions of the good life either. Liberals, in turn, criticise communitarians for not recognizing ethical pluralism (i.e. diversity of conceptions of the good life that exist in society) and for not being able to guarantee minority rights.[34] Wibren van der Burg and Frans Brom argue, however, that 'active state neutrality' offers an intermediate position that recognizes the strengths of both theories. They distinguish between conceptions of the good, the good life, and the good society. Certain goods figure in a specific conception of the good life, but a conception of the good life is more than simply the sum of everything that someone regards as goods. Similarly, a good society is more than the sum of the good lives of its citizens. States cannot remain completely neutral regarding conceptions of the good society; they need to operate according to certain ideals, for example regarding justice. States cannot remain completely neutral regarding goods either. Rather, the question is regarding which goods it should and regarding which it should not remain neutral. Furthermore, if a liberal state values plurality, it should stimulate a wide range of possible ways of life for its citizens to choose from. This entails that states support certain ways of life, such as those of minority groups, in order to provide them with equal opportunities to live their good lives. This is what is meant by active, as opposed to passive, state neutrality.

---

[32] Van den Brink (2000), page 18. Van den Brink draws on the distinction between the moral and the ethical that Habermas develops in 'On the Pragmatic, the Ethical, and the Moral Employments of Practical Reason', in Jürgen Habermas (1993).

[33] Van den Brink (2000).

[34] See Wibren Van der Burg and Frans W.A. Brom (1997), page 53.

3.5 Morality Versus Ethics

Even though the idea of active state neutrality is not blind to the communitarian critique of liberalism, it is still primarily liberal in its orientation. It still rejects a communitarian reading of goods and the good life. As Piercey puts it:

> communitarians reject the liberal conception of the state in favour of one with a richer and more historically specific view of the good. Liberals we might say, want the state to be in the morality business, while communitarians want it to be in the ethics business.[35]

Rawls proposes a 'thin theory of the good', consisting of basic goods that are universally accepted as goods, and argues that the state should remain neutral regarding all other goods. This appears to mean that while states cannot remain neutral on morality, they should remain neutral regarding ethics. Non-basic goods are not universal, and therefore not generalisable, and they should remain in the private realm. The fact that in society we encounter a plurality of conceptions of these non-basic goods and of the good life is, in theory, not politically problematic for liberals, as these conceptions comprise the realm of ethics, and they are not properly the subject of public life. According to Van den Brink, for liberals,

> the *moral* is one while the *ethical* is many. And, at least for normative political theory, the moral must be more *real* than the ethical, because – so the assumption goes – we all share the same capacities for moral action, while we are deeply divided over ethical questions of personal excellence.[36]

This liberal view presupposes, erroneously I believe, that we can neatly separate our moral from our ethical views. Even though ethics and morality can be distinguished analytically, they are not so easily separated in practice, and I think that neither of the two should be accorded more priority than the other. Even though I agree with Rawls that there are basic goods, or shared values (in other words, the stuff of morality), I think these have to be abstracted from people's broader views about the good and the good life (the stuff of ethics), but they are not prior to them in the way liberals suggest. Rather, basic goods are embedded within our broader views and only get shape through them. In practice an interaction between the two always takes place, which makes it problematic to privilege one over the other. Both Rawls and Habermas, however, accord priority to morality over ethics, and thereby to the universal over the particular (or, in Rawls' terms, to the right over the good).[37] Piercey similarly argues that we cannot privilege morality over ethics or vice versa, because the two are not as opposed as is often made out; ethics and morality need each other in order to be what they are. Against Habermas he claims that 'there can be no procedure for testing moral norms that is completely separate from any shared ethical understanding ... Agents can be moral only if they have the good fortune of belonging to a certain kind of ethical community.'[38] We are educated as moral beings in a specific historic and cultural context and we could not participate in a

---

[35] Piercey (2001), page 54.

[36] Van den Brink (2000), page 19.

[37] See John Rawls (1988).

[38] Piercey (2001), page 65.

moral discourse such as Habermas envisages, if we had not learned certain virtues which enable us to test moral norms. These virtues are derived from a particular, not a universal, context.[39]

3.6 Broader Views

When we come together in public life in order to deliberate on important political decisions, we cannot easily separate universal, basic, goods from our context-dependent particularistic conceptions of the good. In many political debates involving moral issues we would be abstracting too much from the real issues that move participants in the debate if we would only allow them to speak about universal moral goods and not about wider ethical contexts. Framing debates only in terms of universally acknowledged goods does not seem to do justice to the way in which people reach moral viewpoints. As I illustrated in Chapter 2 in reference to biotechnology, moral viewpoints are often based on broader beliefs about what constitutes the good life and even on worldviews and metaphysical beliefs. Moving these off the conversational agenda is done at the risk of creating an artificial debate; in other words, if we can only speak in terms of universal goods that everyone can acknowledge, we are missing the essence of disagreement about many moral issues. Separating universal or shared values from so-called personal values discounts the way in which participants in a debate experience the intractable conflict they are in. Indeed, the fact that they experience the conflict as intractable is precisely due to the inseparability of their broader or more fundamental evaluative views from the more generalisable views they hold.

This point is demonstrated regarding the abortion debate by Simona Goi, who argues that people's positions on abortion cannot be understood apart from their wider metaphysical views about what constitutes human life and personhood, or from their views about, for example, the role of the family in society and the role of women in families. If we require these people to argue the issue only in generalisable terms and with reference to universal goods, we are avoiding the real issues. In Goi's words: 'a particular position on abortion cannot be considered apart from the broader set of assumptions that constitute an individual's or a group's world view...we cannot presume to frame the issue in terms that are abstract from the background beliefs of different groups'.[40] Constraining political deliberation by allowing participants to talk only about matters of morality, whilst leaving ethics out of the discussion – as liberals like Bruce Ackerman propose – seems to cut off

---

[39] On the other hand, against Williams, who proposes to privilege ethics over morality, Piercey argues that there are at least some abstract universal obligations (in other words morals) that ethics is subject to. Ethics and morality should not be regarded as opposed categories, but rather as different aspects of life's experience that we cannot disconnect in practice.

[40] Simona Goi (2005), page 65.

3.6 Broader Views

people's reasons from the grounds they have to advance those reasons in the first place. Van den Brink makes a similar point when he argues that the problem with reserving moral reasoning for public matters and ethical reasoning for private ones, is that

> 'it – perhaps unwillingly – presupposes a conception of personhood that borders on the schizophrenic. It expects citizens to be able to largely abstract from personal interests, attachments and purposes in their public lives (the life of autonomous, reasonable and generalisable reasons and actions), while it encourages these same citizens to find their personal fulfilment in substantive and possibly controversial notions of the good in their private lives'.[41]

He goes on to ask 'is it not true that we cannot act autonomously and reasonably without being motivated by personal (ethical) convictions?'.[42]

I argue that through deliberation we might be able to identify shared (though not necessarily universal) values that point to basic goods, but we cannot determine beforehand what these basic goods are and that they alone are open for political deliberation. In summary, I do not believe that citizens can, nor should, leave their comprehensive views of the good life behind when they come together in public life to engage in political deliberations. Broader moral views should be allowed in deliberations in all spheres of civil society and government. State neutrality does not necessarily need to be guaranteed by distinguishing morality from ethics and subsequently moving ethics off the political agenda. State neutrality could also be defined, actively, as providing every citizen, no matter what conceptions of the good life she holds, with the equal opportunity of taking part in public debate.

According to my first set of arguments to favour the deliberative over the political liberal solution, no neutral way exists to determine what should belong to the public and private spheres. In most conflicts, particularly in unstructured ones like the biotechnology debate, state non-interference is not as neutral as it appears. Moreover, in practice we cannot disconnect our ethical and moral convictions and relegate the ethical ones to the private sphere. We would not be doing justice to the importance that people attach to their own views and opinions if we allowed them to only draw on values that could be universally endorsed. I concluded that state neutrality can still be maintained even if comprehensive doctrines enter political debate, if neutrality is defined actively, as providing citizens with equal opportunities to participate in public debate. In other words, rather than constraining debate on comprehensive doctrines, governments should enable debate about conflicts that are based on disagreements stemming from these doctrines. Now I will turn to my second set of arguments.

---

[41] Van den Brink (2000).

[42] Ibid.

3.7 Opinion Transformation

As we have seen, the main reason why liberals want to restrain the use of non-public reasons is that they think that the exercise of political power by the state is only legitimate when it is based on reasons that could be agreed to by everyone. According to Stefan Grotefeld, the underlying motivation for this aim is that according to liberals this is the only way that citizens can show respect for each other as free and equal persons.[43] Grotefeld argues, however, that for citizens to show each other such respect, they need not refrain from referring to their comprehensive doctrines. In his eyes, the liberal condition of exclusion of comprehensive views is based on a narrow view of normative argumentation, which presupposes that citizens have a more or less stable and unified point of view. For if two citizens each have a stable, unchangeable point of view, there seems to be no point for them to provide each other with insight into their views. In that case, they have no hope of ever convincing each other of the correctness of their views and they might as well leave their comprehensive views out of public debate. Grotefeld follows Michael DePaul, who argues that this narrow view of moral argumentation runs along the following lines: every person has a set of more or less stable beliefs, which she draws on when providing others with reasons for a certain position. When she tries to convince someone else of the correctness of her position, she connects to the types of inference that that person accepts. Additionally, she can provide new facts which will change that person's position or she can show that there is an inconsistency within that person's set of convictions. This entails that all changes that one brings about in somebody's opinions are based on a relatively settled set of convictions. However, DePaul argues, correctly in my view, that other legitimate forms of moral argumentation are possible.[44] For example, when an environmentalist's view of the good life consists of a life led in harmony with the species around us and she therefore favours the protection of a nature area that others – whose main concern is with economic growth – would like to see turned into a shopping mall, she can try to argue her case by drawing on her comprehensive views. She could explain what it means to her to live in harmony with nature, for example that it means not giving automatic priority to human interests. By articulating her comprehensive views she is trying to make her opponents understand her reasons for her position and how her different beliefs relate to each other. Even though this might not happen often, it is certainly possible that she convinces other people of her point of view, by exposing them to alternatives to their own values and worldviews. After all, we do at times revise part of our set of convictions after reading or hearing about alternative views or ways of living. We do not only change our convictions after hearing generalised arguments that we already accepted. Moreover, this example again points out that this woman would be abstracting too much from the real content of her arguments

---

[43] Stefan Grotefeld (2000).

[44] Michael R. Depaul (1998).

if she could only use arguments that she already knows are acceptable to her opponents. She would then not be advancing her own point of view, but some abstract generalised point of view.

The woman in our example is appealing to her opponents' reason and there is nothing in this course of action that fails to treat them as free and equal persons. After all, others would also be allowed to appeal to their own comprehensive views. A liberal could object that, of course, simply offering somebody else one's own comprehensive views does not fail to respect others as free and equal, but rather the state would not be respecting them if it would enforce decisions based on one group's comprehensive views and not those of another group. However, if the state has given everyone the equal opportunity to advance their own views, without itself taking a predetermined stand on the issue, we could still say that each citizen has been respected as free and equal. The main concern of this approach for liberals remains the protection of a minority against majority views. However, as I have argued before, on public issues decisions will have to be made and it is not altogether evident that a system that excludes comprehensive reasons succeeds better in protecting minorities than a system that allows them; perhaps even the opposite is true – provided that certain background conditions are met. Considering that the question of what properly belongs to the private and public spheres cannot be settled in advance, it makes more sense to allow comprehensive views in public debate, so that it can be established through this debate what views are and are not allowed. Moreover, if participants in a debate are allowed to explain the broader picture to explain where their viewpoints are coming from, there is more chance that participants will reach mutual understanding. After all, how can we achieve mutual understanding if we do not get to properly explain our views?[45] Furthermore, the static view of people's opinions denies the nature of citizens as independently thinking persons who consider and test arguments before they draw their own conclusions; ironically, it is precisely this independently thinking character of persons that liberals wish to respect.

3.8 Aggregation and Bargaining

The idea underlying the alternative style of moral argumentation proposed by DePaul is that the exposure to other views can influence our own set of convictions; in other words, opinion transformation can take place through the confrontation with viewpoints other than our own. The liberal idea that citizens have stable and unified points of view from which they abstract the reasons they bring into public debate

---

[45] As Mulhall and Swift argue, Rawls takes the burdens of judgment that lead to reasonable disagreement so seriously that he seems to overlook the possibility that reasonable *agreement* can in fact sometimes be reached on conceptions of the good Mulhall and Swift (1996). If it were indeed the case that people could never reach agreement on conceptions of the good, then there would be little motivation to discuss conceptions of the good in public.

does not account for the fact that many people's views are actually quite fragmented and it discounts the potential of opinion transformation. The 'static' view of citizens' opinions has been criticised from a green political theory perspective, because it discounts the influence of social structures in personal opinion (and preference) formation.[46] Political liberals seem to assume that the moral views that people profess in the public sphere are fixed and crystallised opinions and that they have stable interests that merely need to be weighed against each other rationally. This denies the dynamic character of opinion formation regarding unstructured problems such as the biotechnology controversy. In the public arena we need to choose a method that makes it possible for people to engage in social learning and further develop their views, and as I will argue now this is best done by using the mechanism of deliberation.

The liberal static view of citizens' opinions is reflected in actually existing liberal democracies' most common mechanisms to make decisions on regulation in the face of disagreement, namely preference aggregation (or voting) and bargaining. These traditional mechanisms for social choice have come under attack in recent years. Social choice theory demonstrates that the aggregation of preferences through voting is inherently problematic, because when there are more than two options to choose from, there is no clear and non-arbitrary method for aggregating preferences in a way that does justice to the voters' positions.[47] This is, among other reasons, because 'voting cycles' can occur, where option A is favoured over option B, option B over C, but option C over A. Therefore, 'different mechanisms [such as the majority rule or positional rules] will always yield different collective choices from identical distributions of individual preferences'.[48] An ancillary problem with the simple aggregation of individual preferences is that it leaves no room for alternative policies or compromises, as voters can only choose between a predetermined or untested menu of options.

A social choice mechanism that does leave room for compromise is that of bargaining, or 'partisan mutual adjustment'.[49] Traditional liberal pluralistic theory regards political decision-making as a bargaining process between different interest groups that all lobby governments to favour their interests. This model, which is most clearly exercised in American politics, takes all political preferences as expressions of interests. One problem with this assumption is that it treats 'even matters of

---

[46] See Robyn Eckersley: In liberalism 'all citizens/consumers are considered equally free and unencumbered agents and therefore equally capable of making independent choices, that all individuals are antecedently individuated or fully formed prior to making choices, that such choices should be accepted at face value.' Behind this critique lies the familiar communitarian criticism of the liberal unencumbered self. Robyn Eckersley (2004), page 97.

[47] David Miller (1993). Another problem with voting is 'strategic voting, which means misrepresenting your true preferences when you vote with the aim of increasing the chances of your favoured option'. Miller argues on page 79 that voters would be less likely to do this after they have met each other in a process of deliberation.

[48] Dryzek (2000), page 35.

[49] This term originates with Charles E. Lindblom (1965).

principle as though they were conflicts of interest'.[50] This means that fundamental moral views are on a par with material and selfish interests and that the strength of people's convictions does not seem to matter. The degree of a person's convictions rather seems to be measured by how much that person is willing to invest in the issue. This, in turn, is dependent on how much a person can invest in general, in other words, on the person's material resources, and on a person's leverage power, which is dependent on their status in society. According to critics like William Connolly, reaching social choices through bargaining favours powerful elites and disregards the preferences of large segments of society.[51] Moreover, in the case of a persistent or even intractable conflict, bargaining does not do justice to the stakes that are involved. Bargaining on such issues, especially when the conflict is traceable to fundamental value or metaphysical disagreements, would require one to treat one's fundamental convictions as something negotiable. However, if people could do this, there would be no fundamental conflict in the first place.[52] According to deliberative democrats, the process of deliberation is a fairer and more reliable method for reaching social choice, because it offers the possibility of preference or opinion transformation. As the exchange of ideas prior to voting can help to clarify and narrow down the set of options to vote for, deliberation as a social choice mechanism avoids problems like voting cycles. The process of deliberation also meets the mentioned objections against the bargaining model, as it is more inclusive and participatory.

3.9 Deliberation

So far, I have argued that the liberal neutrality thesis is misguided for several reasons. Firstly, it relies on two problematic distinctions, namely between the public and the private spheres and morality and ethics. I have argued that while these distinctions can serve useful heuristic purposes they cannot play the role they have been ascribed by liberals. There is no neutral a priori way of deciding what should belong to the public and what to the private sphere. Moreover, morality and ethics are two inseparable sides of the same coin and we should not give priority to one over the other. The conversational restraint that is based on these distinctions is neither desirable nor necessary in order for governments to retain a neutral position, nor for citizens to show each other respect as free and equal persons. Conversational restraint does insufficient justice to the importance people attach to their own opinions and the way these were formed. This becomes especially pressing when we are dealing with unstructured problems, including controversy about novel technologies, because opinions about these are still in the process of being articulated. A second problem with conversational restraint as a neutrality device is, therefore,

---

[50] David Schlosberg (1999), page 6. Schlosberg quotes Robert Wolff (1965), page 21.

[51] William E. Connolly (1969), page 16.

[52] James Bohman (1996).

that it precludes the possibility of preference or opinion transformation, which is particularly problematic for problems that are still in a dynamic process. In this section I want to elaborate this suggestion by arguing, firstly, that the liberal requirement of conversational restraint gives rise to some problems that could be solved by deliberation and, secondly, that there is a strong link between pluralism and public deliberation.

One problem of conversational restraint is that by not speaking about what divides us we do not learn about other people's points of view and we can only assume that they may have good reasons for their positions, but that we do not share them. We do not have to engage with our opponents' viewpoints, because we will not likely come to agree anyway. This amounts to an attitude of mere toleration of each other's viewpoints and misses the opportunity to build mutual respect. Toleration in the sense I use the term here is something akin to indifference and betrays the attitude that we do not care what another person thinks or believes. Showing respect, on the other hand, involves a recognition that other people might have their own good reasons for their views. In practice, the social choice mechanism of voting also fails to bring mutual understanding, because citizens are not required to understand and identify with their fellow citizens' points of view; something they would have to do in a public deliberation. The simple aggregation of preferences avoids the level at which conflicts actually take place – often the level of values and worldviews – and this may make conflicts even more intractable.

Furthermore, through the requirement of conversational restraint liberals lose the opportunity to identify wrong positions. While it is often not possible to identify the one right solution, it is still possible to identify wrong solutions. In a deliberation it can, for example, be shown that certain arguments are not coherent and that, therefore, certain positions are wrong. The decision to effectively move issues about which disagreement exists off the political agenda precludes the possibility of determining which positions fall in the range of acceptable options, because it cannot distinguish right from wrong positions. Even though by focusing on *reasonable* pluralism liberals claim to distinguish right from wrong solutions to moral dilemmas, their decision to move issues about which moral disagreement exists off the political agenda means that it has to be possible at the outset to establish which positions are reasonable and which are not. It appears that a position is reasonable if it does not conflict with the liberal view of the right, but this solution has a distinct unneutral ring to it.

I have one last, positive, reason for responding to pluralism with deliberation. One way to reach agreement in the face of pluralism is to find values that are shared by the different sides to a disagreement, and work from these. This will not necessarily lead to agreement, but as Gutmann and Thompson argue, at least 'deliberation can clarify the nature of a moral conflict, helping to distinguish among the moral, the amoral, and the immoral, and between compatible and incompatible values'.[53]

---

[53] Amy Gutmann and Dennis Thompson (1996), page 43.

3.9 Deliberation

Deliberation can, in other words, clear up the source of disagreements. Participants in deliberation can, for example, discover whether a disagreement is purely based on an absence of information or a misinterpretation of facts, and therefore is purely a factual disagreement or whether their disagreement is in fact based on value or even metaphysical differences. Deliberation will also help them to understand exactly what values or worldviews they disagree about. Moreover, deliberation might help to find alternative courses of action.

My criticism of conversational restraint and my view that broader issues should be allowed in public deliberation raises the question whether just any views should be allowed in public debate. In Chapter 4 I will devote more attention to this question, but some preliminary remarks are in order here. Firstly, views that hinder public debate itself – such as views that do not consider all participants as free and equal, or those views that would condone the use of threats or violence against other participants – should not be allowed. Secondly, I pointed out above that the paradigm cases of state neutrality concerned those practices or beliefs that can be relegated to the private sphere because they do not interfere with other people's practices or beliefs – religion and sexual conduct between consenting adults being the primary examples. I argued that not many practices or beliefs actually constitute true private cases, because most issues about which moral disagreement exists concern public goods on which state regulation is necessary. In my view those issues that do not need a public resolution – namely those practices and beliefs that do not interfere with other people's practices and beliefs – do not belong to the realm of political debate. Of course, in line with my earlier points about the normative aspects of the terms 'public' and 'private', I do not regard these two categories as either objective or static; rather, the question as to what constitutes public and private matters, again, should be open for public debate. It follows that I do not want to argue that we need political deliberation about the question what comprehensive doctrine, for example what religion, is the correct one. This is not necessary, because citizens can profess their faith without compromising other people's opportunity to profess theirs. I therefore agree with Bruce Ackerman that we should not use public deliberation in order to determine which comprehensive doctrine, such as a religion, is true. Even though citizens can certainly have debate about this question, it is irrelevant to politics, because one person's opportunity to believe in a certain religion does not interfere with that of another, and, therefore, the state does not have to proscribe (nor can it logically proscribe) what beliefs people should have.[54] However, that does not exclude the option of bringing certain religious beliefs into a discussion about something which *is* deemed a public issue.

---

[54] Of course, this is open for debate in cases of religious conflict such as the aforementioned case of Israel and Palestine, but ideally in a deliberative democracy citizens would have no reason or motive to discuss the validity of specific religions in a political context.

3.10 Criteria for Dealing with Intractable Disagreement

In the next chapter, I will elaborate on deliberative democracy and address the question of how we could deal with persistent, intractable, disagreement through public deliberation. But here I must address one issue that precedes this question and that relates to the central question of this book: How should we deal with intractable disagreement in the face of pluralism? From my foregoing account it becomes clear that when we are dealing with intractable problems such as the biotechnology controversy we are dealing with issues that move people deeply and that cannot be simply relegated to the private sphere. We are, furthermore, dealing with many opinions that are still being developed. In order to be able to deal with these in the political arena we need public deliberation that fulfils three criteria. These criteria can also guide us in the next chapter.

- Firstly, the *criterion of non-exclusion* states that no reasonable views should be excluded from public deliberation. More views should be included in the range of 'reasonable' views than traditional liberals allow when they are using this term. My arguments against conversational restraint support this criterion. Moreover, this criterion is based on the respect that governments owe their citizens and this includes respect for their opinions and their intellectual capacities. The confrontation with others and their diverging points of view is necessary to stimulate these intellectual capacities. Furthermore, given the fact that in the real world we are confronted by certain limitations, such as the *condition humaine* or the burdens of reason, it is always possible that certain entrenched beliefs or views turn out to be mistaken. In such a world it is of utmost importance that there are different lines of thought that can be tested against one another. Even if engaging with views opposed to our own does not change our views, at least it helps us clarify and strengthen them. Finally, when we are dealing with new issues about which opinion formation is dynamic it is premature to exclude any arguments at the start; we need all arguments to come to light in order to aid opinion (trans)formation.
- Secondly, the *criterion of absence of power differences* states that in public deliberation decisions should not be forced by powerful groups upon people with less power. This criterion is based on the critiques of voting and bargaining. If we want to stimulate preference and opinion transformation the quality of arguments should be the decisive factor and not external power, as is the case with bargaining or majority power. What we are after is not equal *vote* – like in the mechanism of aggregation – but equal *voice*. In order to achieve equal voice it is important that everyone gets an equal opportunity to take part in the deliberations that are part of the decision-making process. This is only possible when power balances are avoided as much as possible.
- Thirdly, the criterion of *open-endedness* states that the outcomes of a public deliberation should in principle be revisable. This criterion follows from the view that opinion transformation should be possible before decisions are taken. Again, in the real world we are faced with limitations – on human knowledge, interpretation, and the complexity of our problems – and this entails that we are

likely to make wrong decisions at times. Decisions should, therefore, be revisable when new information or arguments come to light. Moreover, if we want to show respect to all persons and their opinions we should keep the possibility open that any group's views may at some point be favoured. In order to give those whose preference was not realised in a specific decision – those who lost the conflict, in other words – the opportunity to revise decisions in future these should in principle be open-ended. As James Bohman argues, deliberation is fair if it makes 'possible continued participation of all groups in a common deliberative framework'.[55] One reason why people in intractable conflicts sometimes resort to violence (for example, killing abortion doctors) or sabotage (for example, destroying test fields of GM crops or liberating test animals) is that they feel their views are sidelined and not taken seriously, and therefore, they have no other means available to be heard. If deliberation is inclusive and open-ended the disaffected will have more reason to accept decisions that they do not agree with as nevertheless legitimate. Of course, the possibility of revision does not necessarily lead to acceptance when people's preferences have not been fulfilled, but they may at least come to understand that they will still have a reasonable chance in the future to influence decisions. I will put these criteria to practice in the last part of the next chapter, when I unfold what version of deliberative democracy I support.

3.11 Conclusion

In this chapter, I have argued that the possibility of meaningful discussion in the face of intractable disagreement implies that it is not necessary to privatise moral conflicts, as political liberals demand. The latter – albeit not all to the same degree – argue that we should only have political debate about what procedures we should employ to ensure people's co-existence in society, and these procedures should be neutral regarding specific conceptions of the good life. Disagreements based on comprehensive notions of the good life should therefore be moved off the political agenda and should only be discussed in private. Only moral viewpoints that are generalisable belong in the public sphere. I have shown that doubt can be cast on the distinction between such moral viewpoints and notions of the good life; rather, these are interconnected. We cannot expect citizens to leave their comprehensive notions of the good life behind when entering the public sphere, because their moral viewpoints rely on these. Conversational restraint demands of people to hold back exactly on those issues that they deem very important. Moreover, where public goods, such as natural or environmental goods, are involved, the state cannot simply refrain from interference, because many conflicts about how to deal with public goods are based on the kind of disagreements about values and worldviews which liberals would

---

[55] Bohman (1996), page 96.

like to relegate to the private sphere. While collective decisions have to be made regarding public goods, citizens would not want their government to simply follow the majority view, because it is always possible that the majority view is mistaken. Dialogue is important in this context, because it brings to light all the different perspectives on a problem and it can provide governments with the arguments that they have to weigh against one another before making a decision.

If the government would simply relegate an issue to the private sphere, and therefore not take a stance, this would effectively mean choosing one value over the other, as the case of abortion shows. The problems with the public/private distinction reveal the failure of liberalism to remain completely neutral; it has an inherent bias towards certain values, such as individual autonomy. It appears, then, that liberalism as it stands now is not neutral enough to accept all relevant positions in a specific problem area. My hypothesis is that, in theory at least, deliberative democracy conforms better to liberalism's own core ideals of liberty and equality in so far as it avoids the inequality in decision-making power that we encounter in bargaining and in so far as it does not exclude groups on the basis of views that are part of their comprehensive notions. Deliberative democracy still favours state neutrality, albeit in a more active sense. In my view, deliberative democracy can be more sensitive to some of the nonneutral consequences of liberalism, especially because it allows people to draw on their comprehensive notions.

I have argued that the political liberal view of what properly belongs to the domain of politics precludes too many issues about which moral disagreement exists. The question as to which issues should be discussed in the public sphere is a topic of discussion in that public sphere itself. Moreover, public deliberation offers a way to find shared values or at least clarify the nature of a moral conflict, or find alternatives. Public deliberation, therefore, provides a better response to the existence of pluralism than voting or bargaining. The idea of public deliberation is central to the theory of deliberative democracy, which is the topic of the next chapter.

References

Ackerman, Bruce (1989), 'Why Dialogue?' *The Journal of Philosophy*, 86 (1), 5–22.

Audi, Robert (2000), *Religious Commitment and Secular Reason* (Cambridge: Cambridge University Press).

Bohman, James (1996), *Public Deliberation: Pluralism, Complexity and Democracy* (Cambridge, MA: MIT Press).

Brom, Frans W.A. (2004), 'Bio-ethiek in een Pluralistische Samenleving. (Bioethics in a Pluralistic Society)', in Theo Boer (ed.), *Ethiek in de Maak. Voordrachten bij het afscheid van Egbert Schroten (Ethics in the Making. Lectures for Egbert Schroten's Farewell)* (Utrecht: Faculteit Godgeleerdheid Universiteit Utrecht (Theological Faculty Utrecht University)), 24–35.

Bromwich, David and Kateb, George (eds.) (2003), *John Stuart Mill: On Liberty* (Rethinking the Western Tradition) (Binghamton, NY: Vail-Ballou Press).

Connolly, William E. (1969), 'The Challenge to Pluralist Theory', in William E. Connolly (ed.), *The Bias of Pluralism* (New York: Atherton Press), 3–34.

References 87

D'Agostino, Fred and Gaus, Gerald F. (1998), 'Introduction. Public Reason: Why, What and Can (and Should) It Be?' in Fred D'Agostino and Gerald F. Gaus (eds.), *Public Reason* (The International Research Library of Philosophy) (Aldershot: Dartmouth Publishing Company), xi–xxiii.

DePaul, Michael R. (1998), 'Liberal Exclusions and Foundationalism', *Ethical Theory and Moral Practice*, 1, 103–120.

Dryzek, John S. (2000), *Deliberative Democracy and Beyond: Liberals, Critics, Contestations* (New York: Oxford University Press).

Eckersley, Robyn (2004), *The Green State: Rethinking Democracy and Sovereignty* (Cambridge, MA: MIT Press).

Galston, William A. (1991), *Liberal Purposes. Goods, Virtues, and Diversity in the Liberal State* (Cambridge: Cambridge University Press).

Goi, Simona (2005), 'Agonism, Deliberation, and the Politics of Abortion', *Polity*, 37 (1), 54–81.

Grotefeld, Stefan (2000), 'Self-Restraint and the Principle of Consent: Some Considerations on the Liberal Conception of Political Legitimacy', *Ethical Theory and Moral Practice*, 3, 77–92.

Guess, Raymond (2001), *Public Goods, Private Goods* (Princeton, NJ: Princeton University Press).

Gutmann, Amy and Thompson, Dennis (1990), 'Moral Conflict and Political Consensus', *Ethics*, 101, 64–88.

Gutmann, Amy and Thompson, Dennis (1996), *Democracy and Disagreement* (Cambridge, MA: Harvard University Press).

Habermas, Jürgen (1993), *Justification and Application: Remarks on Discourse Ethics*, trans. Ciaran Cronin (Cambridge: Polity Press).

Kymlicka, Will (2002), *Contemporary Political Philosophy* (Oxford: Oxford University Press).

Landes, Joan B. (1996), 'The Performance of Citizenship: Democracy, Gender, and Difference in the French Revolution', in Seyla Benhabib (ed.), *Democracy and Difference. Contesting the Boundaries of the Political* (Princeton, NJ: Princeton University Press), 295–313.

Lindblom, Charles E. (1965), *The Intelligence of Democracy: Decision-Making Through Mutual Adjustment* (New York: Free Press).

Miller, David (1993), 'Deliberative Democracy and Social Choice', in David Held (ed.), *Prospects for Democracy. North, South, East, West* (Cambridge: Polity Press), 74–92.

Mulhall, Stephen and Swift, Adam (1996), *Liberals and Communitarians* (Second edn.; Malden, MA: Blackwell).

Nagel, Thomas (1987), 'Moral Conflict and Political Legitimacy', *Philosophy and Public Affairs*, 16 (3), 215–240.

Nash, Kate (1998), *Universal Difference. Feminism and the Liberal Undecidability of 'Women'* (London: MacMillan).

Okin, Susan Moller (1989), *Justice, Gender, and the Family* (New York: Basic Books).

Pateman, Carole (1988), *The Sexual Contract* (Cambridge: Polity Press).

Piercey, Robert (2001), 'Not Choosing Between Morality and Ethics', *The Philosophical Forum*, XXXII (1), 53–72.

Quinn, Philip L. (1995), 'Political Liberalisms and Their Exclusions of the Religious', *Proceedings and Addresses of the American Philosophical Association*, 69 (2), 35–56.

Rawls, John (1987), 'The Idea of Overlapping Consensus', *Oxford Journal of Legal Studies*, 7 (1), 1–25.

Rawls, John (1988), 'The Priority of the Right and Ideas of the Good', *Philosophy and Public Affairs*, 17 (4), 251–276.

Rawls, John (1996), *Political Liberalism* (New York: Columbia University Press).

Rawls, John (1999), 'The Idea of Public Reason Revisited', in Samuel Freeman (ed.), *Collected Papers* (Cambridge: Harvard University Press), 573–615.

Sandel, Michael J. (1998a), *Liberalism and the Limits of Justice* (Second edn.; Cambridge: Cambridge University Press).

Sandel, Michael J. (1998b), *Debating Democracy's Discontent. Essays on American Politics, Law and Public Policy* (Cambridge, MA: Belknap Press/Harvard University Press).

Schlosberg, David (1999), *Environmental Justice and the New Pluralism: The Challenge of Difference for Environmentalism* (Oxford, NY: Oxford University Press).

Schroten, Egbert (1984), 'De "Waarheidsvraag" in de *Theologica Religionum* (The "Truth question" in the *Theologica Religionum*)', *Nederlands Theologisch Tijdschrift (Dutch Theological Journal)*, 38 (4), 310–327.

Van den Brink, Bert (2000), *The Tragedy of Liberalism. An Alternative Defense of a Political Tradition* (Albany, NY: State University of New York Press).

Van der Burg, Wibren (1991), *Het Democratisch Perspectief. Een verkenning van de normatieve grondslagen der democratie (The Democratic Perspective. An Exploration of the Normative Foundations of Democracy)* (Arnhem: Gouda Quint).

Van der Burg, W. and Brom, F.W.A. (1997), 'Eine Verteidigung der Staatlichen Neutralität (A Defense of State Neutrality)', in K.P. Rippe (ed.), *Angewandte Ethik in der Pluralistischen Gesellschaft* (Freiburg: Freiburger Universitätsverlag), 2, 53–82.

Wolff, Robert (1965), 'Beyond Tolerance', in R. Wolff, B. Moore, and H. Marcuse (eds.), *A Critique of Pure Tolerance* (Boston, MA: Beacon Press), 3–52.

Part II
Theoretical Framework

Chapter 4
Deliberative Democracy and Its Limits

> Bernard (Minister's Personal Assistant): "What's wrong with open government? Why shouldn't the public know more about what's going on?".
> Arnold (Prime Minister's Permanent Secretary): "Are you serious?".
> Bernard: "Well....yes sir. It is the Minister's policy after all".
> Arnold: "But that is a contradiction in terms. You can be open or you can have government".
> Bernard: "But surely, the citizens of a democracy have a right to know".
> Humphrey (Permanent Secretary Sir Humphrey Appleby): "No, they have a right to be ignorant. Knowledge only means complicity and guilt. Ignorance has a certain dignity".
> Bernard: "But if the Minister wants open government...".
> Humphrey: "You don't just give people what they want if it is not good for them. You don't give brandy to an alcoholic".
> Arnold: "If people don't know what you're doing, they don't know what you're doing wrong".
> Humphrey: "Can you keep a secret?".
> Bernard: "Yes".
> Humphrey: "So can I".
> Yes Minister (BBC, February 25, 1980): Open Government

> A well-functioning system of democracy rests not on preferences but on reasons.
> Cass Sunstein, 1997, 'Deliberation, democracy and disagreement'[1]

4.1 Introduction

As I argued in the previous chapter, I share liberal democrat Bruce Ackerman's diagnosis of *why* we need dialogue: the only hope people who disagree about the moral truth have to solve their problems of co-existence is to talk about

---

[1] Cass Sunstein (1997).

them.[2] However, I do not share Ackerman's vision of the *how* of dialogue: conversational restraint about the content of our moral disagreements. In contrast, I argued that public deliberation about broad moral views provides a better way of dealing with intractable disagreement than constraining debate about 'comprehensive views'. In fact, we cannot separate our comprehensive views from our political views. In order to test my views in practice, I will shortly make a comparative analysis of two practices in two countries in which public deliberation has been used to reflect on animal and plant biotechnology. But first I need to examine the case for public deliberation, provided by the theory of deliberative democracy.[3] After all, deliberative democracy – sometimes also termed discursive or communicative democracy – is the theory that centers on the ideal of preference or opinion transformation through public deliberation. Before I test the merits of public deliberation in practice, I need to examine what problems have been encountered in the theoretical analysis of deliberative democracy and that is the focus of this chapter. This chapter, in other words, aims to give a political philosophical answer to the question of 'how dialogue?'. Under what conditions can we deal with intractable moral disagreement through public debate, and what obstacles may we encounter?

I start out by providing a framework for this chapter by firstly, introducing three core questions that I think make 'dealing with intractable disagreement' operational. These questions are (1) what should be the goal of deliberation? (2) what type of argument should we deem valid in the debate? and (3) who should participate in the decision-making process? These three questions will structure the remainder of the chapter. Secondly, I introduce the central elements of theories of deliberative democracy by looking at both the practical and the theoretical reasons that have been given for the call for more public deliberation. These reasons reflect the normative assumptions that underlie deliberative democracy. I explain that they demonstrate both an ideal of quality of argument and of inclusiveness and I argue that a tension exists between these ideals. Sometimes one of the two ideals has to yield in order to achieve the other. How one makes the choice between these two ideals has repercussions for the main question of this chapter, namely how public deliberation could deal with intractable moral disagreement. This tension can, then, also be discerned within the three mentioned core questions.

I furthermore elaborate how certain leading proponents of deliberative democracy have dealt with intractable disagreement by investigating how they answer my three core questions. It will become clear that for some deliberative democrats an important goal of deliberation is reaching consensus. This goal, again, forms

---

[2] See Ackerman (1989).

[3] As is pointed out by several deliberative democrats, the idea of deliberation is not new in politics, but goes as far back as the time of Aristotle. According to James Bohman, 'although the idea can be traced to Dewey and Arendt and then further back to Rousseau and even Aristotle, in its recent incarnation the term stems from Joseph Bessette, who explicitly coined it to oppose the elitist or "aristocratic" interpretation of the American Constitution'. See Bohman (1998), page 400.

a tension with the other two goals of quality and inclusiveness. The tensions I describe also motivate some of the critiques that have been mounted against deliberative democracy, in particular those put forward by liberal democrats, difference democrats, and agonistic democrats. I consider (possible) responses to these critiques and elaborate what version of deliberative democracy I support with regards to the three core questions.

4.2 Framework

The main question of this chapter is how deliberative democracy could deal with the existence of intractable disagreement, particularly regarding novel technologies. I will shortly introduce the central elements of deliberative theories of democracy, but first I need to clarify what I understand by 'dealing with intractable disagreement'. In order to make dealing with intractable disagreement through public debate operational, I think three core questions need to be answered: Firstly, what is the goal of deliberation for different proponents of deliberative democracy? Secondly, what type of arguments do they deem valid in the deliberation? And thirdly, who do they think should participate?

4.2.1 Three Core Questions

Why are these three core questions important for the way we deal with intractable disagreement? First of all, 'dealing with' can mean a variety of things. It can mean 'resolving' the conflict, for example by reaching agreement or consensus. It can also mean 'managing' the conflict, for example by redefining the problem area, moving the contentious issues off the agenda, or by agreeing on a neutral decision procedure such as voting, or letting an authority, such as an expert committee, decide. 'Dealing with' can even mean letting the conflict persist. Letting the conflict persist does not have to mean that nothing is won by debate; we can still learn from the conflict, so that even if not all agree on the decision made it is at least a better informed decision. Moreover, when the conflict persists, through debate at least bad decisions or decisions based on purely self-interested motives can be avoided. Also, misunderstandings about the other parties' motives and beliefs can be cleared up.[4] Disagreement is often regarded as something negative, a state we all want to avoid, particularly in a situation where regulatory decisions have to be made.[5] However, it can also be regarded as a catalyst for progress. Without conflict and disagreement there is no change; change is often preceded by a period of disagreement and discontent. 'Dealing with' in this context means creating

---

[4] Hisschemöller and Hoppe (1995).

[5] This is also pointed out by Diana Mutz, who shows that people avoid 'talking politics' with those with whom they know to disagree, in order to avoid conflict. Mutz (2006).

space for disagreement to exist and even flourish, while avoiding possible negative side effects of disagreement, such as violence or the oppression of views. While dealing with intractable disagreement does not require resolving the disagreement, nor even finding shared views – albeit this would be desirable – at a minimum, then, it means finding peaceful ways of avoiding violence. Deliberative democrats argue that dealing with intractable disagreement can best be done through reasoned debate. However, which of the above ways they interpret as best dealing with intractable disagreement has repercussions for the kind of deliberative democracy they envisage. Their interpretation becomes clear when we ask what, in their eyes, is the goal of deliberative democracy.

Secondly, the question of what type of argument we deem valid in the debate is important when we try to deal with intractable disagreement. On a practical level we can argue, with Hisschemöller and Hoppe, that when broader, value-related concerns are excluded from debate, citizens do not feel their viewpoints are taken seriously and this can lead to intractability. In my analysis of the biotechnology debate it became clear that the emphasis of the debate was often on certain quantifiable arguments, while the more fundamental, often value-related arguments, that lie behind these arguments were disregarded. It is my contention that this has led the opposite positions in the debate to become more fixed and inflexible and that it has not led to a greater understanding of different parties' viewpoints and motives. In the previous chapter I argued on a theoretical level as well that broader views should be included in public debate. However, among deliberative democrats different views exist about exactly how far we should expand the scope of arguments into public deliberation in order to still be able to deal with intractable disagreement. Allowing just any utterance in a public debate may make it unworkable or may even increase the intractability of the problem.

Thirdly, a corollary of the question of *how* to deal with intractable disagreement is the question of *who* should deal with the disagreement. Lay persons can contribute their own valuable insight into policy problems. Moreover, when we are dealing with unstructured policy problems no single viewpoint is necessarily superior. Involving a broader group of people in policy decisions can mean different things, however, and the question remains exactly what groups should be included. Should only stakeholders be included or only representatives of certain groups? Should groups who show no interest in public participation nonetheless be stimulated to partake in public debates? The answer to these questions will have repercussions on the capability to deal with intractable disagreement. These three core questions will be guiding in my quest for the 'how of dialogue?' and will structure the three main sections of this chapter. But in order to understand how deliberative democrats have answered these questions, I first need to introduce deliberative democracy more generally and I will do so by describing the reasons deliberative democrats have for proposing public deliberation. As deliberative democracy in fact refers to a group of different but related theories I want to stress that in this book I can only offer a sketch of the 'dominant' account of deliberative democracy.

4.2 Framework

4.2.2 Practical Reasons for Public Deliberation

Public participation in decision-making about emerging biotechnologies has become more and more popular. Several practical circumstances for the call for public participation can be noted. First of all, members of the public are becoming better informed and more highly educated, creating a wish to become involved in political and social issues. At the same time, expert knowledge has become increasingly questioned in the wake of what many see as failures of the technological paradigm, resulting in environmental degradation on a scale never experienced before, and an increased perception of the risks we are exposed to. As most risks are largely invisible and it is often hard to establish the cause of environmental harms, the public relies on expert knowledge in order to discover risks. At the same time, the public is confronted with conflicting accounts of risk assessment by different experts and is aware that some predictions by experts have turned out to be false. Conflicting expert accounts have drawn attention to the fact that science is not value-free and objective.[6] In risk assessment, value choices are made in the problem definition, the research methods, and the interpretation of outcomes. In this context, it is acknowledged more and more that expert knowledge is not purely objective, that it is not infallible, and that lay persons can contribute certain experiential knowledge that is often overlooked by scientific experts.

When we are dealing with GMOs we are not only dealing with risks in the sense of known or quantifiable unknowns, but also with unknown unknowns. In other words, it is such a new technology that we do not even know what possible risks to look for. When doing field trials to establish environmental effects of GM crops, unpredictable consequences are out of necessity left out. Science can only deal with *anticipated* uncertainties and out of necessity discounts unanticipated ones.[7] Scientific risk analysis cannot incorporate such variables and therefore tends to ignore them, while for the public their possibility are an important reason to call for caution. For example, it has been lay people who have drawn attention to the possibility of unanticipated consequences of growing genetically modified crops. As Brian Wynne argues, the value of lay experiential input lies not just in the ability to supplement expert knowledge with factual knowledge, but also in the fact that it tends to be critical of institutional structures and to reflect on broader issues than merely technological ones; in effect, lay expertise can play a role in challenging the validity of present expert knowledge.[8]

Furthermore, public acceptance of gene technology has proven to be influenced, amongst other things, by the amount of trust that citizens have in the institutions that govern the use of biotechnology.[9] Simply providing the public with information, for example about the risks (or lack thereof) of consuming genetically modified food, is

---

[6] Paul Slovic (1999).

[7] Wynne (2001).

[8] Brian Wynne (2003).

[9] Paula and Birrer (2006).

not enough to build trust in the institutions regulating biotechnology. After all, how one interprets this information is already dependent on the level of trust one has in these institutions in the first place. Franck Meijboom et al. convincingly argue that trust is not simply a characteristic of individual citizens or consumers, but that trust is established in a relationship, and government institutions, therefore, need to show they are trustworthy.[10] This calls for what is referred to as 'two-way transformative learning'; rather than a cognitive deficit model where public participation is simply a different word for top-down 'education' of the public, two-way transformative learning assumes that something can be learned in both directions; opinion transformation can take place on all sides of a dialogue between government, stakeholders, and the public.[11]

To sum up, the circumstances of declined trust in experts and governments and the increased perception of risk by the public, together with the acknowledgment that scientific knowledge is not objective and value-neutral, have led to an ever greater call for public involvement in decision-making regarding issues such as biotechnology, in which uncertainty is high and agreement is low. In short, the social position of science is changing and scientific research has to be considered in a broader social context.[12] Increased interest in scientific matters and the acknowledgment of social and normative aspects of science have laid bare fundamental disagreements about how we should deal with issues such as novel technologies.

4.2.3 Theoretical Reasons for Public Deliberation

Deliberative democrats also have more theoretical reasons to propose public deliberation in the context of disagreement. As a general definition, we could say that deliberative democrats hold that a decision has been reached legitimately if it was the result of a procedure in which all those potentially affected by the decision have had a free, equal, and fair chance to influence the outcome through rational deliberation.[13] An important aspect of democracy for deliberative democrats is, then, that all persons should be treated as free and as equal and this amounts to an equal opportunity to participate in the drafting of decisions that will influence their lives.[14] Some deliberative democrats simply pose the principles of freedom and equality because they regard them as universal rights – something they share with traditional liberal democrats. Others argue that these principles are internal to the ideal

---

[10] Franck Meijboom, Tatjana Visak, and Frans W.A. Brom (2006).

[11] Fern Wickson (2006).

[12] Mørkeberg and Porter (2001).

[13] Habermas, for example, states that 'the only regulations and ways of acting that can claim legitimacy are those to which all who are possibly affected could assent in rational discourse'. Jürgen Habermas (1996), page 458.

[14] See Joshua Cohen (1996), page 96.

4.2 Framework

of deliberation itself; participating in 'communicative action', as Habermas puts it, already presupposes an ascent to certain 'rules of discourse'. According to these rules, everyone who is able to do so is allowed to pose arguments in public and question the arguments of others without being coerced.[15]

This definition also entails the view that state power is only used legitimately on citizens when it is based on rules that these citizens could in principle accept.[16] For deliberative democrats this means that a fair procedure has been used to reach decisions and they hold that deliberation is a fairer and better procedure of social choice than the mere aggregation of preferences or 'partisan mutual adjustment' (bargaining). The former tends to treat all opinions as static preferences, excluding the possibility of preference transformation. Deliberative democrats point out that social choice cannot simply be based on the aggregation of preferences, as this assumes that people's preferences are given and coherently ordered, whereas in reality on complex matters people have no such coherent preferences – we in fact often find conflicting views and opinions within one and the same person. Deliberation helps people order their own preferences.[17] Moreover, as we saw in Chapter 3, the mechanism of aggregation runs into problems such as voting cycles. Even though deliberative democrats allow voting[18] when deliberation has failed to produce agreement, the problems inherent in this mechanism are less insurmountable after opinions have been exchanged in deliberation – resulting in an exposure of self-interested arguments, fallacies and misinformation – and after the possibility of preference or opinion transformation.[19]

The procedure of bargaining, as we have already seen in Chapter 3, is deemed problematic, because it gives some people better chances of influencing decisions than others, as a result of inequalities in status and power, and it therefore stretches the norm of equality. Deliberative democrats assume that through public deliberation the influence of power and wealth on political decisions should be decreased.[20] This view is at the core of Habermas' directive which has by now become the mantra of deliberative democracy, that decisions should be based on 'the forceless force of the better argument' rather than on logrolling or bargaining. One important aspect of letting the better argument decide is that people

---

[15] Habermas (1996), page 89.

[16] This is also the basis of the liberal idea of legitimate government based on the consent of the governed; however, in that case a more mediated form of consent is denoted.

[17] Seyla Benhabib (1996).

[18] It should be noted that the mechanism of voting does have inherent value as well and should not only be regarded as a last resort. Thomas Christiano argues that the right to vote is in fact conducive to public deliberation: 'it is because citizens have voting power that they have reason to contribute to public deliberation. A citizen's incentive to listen to another's opinion with which he disagrees strongly diminishes when that other has no power'. Thomas Christiano (1997), page 251.

[19] See Miller (1993). The view behind this is that through deliberation it becomes clear where people's priorities lie and a set of alternative courses of action can be drafted that can prevent voting cycles.

[20] Gutmann and Thompson (1996), page 133.

should be prepared to give reasons for their views publicly. These reasons should be public in two respects: they have to be stated in public, before an audience, rather than be part of deliberation in private, and they should be public in the sense that they should be intelligible for everyone. The latter, of course, does not mean that everyone has to agree with these reasons, but rather that the reasons should be accessible to the scrutiny of others and therefore should be contestable. This entails that points of view have to be put into generalisable terms.[21] The first respect of publicity ensures that decisions based purely on self-interest or unequal power are avoided and the second ensures that no appeals are made to statements that do not properly count as arguments, for example because they are based on divine revelation, that cannot be rationally contested.

Some deliberative democrats, who often base themselves on John Stuart Mill and Hannah Arendt, argue that deliberation is not only valuable instrumentally, because it leads to better decisions, nor that it is only valuable because it is fair from a procedural point of view, but also that participation in a public deliberation in and of itself is valuable for individuals. It is taken to have a beneficial educative effect on them; it enhances their reasoning skills and promotes their moral qualities, because they are confronted with the viewpoints of other participants in the debate.[22] Because participants in the debate have to give reasons for their opinions in public they are forced to identify with others; they have to anticipate other people's opinions so that they can give reasons that other people could in principle accept. Deliberative democrats theorise that this leads to 'social learning' and leads people to adopt a broader vantage point than just their own. This is usually not an independent defence of deliberative democracy, as the educative effects should be seen as a favourable side-effect of deliberation and not the point of starting the deliberation in the first place. Participants in a debate do not deliberate in public merely for the sake of debating – save perhaps in the case of debating competitions – but because they want to reach a decision, want to confront others with their point of view, or want information about possible alternative policy-options so that they can form their own viewpoint. That it is actually the case that public deliberation has beneficial educative effects on participants is often assumed, but needs to be further explored in practice. It has been argued that no independent standard exists that can tell us whether this effect has occurred,[23] but of course it is still possible to ask participants themselves whether they feel they and their co-participants have enhanced their reasoning skills and broadened their moral horizon. This may be a subjective rather than an objective standard, but it is still an independent one. One beneficial effect of 'hearing the other side' does appear to be supported by empirical research. According to Diana Mutz, so-called 'cross-cutting exposure', particularly in those with the capacity to take other people's perspectives, leads to a higher level of tolerance: 'the capacity to see that there is more than one side to an issue, that a political

---

[21] Eckersley (2000).

[22] Maeve Cooke (2000).

[23] Ibid.

4.2 Framework

conflict is, in fact, a *legitimate* controversy with rationales on both sides, translates to greater willingness to extend civil liberties to even those groups whose political views one dislikes a great deal'.[24]

Taking part in deliberations about policy decisions, according to deliberative democrats, does not only have an educative effect, but also leads participants to value civic participation more. When people take part in the drafting of policies they may feel more involved with the social world around them, they may feel like their opinions matter, and they may start thinking more favourably about social cooperation in general. This in turn may lead to higher self-esteem, a feeling of social responsibility, and a willingness to consider social policy questions in future. Moreover, participation in the drafting of a policy is deemed to make people more likely to abide by the rules following from the decision, even if the decision does not match their own opinion. In other words, if citizens participate in the decision-making process they have normative reasons to respect and comply with the resulting policies. This process could counter the distrust and disinterest in politics that is exhibited by certain groups in society, particularly those marginalised groups that through lack of social status feel they have no control over policy decisions whatsoever. If these groups take part in a public deliberation that proceeds in a respectful manner towards all participants and they have had a real chance to make their point, this might lead to more equality and de-escalation of opposition. Preferably, these groups are involved in the decision-making process already at the early stages of agenda-setting and problem definition, because otherwise their viewpoints can be overlooked due to the framing of the debate by dominant groups and they will still feel excluded. According to Hisschemöller and Hoppe, 'the crucial matter is whether a given problem construction takes into account all the differences of opinion about the problem and its possible solutions'.[25] If this is not the case, this eventually backfires and leads to an intractable policy problem with the risk of escalation into protest, sabotage, or possibly even violence.

Deliberative democrats argue that even when participants in a debate do not see their opinions reflected in the final decision, the fact that their opinion has at least been taken into account as well as the possibility of revision stops them from turning their backs on the process and resorting to other, possibly violent, means of making their point. After all, deliberation is an ongoing process and if new reasons are presented or the circumstances change a decision can always be overturned. Decisions in this sense are always provisional endpoints. The fact that in a deliberative democracy a norm that is the outcome of debate can always be revised – when a theretofore excluded group gains access to and reopens the decision-making process or when new arguments or information are offered – should also stimulate the participants to keep an open mind; even when they are absolutely certain they have reached the right conclusion, it can always be challenged.

---

[24] Mutz (2006), page 85.

[25] Hisschemöller and Hoppe (1995), page 48.

4.2.4 Normative Assumptions

The foregoing account of the practical and theoretical reasons for the ever stronger call for public deliberation harbour certain normative presuppositions. Firstly, in complex cases where expert accounts frequently contradict each other, it is desirable that as much relevant information and arguments as possible come to light, and this can be done by listening to the accounts of participants with as many different views as possible on the topic. Deliberation, therefore, is 'a procedure for being informed'[26] and is necessary to bring to light all relevant information, as no individual on her own can possess all information and see a problem from all relevant perspectives.[27] Public deliberation can reveal 'private information', but furthermore, when the problem under discussion is so complex that even if all information is present it is still difficult to make a decision, discussion of different arguments for and against certain policy options may help to reach a better decision. As James Fearon explains, 'discussion might be a means for lessening the impact of bounded rationality, the fact that our imaginations and calculating abilities are limited and fallible'.[28]

Secondly, in cases where normative views to a large extent influence the definitions, research methodologies and interpretations of a certain topic, experts and policy makers alone should not determine policies. After all, in such situations there is no neutral standpoint from which to make decisions.

Thirdly, lay persons should not only contribute their ethical views to the decision-making process, but they also have experiential knowledge that experts may have overlooked and this knowledge should be considered in order for better decisions to be made. Moreover, a valuable contribution that lay persons can make lies in challenging the validity of expert views and methodologies.

Fourthly, trust in governments should be earned by greater transparency not only about what measures governments take, but also by explaining why, and this requires an open discussion with the public rather than a one-way communication strategy.

Fifthly, arguments based on fallacies and self-interested motives should be exposed and, in contrast to bargaining or simple preference aggregation, public deliberation provides an opportunity to do this. As Amy Gutmann and Dennis Thompson put it, 'deliberation helps sort out self-interested claims from public-spirited ones'.[29] This final argument is one of the main reasons why environmental theorists are interested in deliberative forms of governance. As Robyn Eckersley explains, from a green political perspective, the call for 'new discursive designs'

> may be understood as attempts to confront both public and private power with its consequences. . ..[and] to expose or prevent problem-displacement and/or to ensure that the

---

[26] Benhabib (1996), page 71.

[27] B. Manin (August 1987), page 352.

[28] James D. Fearon (1998), page 49.

[29] Gutmann and Thompson (1990), page 43.

4.2 Framework

sites of economic, social and political power that create and/or are responsible for ecological risks are made answerable to all those who may suffer the consequences.[30]

Underlying all these arguments is the broader argument that deliberation, which enables, among other things, critical questioning, social learning, and the correction of mistakes in response to feedback, also acknowledges human fallibilism – that we are not perfect, that we are liable to make mistakes; deliberation forces a society to continually reflect on past mistakes and to learn from them. Of course, deliberative democrats acknowledge that these claims are rarely fulfilled in reality, and therefore they should be regarded as ideal characteristics of deliberative democracy. These arguments all refer in some way or another to the quality of the deliberation – they are about the inclusion of certain types of knowledge and viewpoints and the exclusion of mistakes or self-interest – but in principle this can be reached by deliberation amongst a small group of representatives to ensure that all the relevant views are taken into account. In fact, it is quite possible that 'better' in the sense of 'more informed and thorough' decisions are reached when only a small group of people get together to deliberate; a group big and diverse enough to bring all the relevant viewpoints to light, but small enough to be able to make expedient decisions.[31] Nevertheless, as became clear earlier, many deliberative democrats also make claims about the value of participation itself.

As my overview of theoretical reasons for public deliberation makes clear, deliberative democrats not only take a normative stance on the quality of decision-making but also on the participation of the public, or in other words, on the inclusion of a wide number of people in the decision-making process, with a special emphasis on marginalised groups. This is based on a normative view of legitimacy as well as a view on the desirable effects that public deliberation could have on participants. The view of many deliberative democrats, such as Gutmann and Thompson, is that public deliberation is not only required by the ideal of equality, but will itself lead to more equality between citizens. Moreover, they anticipate favourable consequences of public deliberation for citizens, and thereby for society as a whole. If we assume that participants gain something from partaking in public deliberation – in the sense of moral learning, a broadened mentality, civic education, or even simply self-esteem – it makes sense to want to include as many people as possible or at least a large and diverse set of group representatives, who can subsequently share what they learned with like-minded people. Iris Marion Young even explicitly proposes a principle of inclusion in deliberative democracy, which entails that everyone's opinion and viewpoint should be included in debate, as long as they also accept that other people's viewpoints are included.[32]

---

[30] Robyn Eckersley (2006).

[31] Note that even though the claims I have distinguished here all refer to the public character of deliberation, in principle deliberation can be done in private or at least by a very small group of people behind closed doors. Some even argue that a certain amount of secrecy is necessary to achieve high-quality deliberation. See Simone Chambers (2005).

[32] Iris Marion Young (1999).

4.2.5 Tensions Between Ideals

As I have described deliberative democracy, its normative background is based on two pillars. On the one hand it represents the idea that decisions should be made on the basis of arguments rather than on, for example, lobbying, bargaining power, logrolling or demagoguery, and that, moreover, these decisions are better when the arguments have been exchanged in a deliberative setting. One pillar therefore harbours an ideal of quality of argument. On the other hand, it represents the idea that the deliberation should specifically take place in public and should involve everyone affected, because decisions would then be fairer and more legitimate; in effect, the deliberation should have as many participants as possible. Moreover, including heretofore marginalised groups can have positive effects, because it stops policy problems from becoming even more intractable and gives these groups more sense of control, resulting in a de-escalation of conflict. The other pillar, then, harbours an ideal of inclusiveness.

These two ideals are not always in accord with each other; in fact, we can discern a tension between them. Allowing more participants is likely to make a debate more laborious, time-consuming, and disorderly. A debate with emphasis on quality of argument is more likely to be served by taking place in a small group of people who are not necessarily like-minded, but who at least share common assumptions. When the deliberation is carried out by members of the elite, who are more likely to be highly educated, one can expect that sounder arguments and less argumentative fallacies will be used than if people from 'all walks of life' participate. Even Aristotle, who was the first philosopher to argue for decision-making on the basis of public debate, favoured aristocracy over a complete democracy, because he thought the participants would be more capable and the debate would be of a higher quality.[33] Inclusiveness, even though it is instrumental in bringing to light as many different views as possible, can stand in the way of the quality of the debate, as a larger and more diverse group of participants is more likely to experience miscommunications and constant repetition of the same arguments. Particularly when a complex technological topic is involved, where some background knowledge is indispensable, the larger the group of participants involved, the harder it will become to achieve a shared body of knowledge. In other words, more inclusiveness may lead to less thoroughly formed arguments, and while a diverse set of viewpoints is necessary to bring to light as much information as possible – which would contribute to the quality of the decision – debate of a higher quality is likely to ensue with less participants (above a certain threshold, of course). On the other hand, when debate only takes place in a small circle it is likely to be elitist and it will exclude groups and lead to inequality.

A tension between deliberation and participation was also noted in empirical research. Mutz demonstrates that while political conversation across divides is very valuable for a vibrant democracy, because it leads to tolerance of other people's

---

[33] See Amy Gutmann and Dennis Thompson (2004).

4.3 Deliberative Democrats' Answers 103

views and more balanced judgments, 'cross-cutting exposure' – being confronted with political views other than one's own – tends to lead to a decline in political participation.[34] People tend to associate most with like-minded others and in politically homogeneous contexts people are more stimulated to vote and become politically active than in heterogeneous contexts. This is due to two social psychological mechanisms: Firstly, being exposed to different viewpoints sometimes leads to ambivalence and people who do not have very strong political convictions are less likely to be motivated enough to be politically active. Secondly, political conversation across divides can lead to conflicts, while most people aim to maintain social harmony; 'conflict avoidance was an important deterrent to participation'.[35] Mutz concludes that 'like-minded social environments are ideal for purposes of encouraging political mobilization... but paradoxically, the prospects for deliberative democracy could be dwindling at the same time that prospects for participation and political activism are escalating'.[36]

Perhaps because of these tensions, while all deliberative democrats agree that we need to aim for inclusiveness, they differ in their view of how this should be accomplished and how wide-ranging this should be. Generally, one could say, then, that one pillar puts more emphasis on the 'deliberative' and the other on the 'democracy' in 'deliberative democracy'. Sometimes one of these two ideals will have to yield in order to achieve the other. Different theorists balance the two differently. Later in this chapter, I will describe how I think the balance should be struck and this will inform my evaluation of the empirical case-studies. But first, we should investigate how deliberative democrats propose to strike the balance.

4.3 Deliberative Democrats' Answers

How one makes the choice between the two ideals described above has repercussions for the question as to how public deliberation could deal with intractable moral disagreement. As I explained at the start of this chapter, dealing with intractable disagreement can be analysed with reference to three questions: what should be the goal of deliberation?; what types of argument should we deem valid in deliberation?; and who should participate in the decision-making process? In this section, I will briefly elaborate how leading deliberative democrats respond to these core questions.

4.3.1 Goals of Deliberation

In the literature on deliberative democracy, three broad goals or aims of deliberation can be discerned. One group of deliberative democrats, headed by Jürgen Habermas,

---

[34] Mutz (2006).

[35] Ibid., page 107.

[36] Ibid., page 127.

Jon Elster, and Joshua Cohen, argues that the aim of deliberation is the formation of consensus.[37] Public debate is then a tool for making decisions and could be regarded as a substitute or as a basis for decision-making on a political level (which puts it on a par with voting or bargaining). This view is often based on considerations such as outlined above, that no-one has the sole prerogative to make decisions on complex, value-laden issues about which much disagreement exists, and that people should have a say in their own destiny. Also, it is argued that when people are forced to reach consensus they will display an attitude of openness to each other's viewpoints and will be more willing to be swayed by the arguments of others. If they do not display this attitude, they will become frustrated in their aim and may have to go on deliberating until infinity. It is held that even though it will take longer to make decisions by consensus, the decisions will be more morally legitimate. This, in turn, will lead to better compliance. Habermas and Cohen judge the success of deliberation against the standard of an ideal consensus. For Habermas, this view of consensus is based on an epistemic claim; moral truth or validity is construed by deliberation aimed at a consensus; in other words, in an ideal speech situation without distorting power differences participants to a debate would eventually reach a valid conclusion together. Even if the ideal speech situation is never attained and consensus, therefore, remains a counterfactual norm, the closer we get to consensus, the better. For Cohen consensus should be focussed on the common good; democratic institutions are better the closer they approach the ideal of consensus on the common good.[38]

Other deliberative democrats, such as Benhabib, Dryzek, and Bohman, do not embrace the goal of consensus.[39] They point out that the fact of pluralism makes consensus an unattainable ideal. While Habermas regards this pluralism as an obstacle to reaching consensus, the aforementioned writers hold that consensus is not essential to a theory of deliberative democracy and pluralism is a circumstance of politics, but not necessarily an obstacle.[40] Gutmann and Thompson, while acknowledging that consensus may often not result, do think that public deliberation should at least aim at consensus; in their writing it becomes clear that they value deliberation because they think that it may reduce disagreement or at least increase mutual accommodation of unresolvable differences, even though they do acknowledge that when people are confronted with each other's viewpoints their differences may actually be exacerbated.[41] For Gutmann and Thompson, then, consensus is not a regulative ideal as it is for Habermas and Cohen. Rather, they seem to focus on the aim of consensus as an attitude of openness to other people's viewpoints which may or may not produce actual consensus. They also understand consensus more loosely

---

[37] Joshua Cohen (1997), Joshua Cohen (1989), Jon Elster (ed.) (1998), Jürgen Habermas (1989), and Gerald J. Postema (1995).

[38] See Carol Gould (1996).

[39] Dryzek (2000), James Bohman (May 1995), Seyla Benhabib (1996).

[40] Dryzek (2000).

[41] Gutmann and Thompson (1996).

4.3 Deliberative Democrats' Answers

than Habermas, who treats it as agreement not only on outcomes but also on the reasons and values supporting those outcomes. For Gutmann and Thompson, any agreement seems to be sufficient, either on outcomes, but supported by different parties for different reasons – a position that Cass Sunstein termed 'incompletely theorised agreements'[42] – or even just 'agreement to disagree'. Sunstein points out that agreements can be incompletely theorised when people agree on abstract principles, but disagree on their application in particular cases, or conversely when they agree on decisions in particular cases, but not on the principles behind the decision. Incompletely theorised agreements help people who on some level disagree fundamentally to still co-exist and show respect for each other. [43]

The second goal of public debate could be described as *deepening* understanding of the issue by involving as many different views and sources of information as possible and by discussing which arguments and positions are the most valid or consistent. Deliberative democrats also have more theoretical assumptions, namely that decisions made on the basis of rational arguments focussed on the common good will be better in the sense of more informed and consistent than decisions made on the basis of non-rational or irrational arguments. This goal corresponds to an ideal of quality.

The third goal is that of *broadening* participation. Public deliberation is then a tool to involve more people in the decision-making process, particularly those from marginalised and hitherto excluded groups. As elaborated in the previous section, this goal corresponds to an ideal of inclusiveness. For some deliberative democrats the very legitimacy of the political process is a result of the equal possibility to participate of all citizens.[44] In the context of intractable disagreement, it should also be noted that the possibility of preference or opinion transformation is an important driving force for deliberative democrats. In order for preference transformation to be successful, it is important that many people partake in deliberation. For if only a small group of representatives change their opinion after being exposed to other people's arguments, public deliberation will have achieved only preference transformation on a small scale. We cannot expect those who were represented but did not actually participate in the debate to unquestioningly accept this preference or opinion transformation without being exposed to the relevant arguments.

We have already seen that a certain tension exists between the ideals of inclusiveness and quality of argument; this tension also translates into a tension between the goals of broadening and deepening of the debate. Both of these goals are furthermore in tension with the goal of reaching consensus. The aim of including as many voices as possible is likely to result in a great divergence of opinion and this can obstruct the achievement of consensus. The aim of inclusiveness assumes a role for citizens in decision making; however, when consensus is strived for, but not reached, the influence of this role diminishes and this in turn can lead to a feeling of not being

---

[42] This notion was coined by Sunstein (1997).

[43] Cass Sunstein (1995).

[44] Manin (1987), pages 351–352.

taken seriously. Furthermore, as we will see, critics have pointed out that aiming for consensus can lead to an exclusion of viewpoints and a suppression of dissensus and this would run counter to the goal of inclusiveness. Aiming for consensus could also detract from the goal of deepening, because in complex problems bringing to light more information may actually make it harder rather than easier to achieve consensus. In neither the goals of broadening nor that of deepening it is essential to aim for consensus; it is readily admitted by many deliberative democrats that this may be too much to ask. Public debate in their view primarily leads to a better framing of policy problems and making an inventory of the relevant considerations and values involved without necessarily reaching consensus, although it can lead to an incompletely theorised agreement or an agreement to disagree. As Gutmann and Thompson explain, agreeing to disagree after respectful deliberation differs from agreeing to disagree without dialogue. The latter amounts to toleration, an attitude in which we accept that other people think differently but we have not taken the effort to understand their point of view and in fact do not care; the former, a 'deliberative disagreement', on the other hand, is based on mutual respect and keeps open the possibility that we change our own minds.[45]

4.3.2 What Types of Argument Are Valid?

Decision-making based on public deliberation can take place on three different levels: that of specific policy questions, that of decision-making procedures themselves, and that of the question as to what values should be protected by the constitution. Most deliberative democrats argue that deliberation should take place on each of these levels. Debate about the last two levels appears paradoxical, because some decision-making procedures already have to be in place and some constitutional rights assumed before deliberation can take place according to the deliberative ideal (which, recall, includes freedom and equality).[46] This means that either some rules and values have to be posed as pre-existing or that one has to allow for the possibility that in the deliberative process the procedures are contested and the rights abolished. Habermas would find this latter conclusion inconceivable. In *Between Facts and Norms* he argues that basic civil and political rights and deliberation are mutually constitutive or presuppose each other[47] and he counters the above mentioned problem by including a hypothetical consensus in his discourse theory of ethics, which entails that in an ideal situation everyone would accept certain principles, regardless of the fact whether they actually accept these here and now – a solution which he refers to as a 'rational reconstruction'.[48]

---

[45] Gutmann and Thompson (1996).

[46] See Gould (1996).

[47] This is known as Habermas' co-originality thesis. He disagrees with orthodox liberals, like Rawls, in their claim that rights come first. Jürgen Habermas (2001). For a discussion of the co-originality thesis, see also Alessandro Ferrara (2001).

[48] Frank Cunningham (2002).

4.3 Deliberative Democrats' Answers

A more 'mixed' position is taken by Gutmann and Thompson, who on the one hand argue that those willing to take part in public deliberation must accept some common values and principles, such as mutual respect, and must agree not to use certain arguments, such as sectarian ones and, on the other hand, that the rules of decision-making and the values in the constitution are themselves open to interpretation in public deliberation. Those people who are not willing to show the right attitude for public deliberation, such as fundamentalists or those who are only concerned with their own self-interest, can never be swayed by the case for deliberative democracy anyway.[49] John Dryzek wishes to extend the scope of arguments even further and argues that we should not exclude arguments from any of the three levels from debate a priori. This is because any rules and principles will be interpreted differently in different historic periods and in different places and it is therefore impossible to pin them down beforehand. Moreover, in his estimation the mechanism of deliberation itself makes the stipulation of conditions and principles to limit the debate unnecessary.[50] For example, Gutmann and Thompson's principle of publicity will be generated by deliberation itself, because people will realise that it is more convincing to state arguments in terms of the common good rather than self-interest. Eventually, because they want to avoid cognitive dissonance, these debaters will even come to accept the point of view from the common good.[51]

As I showed in Chapter 3, the fact of pluralism has led political liberals to argue for state neutrality on moral matters, thereby removing them from the political agenda and into the private domain. As was argued by philosophers such as Gutmann and Thompson and Van den Brink, however, state inaction on moral issues is not as neutral as it seems. According to these deliberative democrats, the state can remain neutral while at the same time leaving room for the public discussion of moral issues, including a reliance on comprehensive moral views. Instead of removing issues about which persistent disagreement exists from the agenda automatically, Gutmann and Thompson argue that the only reason not to address certain issues collectively is that they do not properly qualify as moral issues in the first place. In their eyes, an issue does not qualify as a moral one when it is purely self-regarding, when the empirical assumptions behind it cannot be defended by reasons or when these assumptions are 'radically implausible'. By the latter they mean that these beliefs 'require the rejection of an extensive set of better established beliefs that are widely shared in society'.[52] An example that one could think of is the case of abortion. Presumably, the moral acceptability of abortion is an issue that should be debated. Two people who disagree about abortion both argue from an other-regarding perspective and they can each appeal to reasons that are not radically implausible. However, someone who argues that babies that were conceived during a full moon

---

[49] Gutmann and Thompson (1996).

[50] Dryzek (2000). Dryzek also believes that the deliberation would eventually subject all arguments to rational critique.

[51] Ibid., pages 45–47.

[52] Gutmann and Thompson (1990).

should be aborted, cannot defend her point of view with the help of plausible empirical assumptions. A paradigm case of a position that does not qualify as a moral position in Gutmann and Thompson's eyes is one favouring racial discrimination, because arguments in its favour are either purely self-regarding or rest on radically implausible assumptions (or both).[53] It should be noted that while Gutmann and Thompson want to exclude views from public debate that do not fit their criteria, they do not necessarily exclude non-moral reasons, such as reasons of a practical nature. Other deliberative democrats want to exclude arguments that they define as unreasonable, as they are based on prejudice or force.[54] However, the question as to what constitutes prejudice itself may be open to debate.

Gutmann and Thompson do not exclude appeals to religion from public debate completely. However, their principle of reciprocity requires that when participants make certain moral claims in public debate these need to be in principle accessible to other participants. Appeals to 'divine authority' are problematic in this respect, because others can only access these claims if they were to accept the religion in question and live their life accordingly: 'any claim fails to respect reciprocity if it imposes a requirement on other citizens to adopt one's sectarian way of life as a condition of gaining access to the moral understanding that is essential to judging the validity of one's moral claims'.[55] One of the principles that Gutmann and Thompson propose to guide moral argument is reciprocity.[56] Reciprocity demands that participants in a debate try to find a way to deal with their moral disagreement that is mutually acceptable to everyone who participates. It is based on an expectation that others will be fair to us just as we are fair to them. In the context of reasonable disagreement reciprocity means that we have to acknowledge that people whose positions we deem morally wrong might still have reasonable grounds for their position and that we have to respect their position. Gutmann and Thompson propose to deal with intractable moral disagreement through a so-called 'economy of moral disagreement', which in short means that citizens should try to avoid moral conflict with one another while at the same time not compromising their own moral views and principles.[57] They should look for points that they do share with their opponents and also acknowledge the validity of their opponents' views. This could entail, for example, that they support policies that they do not consider to be a priority, but that they know are important for their opponents. Pro-choice activists, for example, could support a policy that allows for abortion but that also includes funding for the promotion of alternative options to abortion. The way Gutmann and Thompson propose to deal with intractable moral disagreement is summed up well in their following statement:

---

[53] Ibid.

[54] For example Bohman (1996).

[55] Gutmann and Thompson (1996), page 57.

[56] The other principles are publicity and accountability. Ibid.

[57] Ibid., pages 85–94.

> By trying to maximize political agreement in these ways, citizens do not end serious moral conflict, but they affirm that they accept significant parts of the substantive morality of their fellow citizens to whom they may find themselves deeply opposed in other respects.[58]

Fundamental disagreement, in their eyes then, can be dealt with as long as participants follow the rules of reciprocity and mutual recognition.

4.3.3 Who Should Participate?

Most deliberative democrats agree with Seyla Benhabib that 'there are no prima facie rules limiting the agenda of the conversation, or the identity of the participants, as long as each excluded person or group can justifiably show that they are relevantly affected by the proposed norm under question'.[59] Nevertheless, as modern day societies are too large and complex to be able to let every citizen take an active part in the decision-making process some form of representation will always be needed. The question is, then, what the criterion for participation should be for the representatives. Many deliberative democrats hold with Benhabib that everyone affected by a certain decision should be allowed to participate. This in effect means that deliberation between stakeholders should take place. Others argue that every citizen has an equal right to participate in the political process, meaning that also lay citizens with no prior knowledge of, or personal interest in, the topic should participate. In this view everyone should be given an equal opportunity and this in turn can entail that such opportunity is actively created or it can simply point to a theoretical requirement that everyone is viewed as equal.

A further question, therefore, is whether only those who actively wish to engage in political debate should be included in deliberations or whether certain groups should be actively stimulated to take part. Some commentators argue that the majority of citizens is simply not interested in politics and that it is useless to try to motivate these people to feel otherwise. Some, such as Val Plumwood, however, argue that particularly those groups that have never shown interest in politics need to be actively included. Often these citizens are part of marginalised groups such as ethnic minorities, people with a low socio-economic status, or women, and often these groups are the worst affected by certain decisions, particularly decisions concerning the environment. Plumwood, in this vain, argues that liberal democracy is not ecologically rational, as it creates the perfect conditions for the occurrence of different types of remoteness.[60] Consequential and spatial remoteness prevent polluters from experiencing the negative side-effects of their actions, because the costs are borne by others, usually marginalised groups. Communicative and epistemic remoteness prevent adequate communication and the exchange of information about

---

[58] Ibid., page 89.

[59] Benhabib (1996), page 70.

[60] Val Plumwood (1998).

these negative side-effects. And finally, temporal remoteness prevents us from being confronted with the future effects of our decisions. In order to counter the problem displacement that often takes place with environmental pollution it is therefore paramount that marginalised groups are involved in policy processes.[61] This entails, amongst other things, that more substantive equality of resources should be strived for as this forms a basis for equal opportunity to participate.[62]

Deliberative democrats, then, do not usually envision a direct democracy Athenian style where each decision is put to an assembly of all those affected. Affected parties can deliberate in a diversity of deliberative fora, such as political parties, citizens' networks and other social movements, and associations such as the consumers association. These together form a 'public sphere of mutually interlocking and overlapping networks and associations of deliberation, contestation, and argumentation'.[63]

4.3.4 Recapitulation

We have seen that just as there is a tension between the ideals of quality and inclusiveness, a tension exists between the different goals of consensus, quality and participation. When we put deliberative democracy to the test in the empirical part of this book we have to be aware of these tensions. As we will see in the next section, some of these goals, particularly the one of consensus, have been criticised, both within and outside of the realm of deliberative democracy. Perhaps, then, we do not need to hold on to each of these goals. As will become clear in the next section as well, a tension also exists on a different level, between these three goals and the demand of expediency in decision-making. The goals of consensus, broadening and deepening of the debate each are likely to demand a longer time frame than is usual in political decision-making. In practice regulatory decisions have to be made (after all, even *not* making regulatory decisions would have important consequences). The decision-making process would run more efficiently and expediently if the goals of inclusiveness and deepening were bypassed. However, this runs the risk that some relevant moral considerations are left out.

4.4 Critiques of Deliberative Democracy

Criticism has been mounted against theories of deliberative democracy from different angles and about many different aspects of the theory. Here, I will focus on those critiques that are relevant for the question as to how deliberative democracy could deal with intractable disagreement. Both liberal democrats and difference

---

[61] About problem displacement of environmental pollution see John S. Dryzek (1987).

[62] This is argued, for example, by Cohen (1997).

[63] Benhabib (1996), page 74.

4.4 Critiques of Deliberative Democracy

democrats fear that deliberative democracy might lead to a tyranny of the majority. Liberals emphasise that certain rights are beyond the reign of public discussion, while difference democrats focus on the possibility of elite domination and the problem of representation. Critics have also pointed out the distorting effects of group dynamics. Agonistic democrats challenge the goal of reaching consensus and both difference and agonistic democrats argue that deliberation alone does not cover the whole terrain of political action. Other theorists have criticised deliberative democracy because of its perceived lack of decision-making expedience and naivety about the willingness of citizens to participate. I will elaborate these criticisms with reference to the three core questions from the previous section.

4.4.1 Goals of Deliberation

The tension between the goals that I have described above also informs some of the critiques of deliberative democracy. It is held, firstly, that the goals of consensus and quality of debate can be counterproductive to the goal of inclusiveness, because consensus is held to suppress minority views and to lead to premature closure of the discussion, while a focus on quality is held to be elitist, as it can be expected that the most highly educated will have most practice at rational deliberation. Critics from another direction, on the other hand, point out that the quality of the deliberation and indeed the very possibility of decision-making under conditions of fundamental disagreement can be undermined by the goal of inclusiveness. Consociationalists, for example, argue that contentious problems are best dealt with by agreement between the leaders of different groups in society, because if ordinary citizens would be involved this could lead to too much conflict.[64] Of the three goals of deliberation that I described above, the goal of consensus has received most criticism. Sunstein argues, for example, that a consensus in the sense of agreement over outcomes as well as the underlying reasons is not automatically just or fair simply because we all agree: 'usually, it would be much better to have a just outcome, rejected by many people, than an unjust outcome with which most or even almost all agree'.[65] This opinion runs counter to Habermas' view that consensus is always right, simply because it is what we would all agree to in an ideal speech situation. Contra Habermas, critics argue not only that consensus is often not attainable in practice – something which Habermas himself concedes – but that it is also an inappropriate ideal.[66]

The focus on consensus has been most fundamentally challenged by a group of theorists named agonists. According to agonists, politics is about contestation; they see contestation as a positive force that can change past or present unjust power relations, and hence lead to emancipation, rather than as a source of conflict to be

---

[64] Dryzek (2000).

[65] Sunstein (1995), page 1769.

[66] Nicholas Rescher (1993).

avoided, as liberals appear to do.[67] They criticise Rawls' more traditional liberal view that while there is a plurality of possible comprehensive doctrines or conceptions of the good, there is but one idea of the right that is based on an overlapping consensus among reasonable people. They also criticise Habermas' deliberative liberal view that political deliberation should be aimed at a consensus, even if this is only a counterfactual ideal. Both of these views presuppose that political coexistence is only possible when consensus – or at least the possibility thereof – on some fundamental level exists. According to Rawls, a mere modus vivendi is not enough; reasonable people with different comprehensive doctrines need to be able to support the consensus for reasons that fit within their own comprehensive doctrines.[68] For Habermas, in an ideal speech situation agreement not only about outcomes, but also about underlying reasons, would be reached. Agonists on the other hand, argue that there is not only reasonable disagreement about questions of the good life, but also about the right and the just, and they argue that a consensus on principles of justice or basic institutions is not needed at all for social and political cooperation.[69] According to agonists, Rawls' and Habermas' views discount the pluralism that they appear to address; they try to restrict pluralism, for example by adding a criterion of reasonableness (in the case of Rawls). Agonists, on the other hand, value pluralism and argue that this means valuing difference and dissensus rather than consensus. A criterion such as reasonableness is too normative in their eyes and excludes groups that do not fit this criterion, which is defined by dominant groups in society. Agonists criticise the liberal distinction between the public and the private; they adopt a very wide conception of the political as 'a form of social interaction characterised by disruption and antagonism' that cuts across the public/private divide.[70] Even arguments between men and women about who should do the dishes are in their eyes political, because they aim at a change in power relations.

According to agonists, deliberative democrats do not take the complex character of intractable disagreements seriously enough. Goi, for example, disagrees with Gutmann and Thompson's treatment of the conflict over abortion as an issue that can be solved by deliberation aiming at consensus and governed by the criteria of publicity, accountability, and reciprocity. In her eyes, this is impossible because we are dealing with incommensurable belief systems.[71] In such a situation a focus on consensus is too premature; we first need to give citizens the opportunity to get to understand their oppositions' viewpoints and to find out the real character of their

---

[67] See, for example, Chantal Mouffe (2000) and Bonnie Honig (1993).

[68] Rawls (1987).

[69] See Bert Van den Brink (2005). It is acknowledged that even though sometimes common views are necessary for making policy, this does not make deep disagreement go away. See also Rescher (1993), page 189.

[70] Thomas Fossen (2008).

[71] Goi (2005). In Goi's analysis, Gutmann and Thompson redefine the abortion debate in terms that everyone can accept as a conflict between the relative values of liberty and right to life, while in reality the disagreement relies just as much on less generalisable views about sexual promiscuity and family values.

4.4 Critiques of Deliberative Democracy

disagreements. We need to create 'agonal spaces' in which the rules of the conversation are less stringent than in deliberative democracy. Rather than requiring participants to frame their arguments in terms that everyone can accept we need to allow them to express their opinions in their own terms. The participants only need to abide by 'basic rules of civility', such as agreeing to listen to other people's viewpoints, and refraining from violence and manipulation. According to agonists, aiming for consensus leads participants to focus too much on reaching a decision and this leads to premature closure of the debate, harbouring the risk that dissensus is suppressed, while we in fact need dissensus and dissent as a way to keep democracy alive. There always needs to be a venue – the agonal space – for those who have been excluded or those who have 'lost' when a decision is made to challenge this decision. If not, the excluded – or 'remainders', in Honig's terms – will either withdraw into sectarian communities of likeminded people or resort to violent means of getting their point across.[72] Moreover, research into actual public deliberations suggests that while people often do seek consensus, this is done at the cost of a more critical stance about the context of the deliberation itself. Existing power structures present in the dominant discourse are often not questioned. In this way, the aim for consensus, again, functions to exclude groups, namely those marginalised groups that do not communicate in the dominant discourse.[73]

There is some empirical basis for the view that the focus on consensus functions to exclude groups and views. Derrick Purdue, who made a detailed analysis of the UK National Consensus Conference on Plant Biotechnology, for example, uncovered mechanisms used by the conference organisers to create conditions favourable to the biotechnology industry's interests.[74] These mechanisms consisted of dividing conference participants into three strictly separated groups of experts, counter-experts from pressure groups such as Greenpeace, and lay people. Anyone who would not fit into any of these groups – because they were 'too lay to be expert, but too expert to be lay' – was excluded.[75] Also, views were excluded by the chairperson, who drew clear boundaries between the different key issues, so that certain, mainly value-related and ecological, concerns were excluded. It is also interesting to note what happened when the lay panel report proved to be more critical of GMOs than some actors had anticipated. Political representatives indicated that the Parliament's decisions would not take the report's recommendations into account, while industry representatives discounted the report as too simplistic to provide clear directives. Others, in contrast, such as the conference organiser, characterised the report as showing 'qualified support for plant biotechnology', while NGO's lauded its precise proposals to curtail biotechnology.[76] Purdue argues that this tendency to

---

[72] Honig (1993); Goi (2005). We can think here of shooting doctors at abortion clinics or threatening animal experimenters.

[73] Kevin Olson (2011). Olson quotes research by Shawn Rosenberg (2007).

[74] Derrick Purdue (September to October 1995).

[75] Ibid., page 171.

[76] Ibid.

114   4 Deliberative Democracy and Its Limits

interpret the outcomes of the consensus conference in various ways is the result of
the assignment to reach consensus and of the directive not to describe any possible
disagreements in the report. This leads him to conclude that

> the goal of manufacturing a "National Consensus" out of a diversity of opinion on such
> a highly-contentious topic as biotechnology was impossible to achieve without silencing
> dissenting voices. In particular, it required the separation of the most articulate criticism
> [the counter-experts] from a less-focused unease among the wider public.[77]

In summary, then, it is held that consensus assumes too much homogeneity and
excludes difference. Consensus can be suppressive of minority views – because it
suppresses dissensus – and it can prematurely close off a discussion. Such closure
can make people disgruntled. In an alternative vision the importance of deliberation
lies in making differences and plurality visible. According to Nicholas Rescher,
for example, even though it makes sense for people to strive for consistency and
uniformity in their own thinking, on a collective level it does not: diversity and
variety of views enrich a culture, because they provide a 'testing ground for the
evaluation of alternatives'.[78]

4.4.2 What Types of Argument Are Valid?

As noted before, some deliberative democrats argue that not only specific policy
decisions, but also the rules and underlying rights of the deliberative process are
contestable through deliberation. This is seen as problematic by liberal theorists
who argue that certain rights, such as equality and freedom from coercion should
not be contestable. These rights need to be guaranteed, because otherwise they fear a
tyranny of the majority and a concurrent trampling of minority rights. The problem
for deliberative democrats, however, is that acknowledging the fact of pluralism they
want to avoid posing substantive principles a priori of the procedure of deliberation.
Nevertheless, it is not sufficient to point out that certain rights are presupposed and
internal to the process of deliberation itself, because as Carol Gould points out, not
all basic rights, for example the right to privacy or to bodily integrity can be justified
by pointing to the process of deliberation.[79] A way out of this problem would be to
argue that rights are not really contestable, but can be interpreted differently and
that this interpretation is open for public discussion. In this case, the existence of
basic rights, albeit a 'thin conception' of rights, needs to be grounded in something
that precedes the deliberative procedure. However, deliberative democrats who take
this line will have to provide a substantive account of the basis of such rights that
precedes public deliberation.

Another line of critique does not so much focus on the question as to what issues
should be open for deliberation, but rather on what form arguments should take

---

[77] Ibid., page 173.

[78] Rescher (1993), page 197.

[79] Gould (1996), page 179.

4.4 Critiques of Deliberative Democracy

in deliberative democracy. This critique is put forward particularly by difference democrats, who constitute a diverse group of thinkers, often with a feminist or an activist background, who argue that in politics we should acknowledge the experience of difference of certain groups whose identity has been shaped by past or present oppression or marginalisation on the basis of sex, gender, social class, race, ethnicity, age, culture, religion, worldview, etc. Difference democrats have uncovered the various ways in which these groups have been excluded from politics and they aim at their emancipation. While some difference democrats embrace deliberative democracy,[80] others oppose it because of its focus on rational deliberation, which in their eyes systematically silences certain voices and excludes groups from participation.[81] They demonstrate that social inequalities influence the course of the debate: 'social power can prevent people from being equal speakers'.[82] One aspect of this problem is that access to education is unequally distributed and that the better educated generally have more experience in formulating rational arguments. Simply increasing access to education does not sufficiently alter the situation, however. A more pervasive problem is that the standards for convincing argumentation are set by the dominant group in society, thereby automatically disadvantaging marginalised groups. These standards tend to favour formal, dispassionate, competitive, confrontational styles of argumentation; all styles that are more common for men than women, who tend to argue in a more personal and passionate way and tend to be more cooperative rather than competitive and confrontational.[83] Men often do not take women's argumentation styles seriously: 'because women are not treated as men's equals in daily life, it is impossible for men to recognize them as free and equal in deliberation'.[84]

So, while deliberative democrats focus on the giving and testing of reasons in debate, what counts as a good reason in practice depends not only upon the logical consistency of arguments, but also on external influences. This is shown convincingly by Sanders, who analysed group dynamics of deliberations by American courtroom juries and concluded that 'when Americans assemble in juries, they do not leave behind the status, power, and privileges that they hold in the outside world'.[85] Sanders found that those who held a privileged position in the outside world were more likely to behave in a way that gained them respectability. They were more likely to take the lead in discussions and generally talked more. Those who talked most and loudest and often repeated their arguments were most likely to be supported by the other members of the jury, irrespective of the force or

---

[80] For example, Anne Phillips, Ibid.

[81] Dryzek (2000).

[82] Iris Marion Young (1996), page 122.

[83] Ibid., page 123.

[84] Andrea Hickerson and John Gastil (2008), page 286. Hickerson cites empirical research that demonstrates this claim. Interestingly, in studies of mock jury deliberations gender did have an effect on outcomes, whereas in actual jury deliberations this did not appear to be the case.

[85] Lynn Moss Sanders (1997), page 364.

logical consistency of their arguments. White, middle class men were most often selected as jury leaders and their argumentation style, which was competitive and verdict-driven, was most successful.[86] The arguments of women and racial minorities were often ignored, as they did not show the same debating tactics. Based on her research, Sanders claims that if members from dominant groups give bad reasons for their position we cannot rely on deliberative processes to uncover these; in practice bad reasons often remain concealed due to prejudice and elitism.[87] Young argues, similarly, that

> the social power that can prevent people from being equal speakers derives not only from economic dependence or political domination, but also from an internalized sense of the right one has to speak or not to speak, and from the devaluation of some people's style of speech and the elevation of others.[88]

This finding could pose real problems for deliberative democracy, because it calls into question the very possibility of a process of deliberation 'governed by the norms of equality'. While many deliberative democrats theorise that public deliberation has an emancipatory effect, critics argue that they do not sufficiently take into account that social differences are present within different forms of speech and that deliberation can therefore perpetuate these differences. Kevin Olson, for example, points out that group identity becomes manifest in styles of speech – including body language, accents, and attitudes – distinguishing members of different groups from each other.[89] Arguments made in deliberation are not only appraised on their content, but also on styles of speech as an implicit marker of identity. Dominant groups in society tend to set the norm for what qualifies as correct argumentation, style of communication, and deliberative competence of the person who uses them. The dominant discourse thus sets the standard for 'legitimate speech', and according to Olson this includes who is seen as an expert – women and ethnic minorities are less likely to be taken seriously as an expert witness – and what type of opinions are judged as valid. If deliberative democrats focus too much on a particular kind of rational deliberation as valid this could result in more rather than less marginalisation. A related aspect of public deliberation that may work to exclude marginalised groups is that arguments have to be given in generalisable or universal terms that everyone can understand. This requirement puts a greater burden on those deviating from common opinions. If minority groups do not state their arguments with reference to dominant values, their arguments are more likely to be disregarded.[90] Moreover, the focus of many deliberative democrats on appeals to the common good is thought to overlook differences between people; the idea of the common good tends to be cast in elitist terms.[91]

---

[86] Ibid.

[87] Ibid.

[88] Young (1996), page 122.

[89] Kevin Olson (2011).

[90] Will Kymlicka (1989), page 901; Sanders (1997).

[91] Gould (1996); Young (1996), page 126.

4.4 Critiques of Deliberative Democracy

Difference democrats such as Sanders and Young do not want to do away with public deliberation completely, however. Instead they want to curb its elitist tendencies by supplementing rational deliberation with other forms of communication, in particular storytelling or testimony, greeting, and rhetoric. Storytelling in their eyes is a more egalitarian form of communication, because it is about sharing experiences with others in order to get them to understand other people's perspectives and it does not specify a particular style: 'because each can tell her story with equal authority, the stories have equal value in the communicative situation'.[92] It is not immediately self-evident how greeting can play a role in deliberation, but it is important to difference democrats, because by greeting in an elaborate way one acknowledges the presence of others and accepts the identity of the 'other'; the other is deemed trustworthy and is welcomed into the debate. Greeting is also a sign of respect and reciprocity and prepares the ground for discussion.[93] Rhetoric is often regarded as an unfair way of manipulating people by persuading them on the basis of emotions rather than arguments. Some deliberative democrats, including Habermas, therefore want to rid public deliberation of rhetoric. Young, however, argues that rational discourse and rhetoric are not as opposed to each other as is often made out. Both are contextual in that they are shaped for the particular audience towards which they are directed. Both try to persuade the audience, but rhetoric uses more different forms to achieve this aim, including humour, flattery, and ridicule.[94] It is not surprising that difference democrats propose the use of rhetoric, because as I pointed out in Chapter 2, 'critical rhetoric' can expose the 'discourse of power' in a particular debate.

4.4.3 Who Should Participate?

Even though deliberative democrats argue at length that all the affected groups, or even all citizens, including especially marginalised groups, should be included into the decision-making process, some critics argue that this is merely a theoretical supposition. In their eyes, proponents of deliberation do not sufficiently take into account the differences in wealth and power that permeate the real world. Some groups have so much power that they can 'shape the terms of public debate'.[95] This poses a real problem for deliberative democrats who view public deliberation as a way to include marginalised groups. Difference democrats argue that in order for more equal participation to ensue, specific measures will have to be taken; we cannot wait for the beneficial effects of public deliberation to bring about equal access to decision-making processes. In practice it is much easier for some groups to be heard than for others, if only because some people have the money to buy 'air time' on

---

[92] Young (1996), page 132.

[93] Dryzek (2000), page 66.

[94] Young (1996).

[95] Ian Shapiro (1999).

TV or radio or have the means to attend organised public hearings (including getting time off work, arranging child care, reading the newspapers in which the hearings are published, etc.).[96] So how can we ensure equal access of all the different groups that might be affected by a decision?

Anne Phillips notes that in the past it was believed to be sufficient to represent the views of different, often marginalised, groups in politics; as long as the whole spectrum of different beliefs and opinions were taken into account in a decision then equality had been achieved.[97] She terms this view the 'politics of ideas' and contrasts this with a 'politics of presence', in which it is held that the physical – perhaps even proportional – presence of members of all the different groups is necessary in order to stimulate real equality. After all, we cannot easily divorce ideas from the persons who hold these ideas; ideas are shaped by experiences and differently situated groups in society will have different life experiences. Through past experience it has become clear that having an elite of highly educated white middle-class men represent the ideas of all groups in society leads to too much homogeneity and an exclusion of difference. As Phillips and other difference democrats are aware, however, simply using quota systems in order to ensure the presence of marginalised groups appears to rest on a false essentialist notion of identity: one woman cannot represent all women and one black man cannot represent all black men and women or all ethnic minorities.[98] A person's identity is made up of many different experiences, not just that of one's gender, race or culture; a 'diversity of subject positions' exists.[99]

The tension between inclusiveness and quality of argument reverberates in the problem of representation, namely in the tension between ideas versus presence. If the focus is on the quality of debate it would be sufficient to have one representative speak for each (or several) group(s), as long as all the relevant information and arguments are brought to the table and a reasoned choice is based on these. If, on the other hand, the focus is on inclusiveness, just one representative would not be enough, because this would discount that some groups are more powerful than others. It is argued that marginalised groups need to have full representation in order to have sufficient influence and to be able to account for the range of ideas present within the group.[100] The problem of essentialism can be countered to a certain extent by focussing not on the biological traits that define a group, such as sex or skin colour, but rather on contingent characteristics, such as the shared experience of oppression or exclusion – for example by having been denied voting rights in the past.[101] Moreover, groups should not be regarded as fixed, but as dynamic, just

---

[96] Iris Marion Young (2003).

[97] Anne Phillips (1996).

[98] Ibid.

[99] Phillips quotes Stuart Hall (1992), page 254.

[100] Jane Mansbridge (1999).

[101] Ibid.

4.4 Critiques of Deliberative Democracy

as identities are always changing.[102] It has to be noted, however, that as long as the speech styles of certain groups are undervalued, representation may still not have the desired effect of giving everyone an equal chance to influence the decision-making process; in fact, representation by members of marginalised groups themselves can in that situation even work against them. Representation will, therefore, have to be coupled with inclusion and valuation of alternative speech styles.

Some critics argue that if measures are not taken to ensure inclusion of marginalised groups in a meaningful way, participation can degenerate into 'co-option of potential troublemakers'.[103] Groups that are disgruntled because their views have been excluded can be given symbolic entrance to official political structures in order to avoid threats to the powers that be. These groups will not have real access to political participation, however. According to this criticism, participation in a deliberative democracy will be merely symbolic and, therefore, marginalised groups would be better off to use other means to get their point across, such as protest and activism. While activism may be criticised for pushing sectarian interests rather than the common good, activists themselves usually argue that they serve a universal interest in restoring past injustices or inequalities.[104] Moreover, even though activists do not take part in a specific public deliberation, their methods do include communicating certain ideas to a broad public. The effect that activists aim for with their methods of protest, boycott, civil disobedience, and disruption is getting people to question dominant views and to transform their preferences and opinions, just as deliberation does. However, activists refuse to take part in deliberations with whom they regard as oppressors, because through their cooperation they would be legitimating the unequal power relations that characterise current institutions.

Deliberative democrats, then, are criticised for not taking seriously enough the obstacles to deliberation in real life democracies.[105] One of the problems that can particularly be overlooked is that the framing of specific policy problems tends to be done within the dominant discourse and tends to conform to dominant interests. Activists do not want to rely only on rational argumentation, because it is very difficult to challenge the dominant discourse or ideology with reasoning alone. So, again, they argue that recourse sometimes needs to be taken to other means of communication, not only rhetoric, greeting, and storytelling, but also non-discursive means such as artistic expression. Taking part in rational debate, in other words, is not the only way in which citizens can show civic participation and try to transform preferences.

According to one last criticism, deliberative democrats are not sufficiently attentive to the possible adverse effects of group dynamics. Basing themselves on research in social psychology, Blamey et al. list a number of ways in which group

---

[102] Phillips argues for mechanisms that 'address the problems of group exclusion without fixing the boundaries or character of each group'. Phillips (1996), page 151.

[103] Dryzek (2000). See also Bill Cooke and Uma Kothari (2001).

[104] Young (2003).

[105] Ibid.

interactions can influence outcomes of deliberations and which can run counter to the deliberative democratic ideal of deliberation as equals.[106] As was also put forward by difference democrats, group members with a higher status stand a better chance of influencing decisions. Furthermore, individual participants of group discussions often change their own views in order to conform to the majority view. However, when a facilitator manages to get more attention for minority views – without appearing to be biased in favour of the minority – this can enrich the deliberations, because more information comes to the table and participants are forced to reflect more on their own presuppositions. Another potentially distorting factor can be the effect of group polarisation. This is described in more detail by Cass Sunstein, who argues that the composition of the group greatly influences the participant's final views.[107] When people with a similar background deliberate, their views tend to become more entrenched and more extreme than they were prior to deliberations (this is called polarisation). On the other hand, when the group is more diverse, the participants' views tend to become more moderate. Group polarisation, according to Sunstein, is the result of social conformity and the fact that the pool of arguments is smaller than in a diverse group. A final distorting factor has been dubbed 'groupthink', which is what happens when people become so involved in a group that their 'strivings for unanimity override their motivation to realistically appraise alternative courses of action'.[108] This could lead amongst other things to self-censorship in order to avoid conflict, the illusion of agreement when there is in fact none, creating a stereotypical image of an adversary, such as government, industry, or environmentalists as untrustworthy (and associated to this an 'us versus them'-mentality), and pressure on disagreeing minorities. Whether such distorting group dynamics get a chance to supervene in practice will rely to a large extent on the skills of the facilitator or chairman of the deliberation.

4.4.4 Recapitulation

A large part of the criticism of deliberative democracy is based on the view that rational deliberation under the conditions specified by its main adherents does not lead to more equality and the inclusion of hitherto marginalised groups, but rather exacerbates marginalisation and leads to a tyranny of the majority. Deliberation does not take place in a social vacuum, but is influenced by power differences in the outside world. Difference democrats and agonists argue that alternative styles of communication should be allowed in deliberation and that alternative forms of political activity are necessary in order to emancipate marginalised groups. We cannot rely on deliberation alone to create political equality, but need more substantive action to this effect. Finally, according to critics, deliberative democrats do not have

---

[106] R.K. Blamey, P. McCarthy, and R. Smith (2000).

[107] Cass R. Sunstein (2002).

[108] Blamey, McCarthy, and Smith (2000). This term was coined by I.L. Janis (1982).

enough eye for the adverse effects of group dynamics, such as conforming to the majority view, group polarisation, and groupthink.

Regarding the different goals of deliberation, it appears, then, that most critics share deliberative democrats' goal of inclusiveness, but question deliberative democracy's ability to deliver. Besides the tension between the three goals, we can discern another tension, namely between deliberation and decision-making expedience. Making decisions by aggregation is a lot more efficient (at least in the short term) than by the time-consuming process of deliberation. This tension is pointedly formulated by Graham Smith: 'a decision implies the end of a discursive process. But deliberation is in principle ongoing...On the other hand, politics requires decisions: there is a temporal limit to debate'.[109] Simone Chambers even points out that the closer a debate approaches the ideal view of public deliberation the less efficient it will be, because a tension exists between mutual understanding and efficiency.[110] She therefore claims that 'a realistic model of deliberative democracy must concede that decision rules in large democracies will always place constraints on constraint-free dialogue'.[111]

A final often-heard critique of deliberative democracy is that citizens either lack time and interest in participating in democratic decision-making, or have lost trust in politics and have become too cynical to believe that their input would matter. In Goi's diagnosis of the conditions of late modernity the emphasis in politics has been on strategic rationality and efficiency and this has led citizens to respond in one of three ways: either they have lost interest in politics and focused solely on increasing their own wealth, or they have withdrawn from the public arena into communities of like-minded people, or they have tried to make themselves heard and gain political power by using violence. The unspoken assumption of deliberative democrats that citizens are willing to participate in rational deliberation appears naive in this context.[112] All these criticisms pose important challenges to the theory of deliberative democracy and they need to be dealt with before we can investigate how deliberative democracy holds up in practice.

4.5 Deliberative Responses

In this section, I will examine what responses could be given to the critiques elaborated above. It should become clear in the process where I stand with respect to the three core questions. In the previous chapter, I proposed three criteria for dealing with intractable moral disagreement: (1) we should include as many views as possible, (2) there should be no power differences large enough to distort the decision-making process, and (3) the outcomes of the debate should be open-ended or open to revision. These criteria form the background that informs my position.

---

[109] Graham Smith (2000), page 37.

[110] Simone Chambers (1995), page 241.

[111] Ibid., page 255.

[112] Goi (2005).

4.5.1 Goals of Deliberation

In Chapter 2, I argued that the controversy over biotechnology is a good example of an unstructured problem: deep disagreements exist about facts and values involved and there is conflict not only about policy outcomes, but already on the level of problem definition and possible solutions. I shall argue that in the case of unstructured problems more emphasis should be placed on the underlying moral considerations that are involved. This should be done in the first stages of problem definition and agenda setting, rather than at the end of the pipeline, when the problem has already been framed and possible solutions demarcated, because the latter strategy harbours the risk of excluding opinions and insights. After all, it is the normative problems that usually need most time to be mulled over and as I argued in Chapter 2, most intractable disagreements can be traced to underlying disagreements about values and worldviews. Particularly in the case of novel technologies it is important to include normative issues in public debate at an early stage, because here it is especially important to anticipate potential future problems. The further the development of a novel technology has progressed, the harder it will become to call this development to a halt or to take alternative directions on the basis of normative reasons.[113]

Of course, it could be argued that when disagreement exists at the moral and metaphysical level a problem such as the biotechnology controversy will always amount to an intractable policy problem. Indeed I argued that biotechnology is an unstructured problem, because there is disagreement at the moral and metaphysical level and hence that there is no consensus on values, but also because it exists at the scientific and factual level, and hence that there is no consensus on facts. Drawing on Hisschemöller and Hoppe's account of how policy problems tend to become intractable, I argued that under these circumstances, if these types of unstructured problems are not dealt with adequately they are likely to become intractable policy problems. Whether this actually happens for any specific policy problem can, of course, not be prejudged, but will come out of deliberation. However, at a different level an intractable policy *problem* should be distinguished from intractable *disagreement*. It may be true that the moral and metaphysical nature of the disagreement over GM makes it an intractable disagreement by its very nature, but this does not necessarily have to translate into an intractable policy problem if the parties to the debate can agree to disagree or reach a compromise or an incompletely theorized agreement.

When we are dealing with unstructured problems a focus on consensus is premature. This is because, as critics of deliberative democracy pointed out, too much focus on consensus building through deliberation harbours the risk that certain voices will be excluded. Consensus can be suppressive of minority views and it can close off the discussion (or at least give the appearance of closure) and these effects run counter to my criteria of inclusiveness of views and open-endedness.

---

[113] See Marta Kirejczyk (2005).

4.5 Deliberative Responses

Consensus is not simply 'too much to ask' – an unapproachable ideal – but it should not be the aim in the first place. As we are dealing with an unstructured problem, in which the problem definition and possible solutions themselves are already disputed, we benefit from fostering diversity of opinion, while a focus on consensus can function to prematurely narrow down the debate.[114] I agree with critics that the aim of consensus assumes too much homogeneity and therefore creates a blind spot for difference and plurality. When we abandon the aim of consensus, we can see that the importance of deliberation can also lie in making differences and plurality more visible. As John O'Neill forcefully states it: 'The virtue of deliberative democracy may lie not in claims that it resolves conflicts, but in its tendency to reveal them'.[115] I am not arguing that consensus can never be a justified aim. An aim for consensus can be beneficial in certain circumstances, as it requires participants to keep an open mind and to truly listen to one another and this in turn is more likely to lead to opinion transformation. When the problem definition is well demarcated (and there is little disagreement about it), when all relevant information and opinions have come to light, all views have been extensively discussed in previous public debates and the stage of decision-making has been reached, striving for consensus is a worthwhile aim. At this stage of the process, however, the problem is no longer unstructured.

While I agree with the critique of consensus, I do not think it is debilitating for deliberative democracy. In fact, many deliberative democrats do not uncritically embrace this goal anymore. Dryzek and Niemeyer, for example, argue that in political practice a balance has to be struck between consensus and pluralism.[116] Sometimes deliberation can expose shared views underneath a plurality of viewpoints that would have gone unnoticed without deliberation, suggesting that consensus can be possible on one level, while there is disagreement on the other. Besides consensus, two other possible outcomes of deliberation spring to mind. Firstly, agreement can be reached on a policy even if no agreement exists on the underlying values that support the policy. The policy can be supported by two or more opposing comprehensive views, for different reasons. While Sunstein terms it 'incompletely theorised agreements on particular outcomes' perhaps this should be termed 'convergence' rather than agreement.[117] Secondly, just as deliberation can sometimes uncover self-interested or opportunistic arguments, at other times through deliberation participants can learn that their opponents' viewpoints are held

---

[114] This view has been extensively argued in feminist political theory literature, such as in the idea of the Rainbow Coalition by Iris Marion Young (1990).

[115] John O'Neill (2006), page 276.

[116] Dryzek and Niemeyer (2003). Dryzek and Niemeyer try to reconcile consensus and pluralism by introducing the notion of 'meta-consensus', which they think should also be embraced by agonists and difference democrats, because 'meta-consensus implies reciprocal understanding and recognition of the legitimacy of the values held by other participants in political interaction' (page 9).

[117] Sunstein (1997), page 96. I use the term convergence here similarly to Bryan Norton, who proposed the 'convergence hypothesis'. Bryan Norton (2003), page 11.

sincerely and are not just based on self-interest. [118] These are all ways to deal with intractable disagreement that do not require consensus.

As long as deliberative democrats do not hold on too strictly to the goal of consensus, and accept that sometimes a (partial) consensus is possible and sometimes it is not, but that it is by no means a necessary pre-condition of deliberative democracy, agonism in my view is not necessarily opposed to deliberative democracy, as long as deliberation is supplemented by alternative forms of political action, such as protest and rallying.[119] In my view, these alternative forms of political activity are sometimes necessary, depending on the stage a policy problem is in and on the actual opportunity groups have to equal access to the decision-making process. If in a deliberation about a particular policy problem the dominant groups in society are still found to exclude other groups – for example by biased ways of framing the policy question – other ways of getting heard are legitimate. It should be kept in mind, however, that there can be a tension between deliberation and alternative forms of political activity; it is difficult to both protest, rally, or perform some other form of activism and at the same time take part in rational deliberations with those whose position one is challenging.[120]

Regarding the goals of inclusiveness and quality, the question whether one should emphasise one or the other also depends on the type of problem involved. In unstructured problems as many voices as possible need to be heard, because the disagreement already exists on the framing of the policy problem and, following the criterion of inclusiveness of views as many different viewpoints as possible should be included in the agenda-setting and problem defining stages of deliberation. Unlike structured problems – which can mostly be dealt with by experts – in unstructured problems no group has the sole prerogative of framing the debate. Another important reason to emphasise inclusiveness is that participation in itself could have beneficial effects on the participants. Deliberation between a wide variety of people has more educative potential than deliberation among similar people, because of the confrontation with different people's viewpoints; moreover, participation can lead to empowerment, part of which is the feeling that one's opinion matters and that one has the competence to take part in political decision making. In due time, these beneficial effects may also lead to an increased quality of argument, because of the possible educative and empowering effects of deliberation. When a problem has become moderately structured it may be time to place more emphasis on the quality of the argumentation.

Rational public deliberation can, then, be regarded as one stage in a decision-making process. Perhaps we could envision an approach along three co-existing tracks, of which one track consists of alternative political activities and agonal spaces in which the main goal is agenda-setting and finding out where people's

---

[118] See also Mark E. Warren (2002).

[119] I therefore regard agonism as an immanent rather than an external critique of deliberative democracy.

[120] Young (2003). There is empirical evidence for this claim. See Mutz (2006).

4.5 Deliberative Responses

differences lie. This 'agonal' track could have an emancipatory function and could provide an avenue to constantly challenge dominant views; those who do not agree with the terms of the debate in other tracks can express themselves here. In a second 'deliberative' track more organised public debates between representatives of all the different affected groups could be used in order to deepen debate about the issue and also try to find shared values and perhaps reach incompletely theorised agreements, albeit with no requirement that consensus must be reached. This process needs a longer time-frame than is often available for regulatory decisions and is likely to be open-ended.

Of course, the decision-making process would function more efficiently and expediently if the goals of inclusiveness and deepening were bypassed. However, this runs the risk that some relevant considerations, particularly value-related ones, are left out. Without formal status for public involvement in the decision-making process, broader moral considerations may easily be bypassed and as I explained before, these should not be bypassed, because they are the fundamental issues about which disagreements exist in unstructured problems. It is fair to say that there will always be a tension between decision-making expediency and deliberation. On the other hand, public deliberation could also be regarded as an expedient way of bringing to light relevant information and creating better compliance with policy decisions, which is also a form of expediency. For reasons of decision-making expedience a third track is needed of decision-making institutions. The decision-making process in these institutions should be as transparent as possible and decisions should always be revisable in future to fit within a deliberative democratic framework. However, the goals of broadening and deepening have to give way to a certain extent to decision-making expedience in this third 'institutional' track. In order for deliberation and participation to have influence on actual decision processes there somehow needs to be a connection between the three tracks, for otherwise these tracks would be purely symbolic. The institutional track should 'feed off' the other two tracks, so to speak; it can receive input in the form of arguments from the other tracks, and it has to be accountable in the sense that decision makers should explain why they favour certain solutions over others. The debate in the other two tracks can take place on a more general level as decisions do not have to immediately follow from them; their influence would be felt more in the long run.

In the three-track approach deliberation would then be separated from decision-making. In the first track power structures and discourses and terms of the debate in the other tracks can be exposed and challenged, while in the second track broader views about a particular policy problem can be discussed. Unstructured policy problems should primarily be debated in the first two tracks, as it is often too premature to make a decision. If a regulatory decision must be made it has to be a very provisional one that is revisable. How exactly this three track proposal should be institutionalised is a matter for further investigation. Some empirical research has already been done in this regard. Archon Fung, for example, has argued that the goals of a specific deliberative practice should determine its specific design choices.[121] For example,

---

[121] Fung (2003).

if the goal of a particular organised debate is public learning, the ideal set-up would be to have frequent meetings and to create a monitoring function. These would also contribute to accountability and legitimacy. If the empowerment of marginalised groups is the goal, emphasis should be placed on presence of these groups and on 'accessible modes of deliberation'; also the outcome of the deliberation should have important status, because 'a mini-public cannot advance justice without power'.[122] It should be mentioned how my three track approach differs from Habermas' view of democracy which some authors have labelled 'two track', in which the first track consists of informal deliberation in the public sphere and the second of more formal deliberation and decision-making by government institutions.[123] In my approach a track is added by sub-dividing the deliberation that should go on in the public sphere into one – more informal – agonal and one – slightly more formal – deliberative track.

In the three-track approach that I envisage, how should we regard the role of expert committees and consensus conferences? Are they instances of depoliticisation or not and how should we evaluate these? Many proponents of deliberative democracy assume that the public will be more likely to deliberate in terms of the common good and that in public deliberation self-interested arguments will be curbed. In this context, commentators like Pettit and Fishkin and Luskin regard public involvement as an instance of depoliticisation.[124] They view depoliticised bodies such as committees and consensus conferences as checks on government control, as bodies that foster impartiality. This appears to be based on a view of current politics as primarily located in the arena of official government institutions, which are heavily influenced by interest group lobbying. The underlying assumption is that in public deliberation interest group lobbying will not be as influential. Depoliticisation in this view means taking power away from the formal political institutions and including the public. This is exemplified in the following quote by Pettit: 'democracy – deliberative democracy – is too important to be left in the hands of the politicians. No democratization without depoliticization'.[125] This view does not take account of other functions of the political arena – including civil society – such as facilitating debate about what is in the common interest. Hisschemöller and Hoppe, on the other hand, do not see a democratising potential for depoliticised bodies. In contrast, they view depoliticisation as a method for governments to remain in control of controversial policy issues, carefully keeping these issues away from the public. They view depoliticisation as something negative, as a method of stifling the public debate necessary for dealing with unstructured problems. Their view should be understood against the background of technocratic governments. In such systems, important decisions regarding novel technologies are left up either to the industry or solely to scientific experts, and this could be regarded as depoliticisation. Public

---

[122] Ibid., page 351.

[123] See for example Joshua Cohen (1999).

[124] Pettit (2004); Fishkin and Luskin (2005).

[125] Pettit (2004), page 64.

4.5 Deliberative Responses

influence is seen as a way to counter technocracy, as a way to bring controversial policy issues out in the open, so that they do not remain the prerogative of technical experts and the governments who use these experts. The do not see public influence, then, as a way of depoliticisation, but rather of *re*politicisation; it counters the depoliticisation that took place under technocratic governments. This is more congruent with views on deliberative democracy that see no role for depoliticisation, but rather for more politicisation. Porsborg Nielsen et al., for example, write that 'the development and spread of the participatory consensus conference is, in effect, an attempt to "expand the political possibilities of action" through public involvement;.... it can be seen as an attempt to ... "put politics in command", that is, expand the political possibilities of action vis-à-vis the growing dynamics of scientific and technological developments'.[126] Understood this way, deliberative democrats in fact want to politicize more issues rather than depoliticize them.

These two meanings of depoliticisation in relation to the role of the public, then, seem diagonally opposed. This is because the two meanings are based on different views of politics. In the first 'Pettit-meaning', a contrast is sketched between the sectorial or personal self-interest that cannot escape formal political institutions, on the one hand, and the common good that is advanced by the public, on the other hand. The public plays a depoliticising role because it counters the self-interest and lobbying that characterises politics. In the second 'Hisschemöller-meaning' technocratic government control via depoliticisation is contrasted with public deliberation. Depoliticisation in this view entails containment of debate and containment of contestation. It appears to be based on a view of politics as contestation, similar to that of agonists. My three-track approach is closer to the Hisschemöller-view, because it wants to involve the public in all three tracks, but not as a way of taking control out of the hands of government. After all, the final decisions are made in the institutional track. This track should be open to input from the agonal and deliberative tracks. Like Porsborg Nielsen et al., I view public input as a form of repoliticisation; public debate should not be confined to specific closed committees or depoliticised bodies, but should take place on several different levels. On the other hand, I view the role of public debate as both countering self-interest and lobbying, and countering the primacy of technocratic decision-making by experts, and I therefore incorporate at least part of the 'Pettit-meaning'. Through *re*politicisation citizens in fact become more critically aware of, and more likely to be involved in solving, collective problems and this could help to counter decisions made on the basis of politicians' self-interest. The function of public participation in policy regarding novel technologies in my eyes, then, is not an instance of depoliticisation, but rather of repoliticisation. As will become clear in the empirical part of this book, what is seen as depoliticisation and its potential is dependent on the context of the political culture under examination.

---

[126] Annika Porsborg Nielsen, Jesper Lassen, and Peter Sandøe (2007), page 17. The authors here quote Leonhard Hennen (1999), page 304.

4.5.2 What Types of Argument Are Valid?

For reasons alluded to before, I agree with deliberative democrats such as Gutmann and Thompson, and Dryzek that all views, including comprehensive views, should be allowed in public debate. Whereas Ackerman argues that individuals should discuss comprehensive views in private, outside of the political arena, I hold that there is actually more reason for individuals to practice conversational restraint in personal relationships than in politics. While we need to discuss our comprehensive views in order to understand better what divides us and to make better political judgements, so that we will be able to co-exist in society, we do not need (although of course we are free) to discuss our views on religion, technology, animal rights, etcetera, in our personal relations. In fact, conversational restraint could save friendships.[127] When I say that we should discuss our comprehensive views I do not mean that we need to discuss what religion or other comprehensive view is best and should be adopted by everyone. This is something that every individual should be able to decide for herself. I merely mean that since our positions on political matters are often based on fundamental values drawn from our comprehensive views, it would be better to acknowledge the views that inform our positions than to let them silently inform our judgments in the background. In fact, we would get an artificial conversation if we could not refer to these comprehensive views at all. Allowing comprehensive views at least opens up the possibility of preference or opinion transformation to occur and this is not likely under the liberal condition of conversational restraint.

Theorists who propose a procedural model of deliberative democracy argue that in the context of intractable disagreement as a result of value pluralism we need to look for agreement not on substantive issues but on procedures. One can wonder whether implicit in this view is the kind of separation between morality and ethics or the right and the good, which I have problematised in Chapter 3. Seyla Benhabib, for example, states that

> The challenge to democratic rationality is to arrive at acceptable formulations of the common good despite this inevitable value-pluralism. We cannot resolve conflicts among value systems and visions of the good by re-establishing a strong unified moral and religious code without forsaking fundamental liberties.[128]

Benhabib seems to assume that we either discuss the truth of substantive beliefs or only discuss procedures that are completely neutral vis-a-vis substantive beliefs. However, if we accept my point that comprehensive views of the good life cannot be neatly separated from universal moral goods this is problematic. I agree that we need not discuss the truth of religious views in order to be able to co-exist peacefully. However, most political issues on which decisions need to be made will need to

---

[127] Empirical research also shows that 'people tend to care more about social harmony in their immediate face-to-face personal relationships than about the larger political world'. See Mutz (2006), page 106.

[128] Benhabib (1996), page 73.

4.5 Deliberative Responses

rely on discussion that may involve an appeal to substantive beliefs. Recall that we are dealing with public goods. Neutral procedures should therefore be the *basis* of public deliberation, but not the endpoint. The consequence of this view is that we will just have to accept that no neutral outcomes are likely to ensue; this is one reason why I proposed the criterion of open-endedness. Of course, some set of procedures and rules will have to be agreed upon before public deliberation can be held; we cannot allow behaviour that undermines the process of deliberation, for example. These procedures are internal to the process of deliberation, however, and do not rely on the distinction between ethics and morality. Moreover, these rules are open to change through the process of deliberation itself. I agree with Benhabib in this respect; she likens the conditions of deliberation to those of a game 'that can be contested within the game but only insofar as one first accepts to abide by them and play the game at all'.[129]

My view that broader issues should be allowed in public deliberation raises the question whether just any views should be allowed in public debate. The logic of deliberative theory entails that views that hinder public debate itself – such as views that do not consider all participants as free and equal, or those views that would condone the use of threats or violence against other participants – should not be allowed. Furthermore, as pointed out before, Gutmann and Thompson argue that a position should only be excluded from political dialogue when it does not qualify as a moral position in the first place. In the context of my criticism of the ethics/morality distinction I would not use the word moral, but for Gutmann and Thompson at least, a position fails to be moral when it does not meet either of the following three requirements: (1) it 'presupposes a moral point of view'. . ., (2) 'any premises in the argument that depend on empirical evidence or logical inference should be in principle open to challenge by generally accepted methods of inquiry', (3) if the latter is not applicable, then at least the evidence 'should not be radically implausible'.[130] Racist views do not meet any of these conditions and, therefore, policies favouring racism should not be allowed on the political agenda. These grounds for preclusion from the political agenda are different from traditional liberal grounds. A position is not precluded because it is a moral position on which reasonable disagreement can exist, but because it does not even qualify as a moral position. In other words, the liberal neutrality thesis is not adequate, because it removes racist policies from the political agenda for the wrong reason – namely, because no agreement on it can be reached – and not because it is not a reasonable position in the first place. If anything, lack of agreement is the reason par excellence why issues should be on the political agenda.

While I agree with Gutmann and Thompson that liberals move issues off the agenda for the wrong reason, I do not wholly support their grounds for preclusion from the political agenda either. In my view even less restrictions to the domain of debatable issues are necessary than they propose. The requirements mentioned

---

[129] Ibid., page 80.

[130] Amy Gutmann and Dennis Thompson (1990), page 71.

above could be used to limit the debate more than necessary. For example, their first requirement may exclude positions that are based on specific situations or interests that are quite legitimate, such as feminist points of view. Their second requirement raises the question as to what should be understood by 'generally accepted methods of inquiry'. If interpreted literally, this could mean that positions that are based on alternative scientific paradigms are not allowed. Nevertheless, paradigm shifts have taken place in the past and what was generally accepted has consequently also changed. If 'generally accepted methods' is taken to mean what the majority of people in a given country accept this could also be problematic. What if, for example, the majority of citizens in the United States reject methods of inquiry that support the theory of evolution?[131] This would rule out a lot of positions that I think are quite acceptable in a public debate.

I do agree with Gutmann and Thompson's recommendation that we determine through public deliberation which specific moral issues that people disagree about should properly belong to the public realm of politics. They hold that 'democratic deliberation addresses the problem of moral disagreement directly on its own terms. It offers a moral response to moral conflict'.[132] While Gutmann and Thompson argue that views that do not conform to their criteria for constituting a moral position should be excluded, I agree with Dryzek that such criteria are not necessary. Only behaviour that jeopardizes the deliberative process itself, such as violence or threats should be excluded and the deliberation itself should be able to sort good from bad positions. Likewise, Gutmann and Thompson's view of moral accommodation in the face of disagreement in my view unnecessarily narrows the debate because of the types of argument it excludes. Their emphasis on mutual respect as a condition for fair deliberation seems to have led them to assume that all participants in a debate will participate with the common good in mind and without self-interested motives. Therefore, they propose that in a debate we should not question the moral status of the views of our opponents by expressing a distrust of their motives, but solely discuss the content of the arguments put forward.[133] This restricts the opportunity for exposing self-interested motives that is in fact an important reason to hold a debate publicly in the first place.

Dryzek points out that Sanders' criticism of deliberation as elitist fails to separate deliberation as such from deliberation in a particular context, namely courtroom juries in American style liberal democracy. It is therefore not clear whether her criticism would hold in other contexts of deliberation.[134] Young's criticism, on the other hand, may be more damaging for deliberative democracy, as it points its arrows at deliberation as such, particularly when deliberation is taken to require common or shared premises. She argues that due to the existence of pervasive difference

---

[131] I am endebted to Betsy Postow † for providing me with this example.

[132] Gutmann and Thompson (1996), page 41. Gutmann and Thompson appear to use the term 'moral' not in the Habermasian sense, but in a sense encompassing both ethical and moral.

[133] Ibid., page 83.

[134] Dryzek (2000).

no neutral premises exist on which to build the deliberation. Supposedly neutral premises tend to reflect the political status quo.[135] However, as Dryzek correctly points out, we will only know whether this criticism is really sustained by investigating deliberation empirically. And even then much will hinge on how the deliberation is set up and on the context in which it takes place. Claims of difference democrats can be somewhat toned down by empirical research into jury deliberations. Research by Andrea Hickerson and John Gastil, for example, demonstrated that gender was not a clear determinant of dominance in jury deliberations or its outcomes, nor that women's speech styles were systematically undervalued.[136] The deliberation, in general, displayed an equalising effect between men and women. This supports the view that over time marginalised groups and their styles of argumentation may become more accepted and that the dominant discourse can change. This would not only involve a learning process on the part of marginalised groups about how to make themselves better heard, but also on the part of dominant groups about how they can listen better.

This latter view entails that forms of communication other than rational argument should be allowed in public deliberation, like Young and Sanders propose. As a response to difference democrats' critique of overly rational deliberation styles, Benhabib argues that in the public sphere there are many associations, groups, and networks where difference is stimulated. However, the problem with other modes of communication, such as storytelling, greeting, and rhetoric is that they cannot be the bases of democratic institutions and the rule of law, as they are not commonly shared. People might not understand each other's stories and greeting can only get you so far.[137] I do not agree with this position, because the same could be said about arguments; the form arguments take are not necessarily commonly shared or understood either. On the face of it there is no reason why deliberative democracy could not encompass greeting, storytelling, and rhetoric, but I agree with Dryzek that these forms of communication have to fulfil certain conditions in order to avoid abuse. Storytelling particularly could be a very effective tool when we want to confront people with different viewpoints. However, I doubt whether storytelling will really avoid elitist tendencies. As Dryzek points out, storytelling can just as well become coercive because of 'normalising' behaviour in groups.[138] He, rightly, proposes to allow only forms of communication (including rational argumentation) that can withstand two tests: it should not be coercive and it should be able to rise above the particular situation and be applicable to more general views: 'if an individual's story is purely about that individual then there is no political point in hearing it. . .a truly effective story about a particular repression will also involve implicit appeal to more universal standards'.[139]

---

[135] Ibid.

[136] Hickerson and Gastil (2008).

[137] Benhabib (1996), pages 83–84.

[138] Dryzek (2000), page 68.

[139] Ibid., page 69.

Some deliberative democrats argue that the legitimacy of certain types of communication and its effects on marginalised groups can and should be questioned within the deliberative forum itself. This is no simple feat, however. As Olson points out, in order to be taken seriously, challenges to the dominant discourse or the terms of the debate must be formulated in the very idiom that one is trying to challenge.[140] Olson admits that this is not entirely impossible, but that the success of this strategy is relative to the context of the political culture in which it takes place. It depends, among other things, on the question how reflexive a deliberative democracy is – how much self-reflection it stimulates about its own practices – and how open it is to alternative styles of speech. This in turn depends on the willingness of elite groups in society to accede challenges to their hegemony. Olson furthermore pointed out that the question of who is regarded as an expert is also based on the standards of dominant groups in society. However, as we saw at the start of this chapter, one of the reasons for calling for more deliberation was that the status of expert knowledge has come under attack. A viable deliberative democracy will, therefore, have to be structured in such a way that official expert knowledge can be challenged and hidden assumptions about expertise identified.

Finally, taking into account the law of group polarisation, it is especially important to ensure that deliberation takes places across divides. Groups that deliberate should not be formed on the basis of common identities or opinions, but rather should have a diverse composition. So, for example, if a debate is organised about genetically modified animals, we should not put all members of animal liberation societies and vegetarians together in one group, but make sure that both proponents and opponents are represented.

4.5.3 Who Should Participate?

Because one of the educative effects of public deliberation is taken to lie in the confrontation with other people's points of views and experiences, a case can be made for physically including representatives of groups and not only representing their ideas. Broad participation is therefore an important tenet of deliberative democracy. However, as difference democrats have pointed out, this interpretation of representation rests on false essentialisms. This is a tension that is not easily overcome, however, and we might have to run the risk of false essentialisms at least in the short run, until more equal participation has been reached and we can diversify identities. As we have seen, difference democrats argue that equal participation is merely a theoretical supposition that does not sufficiently take into account the differences in wealth and power that permeate the real world. Against these critics we could argue that the current emphasis on bargaining tends to disadvantage women and marginalised groups more than deliberation does and deliberative democracy is therefore already a step in the right direction. However, they are right to point

---

[140] Kevin Olson (2011).

4.5 Deliberative Responses

out that a mere formal requirement of equality will not be sufficient.[141] Substantive equality will need to be ensured, for example with the help of quotas, education and positive discrimination. The focus on inclusiveness and equal participation in deliberative democracy should also function to avoid tyranny of the majority; again, this will only be true if substantive measures are taken to ensure political equality. Even though intelligence and knowledge is not distributed evenly among citizens and the risk therefore exists that the highly educated elite will take a dominant position in deliberations, as Manin points out, among the elite or experts disagreement exists as well and being exposed to such disagreements provides a learning opportunity for those less knowledgeable and at the same time mitigates the dominance of the elite.[142]

Whether the criticisms described above are warranted depends to a certain extent on the specific set-up of the public deliberations. For example, in a deliberative poll, in which an organized form of public deliberation on a relatively broad scale precedes an anonymous vote by the participants, the distorting effects of group dynamics play less of a role. Because the participants vote anonymously after the deliberation the problem of conformity or group pressure does not occur and because consensus is not strived for, deliberative polls tend to avoid the problem of suppression of dissent.[143] Moreover, the problem of group polarisation occurs less with deliberative polls than in many other political discussions. Research has shown that after participating in a deliberative poll the participants often have less extreme and more nuanced and better considered views on the issue under debate and they have acquired more understanding of their opponents' viewpoints.[144]

Finally, difference democrats make a good case for supplementing deliberation with other forms of political activity, most notably activism. In fact, if we recall the results of Mutz's empirical research it appears that a tension exists between participation and deliberation and this suggests that a combination between deliberation and political activism is necessary. Maximum participation would lead to enclaves of like-minded people who only reinforce each other's views and this in turn leads to polarization. On the other hand, in some contexts 'like-minded networks' can stimulate very important forms of political action, such as social movements against slavery or sex discrimination. At the same time we need people who are willing to engage with opposing viewpoints, as this open-mindedness is more conducive to tolerance, well-balanced judgments, and compromise. According to Mutz, 'a "mixed" political culture would result in greater political stability than one with maximal levels of participation'.[145] Deliberation, in other words, is just one aspect of politics; while more emphasis should be put on deliberation and on making those instances in which deliberation takes place more successful, there should always be other avenues of political action as well.

---

[141] Young (1999), page 156.

[142] Manin (1987), page 354.

[143] Fishkin and Luskin (2005).

[144] Dag Elgesem (2005).

[145] Mutz (2006), page 132.

4.5.4 Recapitulation

In this section, I argued that when we are dealing with unstructured problems aiming for consensus is premature and can function to exclude viewpoints. I proposed a three-track approach in which deliberation and decision-making are separated, but in which the institutional track would receive input from the agonal and deliberative tracks. I furthermore argued that no views should be excluded from public debate at the outset, unless these views include threats of violence or counteract the possibility of holding a debate in the first place. Finally, I argued that if special care is not taken to create actual conditions of equality of voice, deliberative democracy could in fact lead to more inequality by privileging already dominant groups in society. Especially in a situation of transition to more deliberative forms of governing other political measures, such as redistribution and affirmative action, and other political activities, such as protest and activism, will be necessary to create more equal conditions for deliberation. Under these conditions, the deliberative process, in turn, could lead to more political equality. The proposed measures can, of course, at a later stage always be questioned through deliberation, and this is where the criterion of open-endedness plays a central role. Whether deliberative democracy is successful in including all affected groups and whether distorting group dynamics can be avoided is something that in the end will have to be tested in practice. In the case of 'groupthink', for example, research shows that deliberation among a group of people tends to lead to results closer to the truth: 'collectively formed group opinions are generally more closely 'on target' than the mere mechanical average of individual opinions'.[146]

4.6 Conclusion

In Chapter 3, I argued that deliberative forms of democracy are better able to deal with intractable disagreement than traditional forms of liberal democracy, as they focus on the possibility of preference or opinion transformation rather than voting or bargaining. The emphasis on preference transformation does not make deliberative democracy necessarily an anti-liberal theory, however.[147] In fact, many of the core principles of liberal democracy are compatible with deliberative ideals: deliberative democrats 'value and see as essential the liberal legacy of rights – not so much rights of private property, but rather the rights that attach to individuals, such as rights of security, citizenship, due process, equal protection, political participation,

---

[146] Rescher (1993), page 35. See also Herman Van Gunsteren (2006).

[147] While deliberative democracy can be regarded as a form of liberalism, its more radical versions distance themselves from liberalism, particularly from the condition of conversational restraint, but also from the close alliance between liberal democracy and capitalism in practice. See for example John S. Dryzek (1996); Dryzek (2000). Whether this is warranted is an issue for further exploration, for which I unfortunately do not have space here.

4.6 Conclusion

speech and association'.[148] The liberal basis of deliberative democracy, I would argue, is that everyone should have an equal starting position in the debate. Because deliberative democracy tries to counter inequality in decision-making influence – as decision-making should not be based on coercive power but on arguments – it in fact conforms better to the liberal principle of equality than the mechanism of voting. In other words, I hold that the liberal democratic principle of equality should be broadened from equality of voting to equality of deliberation. Therefore, I do not wish to present deliberative democracy as an alternative to liberal democracy, but merely wish to broaden, and indeed strengthen, liberal democratic theory and practice by placing more emphasis on deliberation.

In this chapter, I have examined how different deliberative democrats have dealt with the existence of intractable disagreement by considering three core questions and the criticisms that have been mounted against deliberative democracy. An awareness of these criticisms helps us to focus on what obstacles we might encounter in the practice of public deliberation. These criticisms have really driven home the idea that the criterion of inclusiveness has to be more than a mere theoretical presupposition, more than an assumed condition, for deliberation to succeed. For deliberation to be truly inclusive, differences between groups have to be recognized more and the demand for consensus has to be less stringent than in the models of Habermas and Cohen. Wibren van den Burg puts it eloquently: 'Even if we aim to build consensus, we should cherish plurality and dissensus in the process and take the dissenting opinions seriously, since they may be the heralds of a newly emerging morality'.[149]

Another problem for deliberative fora such as consensus conferences is that while the distinction between experts and lay persons has been challenged a tendency exists to resort to expert knowledge. The status of expert knowledge and scientific research in general should be the subject of public debate and we should strive for a two-way relationship of information exchange between experts and lay people with experiential knowledge.

My response to the core questions can be translated into three conditions that need to be met for a theory of deliberative democracy to be able to deal with intractable disagreement. My responses to the core questions and the conditions that follow from them are related to the criteria for dealing with intractable disagreement that I elaborated in Chapter 3. The conditions that I will formulate each lead to a more specific set of criteria. These criteria will be applied in the empirical part of this book, where I will investigate how two organised deliberative practices – biotechnology ethics committees and consensus conferences – fared.[150] I use these case-studies in order to seek the implications of

---

[148] Warren (2002), page 176.

[149] Wibren Van der Burg (2003), page 27.

[150] As I have argued elsewhere, these conditions and criteria could be applied to evaluate different deliberative practices, such as planning cells or citizens' juries as well. See Bernice Bovenkerk (2009).

my conditions and criteria. In other words, the aim of my empirical analysis is not only to find out to what extent real life deliberative fora conform to the ideals of deliberative democracy, but also to examine the feasibility of my conditions.

Regarding the goals of deliberation, my condition is that the aims of the deliberation should match the type of policy problem. When we are dealing with an unstructured policy problem the broader aim should be opinion transformation – as thinking about such problems is still in a developmental stage – and this entails firstly inclusiveness, or broadening, and in a later stage deepening. When we are dealing with more structured problems, in which broad agreement exists about problem definition and demarcation and solving the problem comes down to applying a set of standards or techniques, consensus may well be a plausible aim. However, public debate is in itself less necessary in such cases, and may primarily be about the question of how to interpret certain standards. In the case of moderately structured problems, consensus may be a worthwhile aim in the later stages of the decision-making process, after the problem has been more clearly defined and demarcated and there is less risk of the suppression of dissensus. The benefit of aiming for consensus in moderately structured problems would be that it forces the participants to come up with a solution, which would both improve decision-making expedience and force participants to take each other's points of view. However, aiming for consensus is only desirable when the strategies (or biases) of excluding participants or narrow problem framing can be avoided. While I argued for a three-track approach I emphasised that the first and second tracks should somehow influence the decision-making process, in order to be more than purely symbolic. This entails, among other things, that policy outcomes are revisable in future; I argued for revisability in Chapter 3 regarding my criterion of open-endedness.

Regarding the type of arguments, my condition is that we should strive to exclude as few arguments and argumentation styles as possible. This condition is based on my criterion of non-exclusion from Chapter 3, which is supported by my arguments against conversational restraint and by the view that governments owe citizens respect for their opinions and intellectual capacities. Furthermore, this condition is based on the view that given the limitations we are faced with in this world, it is always possible that certain entrenched beliefs or views turn out to be mistaken. In such a world it is of utmost importance that there are different lines of thought that can be tested against one another. In this chapter, I also highlighted the importance of lay person input in the debate and one more specific criterion is, therefore, that lay persons should be able to bring in additional arguments or experiential knowledge and have a say in determining the framework for deliberation. The dominance of formal expert knowledge and scientific methods should be open for discussion, and this is enabled, amongst other things, by open expert contestation. Finally, a specific criterion is that it should be possible to uncover fallacies and self-interested motives through the deliberation.

4.6 Conclusion

Regarding the question of participation, my condition is that we should aim to be as inclusive as possible, particularly when dealing with unstructured policy problems, and this means paying close attention to mechanisms of exclusion and obstacles to equal participation. As I argued in Chapter 3, in public deliberation decisions should not be forced by powerful groups upon people with less power. This leads to some more specific criteria: women and marginalised groups should not be disadvantaged in the debate; mechanisms should be in place to counter power differences; distorting group dynamics should be avoided; people with a wide variety of backgrounds and opinions should be included. In practice this means that quotas or other mechanisms should be used in order to achieve representation of marginalised groups. I also argued that the status of expert knowledge should be relativised. Finally, I argued that we should aim for as many participants as possible, amongst other things because of deliberation's educative effects. This leads to the following criteria: participants should learn from the debate; the arguments and outcomes of the debate should reach people other than the direct participants. Because of the tensions between the different pillars of deliberative democracy that I noted, it is not expected that all of these criteria could be fulfilled at the same time.

These conditions and criteria are displayed in Table 4.1:

Table 4.1 Conditions and guiding criteria

Goals
Condition: the aims of the deliberation should match the type of policy problem
Criteria: (1) opinion transformation should take place
        (2) deliberation should influence the decision-making process
        (3) outcomes should be revisable

Arguments
Condition: no relevant arguments or viewpoints should be excluded
Criteria: (1) there should be lay person input
        (2) it should be possible to uncover fallacies and self-interested motives

Participants
Condition: we should aim to be as inclusive as possible
Criteria: (1) women and marginalised groups should not be disadvantaged in the debate
        (2) mechanisms should be in place to counter power differences
        (3) distorting group dynamics should be avoided
        (4) people with a wide variety of backgrounds and opinions should be included
        (5) there should be open expert contestation
        (6) participants should learn from the debate
        (7) the outcomes of the debate should reach people other than the direct participants

---

We will now move to the empirical part of this book, in which I will examine the implications of these conditions and guiding criteria when they are applied to real-world contexts.

Intermission: Between Theory and Practice

Up until now I have argued that the debate about animal and plant biotechnology has reached a state of intractable disagreement, due to all the different dimensions of disagreement that it involves. I have defined biotechnology as an unstructured problem, meaning that no consensus exists on either the facts or the values involved; indeed, facts and values are often interwoven. In this context, it is increasingly acknowledged that expert knowledge is not purely objective, that it is not infallible, and that lay persons can contribute certain experiential knowledge that is often overlooked by scientific experts. On a policy level unstructured disagreements can become intractable when the problem is inappropriately treated as structured or moderately structured. This happens when one or both of the dimensions of disagreement are not acknowledged. My next question was how governments of democratic societies should deal with the existence of such intractability. I argued that on a theoretical level, deliberative democracy could deal better with intractable disagreement than traditional variants of liberal democracy, most notably political liberalism, because the emphasis on deliberation would enable citizens and governments to address the real sources of disagreement and would enable opinion transformation. Governments, therefore, should enable rather than constrain public dialogue.

Behind the core aim of deliberative democracy – deliberating together in public about the common good – lies an inclusive ideal of the direct democracy. This ideal is unattainable in modern complex states. Governments that nevertheless wish to implement this ideal have put their hopes on concrete deliberative fora. The purpose of Part III, is to examine whether deliberative fora actually contribute to the inclusive ideals of deliberative democracy, or whether they give cause to revise this political philosophy. In this chapter, I have defined three conditions and a set of guiding criteria to aid my examination. My general conditions are: (1) the aims of the deliberation should match the type of policy problem, (2) no relevant arguments or viewpoints should be excluded, and (3) we should aim to be as inclusive as possible.

In order to determine what circumstances are favourable to an open and inclusive debate, in the empirical part of this book, I will analyse those two fora that have been used most by governments and social organizations to give shape to the wish for more public deliberation regarding biotechnology – namely biotechnology ethics committees and consensus conferences. I will investigate these two in two different political cultural contexts, the Netherlands and Australia, because my assumption is that these contexts can make a difference. This double comparison should particularly shed light on the question to what extent the political culture within which the deliberative fora are embedded influences their success or failure. With a few notable exceptions,[151] this link between political culture and the merits of deliberative fora has not been studied before. Evaluations of the committee system and consensus conferences tend either to focus only on their functioning in one country, assuming that they are so bound to one country that no general conclusions can be drawn, or address their functioning more generally, thereby discounting cross-cultural differences. My research, in contrast, compares two institutional designs in two different political cultural contexts, because my intuition was that political culture is influential for the functioning of deliberative fora. This intuition was confirmed by research carried out by Porsborg-Nielsen et al.[152]; however, the difference between their and my research is that these authors concentrate on the political philosophical differences between countries that are implicit in these countries' political cultures, while I make the latter explicit. While one might think that the different subject matters of the Dutch and Australian consensus conferences and committees will make a comparison difficult, I do not so much focus on the content of the recommendations as on the way each deals with intractable disagreement. Moreover, all studied deliberative fora involve animal and/or plant biotechnology.

Before explicating the differences between consensus conferences and the committee system, we need to pause and consider the rationale for and the status of my empirical research. As I already argued in the Introduction to this book, the question about the merits of deliberation cannot be answered solely on the basis of theoretical analysis, because despite its theoretical attractions, a political theory ultimately gains its force from practical application. In order to judge whether the normative assumptions and abstract claims of deliberative theories are justified we will ultimately need to look at deliberative practice. In turn, this practice can be criticised for not meeting certain normative criteria or at least for being unclear about what normative aims are strived for. In the words of Graham Smith,

> by studying the actual practice of institutions, we are in a better position to interrogate the evaluative claims of democratic theory and to open up areas of theoretical inquiry that may have been overlooked at a more abstract level of analysis.... Focusing on the practice of mini-publics can highlight overlooked issues deserving of theoretical elaboration.[153]

I will argue, amongst other things, that one issue that has been overlooked is the difference that political cultures make to the success or failure of deliberative fora. Moreover, in this chapter I identified a set of tensions between the goals of

---

[151] As will become clear in part III, the following authors do specifically look at the effects of political context in the evaluation of consensus conferences: John S. Dryzek and Aviezer Tucker (2008); Porsborg Nielsen, Lassen, and Sandøe (2007). Jan Gutteling has put the Dutch committees and consensus conferences in a political perspective. See Jan M. Gutteling (2002).

[152] Porsborg Nielsen, Lassen, and Sandøe (2007).

[153] Graham Smith (2008), page 19.

consensus, quality, participation, and expediency. An appreciation of these tensions could help in the evaluation of different deliberative practices, because the tensions in deliberative theory may be mirrored in the practice of deliberative fora. For example, the tension between decision-making expediency and participation becomes visible in the discussion about the influence that lay panel reports should have on political decision-making.

To reiterate, the aim of my empirical analysis is to test deliberative democracy – and in particular the conditions and guiding criteria – through case-studies. This test will move in two directions; on the one hand, I aim to develop insights about the strengths and weaknesses of specific deliberative fora by examining to what extent they conform to deliberative theory while, on the other hand, I aim to refine my normative and sociological insights in the light of my comparative, empirical investigation and make them more practically relevant. However, in the Introduction, I already pointed out the inherent difficulty in any 'attempt to bridge the disciplinary gap between normative theory and empirical political analysis'.[154] I argued that on the basis of limited empirical research one cannot draw sweeping conclusions about deliberative democracy in general. We can only test partial aspects of specific deliberative theories in practice in order to understand whether certain claims that deliberative democrats make are feasible. I investigate one such partial aspect, namely the ability to deal with intractable disagreement. Moreover, many different theories of deliberative democracy have been proposed and, as has become clear, each makes its own normative assumptions and has its own emphases. Empirical research can be helpful in relating deliberative theories to institutional design choices. For example, as we saw in this chapter, in response to criticism by agonists and difference democrats some deliberative democrats have now accepted that we should not aim for consensus. I supported this move particularly in the case of unstructured problems. In order to test this claim we can investigate the practice of consensus conferences to see whether consensus in fact does work to suppress minority views. If so, one could argue that a different type of deliberative forum should be used for unstructured problems. To mention another example, the claims of critics about pernicious group dynamics may be confirmed by some deliberative practices and not by others. Deliberative theorists may, therefore, need to qualify some of their claims by relating them to specific practical and institutional contexts.

Of course, one can wonder whether the two types of deliberative forum that I study in fact correspond to deliberative democracy's picture of public deliberation. Most theoretical accounts of deliberative democracy do not specify exact methods and designs of public deliberation, although Bohman has argued for a revival of 'deliberation-promoting conventions, such as delegating certain authority to committees in order that they may acquire more informed and well-reasoned opinions', and points out that deliberative polls offer one procedure to increase public deliberation.[155] Moreover, many authors who advocate consensus conferences

---

[154] Ibid., page 19.

[155] Bohman (1996), page 189.

do so on the basis of deliberative democratic theory. Still, one important difference between committees and consensus conferences and the theory of many deliberative democrats is that the latter envisage the deliberating public as made up of voluntary associations in civil society between citizens with a special interest in the topic under discussion, whereas committee deliberations take place between experts while consensus conferences take place between lay people with no pre-existing interest in the topic. Nevertheless, as ethics committees and consensus conferences are two major ways in which governments and social organizations have given shape to the call for more public deliberation they merit analysis. Perhaps more importantly, one can wonder whether associations in civil society actually conform to the ideals of deliberative democracy. As Smith argues, while deliberation certainly goes on in voluntary associations, the democratic character of these is questionable and neither is it clear that the reasons given in these deliberations are focussed on the common good: networks in civil society 'tend to be populated by fairly like-minded individuals and organizations and their interaction with other organizations and networks with very different perspectives would typically be cast in terms of strategic rather than communicative action'.[156] Of course, certain obstacles to a genuine open and inclusive debate that we will encounter in the committee-system and in consensus conferences may not appear in other deliberative settings, while those aspects of my deliberative fora that work well may not be easily replicated in other deliberative settings. In future research it would be desirable to put my criteria to the test in the context of other deliberative fora with different designs.

Differences Between Committees and Consensus Conferences

Before I proceed to compare my different case-studies it should be noted that there are differences in form and function between the committee system, on the one hand, and consensus conferences, on the other. Of course the most obvious difference is that committees are composed of experts and consensus conferences of lay people. One problem that I will discuss in this context is that this leaves out the group in between that is 'too expert to be lay and too lay to be expert'.[157] One other main difference that I already discussed is that expert committees primarily focus on factual and technical disagreement and consensus conferences on value disagreement, while I argue that both need to focus on both types of disagreement, because we are dealing with an unstructured problem. Establishing an ethics committee amounts to treating biotechnology as a moderately structured/ ends problem and organising a consensus conference as a moderately structured/ means problem. Considering the points I have made in Part II about political liberalism and deliberative democracy, one should perhaps position committees as belonging more in the liberal democratic paradigm and consensus conferences more in the deliberative

---

[156] Smith (2008), page 5.

[157] Purdue (1995), page 171.

paradigm (albeit one open for improvement). One of my main arguments was that political liberalism unjustly constrains debate. The committee system fits well in this paradigm, as I will argue it too constrains debate by relegating more fundamental moral and metaphysical disagreements to the private sphere. In traditional liberal democracies policy problems tend to be cast in terms of factual disagreements and decision-makers tend to rely on expert accounts, because if they would focus more on value disagreements they would be perceived not to be neutral between different citizens' conceptions of the good. I will argue that while there was some room in committees to discuss value disagreements, the main normative decisions had already been made before the instalment of the committee and the conclusions of discussions about values were not given much weight anyway. Also, in the functioning of both committees it became clear that most members wanted to relegate fundamental moral or metaphysical views to the private sphere, to the frustration of a small minority of the members. Consensus conferences conform more to the image of genuine public deliberation, because they involve lay persons deliberating together in public about the common good. The reason why I do discuss both types of deliberative forum in the context of deliberative democracy is that both committees and consensus conferences are meant to play a role in stimulating public debate, although I will argue that both only achieve this function to a limited extent – as committees in practice substitute public debate and consensus conferences do not reach quite as large an audience as is aimed for. Still, a consensus conference could be termed a mini-public, because it includes representatives from the lay public, while committees could not, because their role in public deliberation is more indirect.

Another difference between the two is that committees convene on a regular basis and the members need to be able to develop a working relationship for a longer period, while consensus conferences are usually one-off affairs and, although they may last for a couple of weekends, the participants may only meet each other a few times. This difference will influence the form the discussions will take. The status of the recommendations is a further difference; a committee's recommendations have a direct advisory function to government (particularly in the case of the Dutch Committee for Animal Biotechnology, where the Minister nearly always adopts the recommendation), while the status of consensus conferences is more unclear and non-committal. This leaves the latter more free to explore fundamental disagreements, while the former will be more likely to meet other demands such as decision making expedience, efficiency and influence on the decision-making process.

Once could argue on the basis of the foregoing that committees and consensus conferences occupy two different political spaces. Committees seem closer to the formal institutions of government while consensus conferences seem to belong primarily to the sphere of civil society. Nevertheless, if I translate the role of both fora to my three track approach I think that both committees and consensus conferences belong to the deliberative track. Committees are meant to give independent advice on the basis of non-political deliberation and as such they do not belong to the institutional sphere of government. Despite my criticism of the committee-system in the following chapter, I will argue that committees are still necessary, but that the status of their recommendations has to be toned down; they should be seen as one input into a public deliberation, while consensus conferences could form another input. Both committees and consensus conferences, on the other hand, are too formal and restricted to belong to the agonal track, which also involves political participation such as protest and activism. The differences I have pointed out between the two types of deliberative forum could lead one to question the very possibility of comparison. However, I do not simply compare consensus conferences and committees as a whole, but I examine how well each different institutional design conforms to my criteria for dealing with intractable disagreement. It will become clear that one conforms better than the other and this is not surprising if we consider that committees could be regarded as fitting more in a liberal democratic model and consensus conferences in a deliberative democratic one.

References

Ackerman, Bruce (1989), 'Why Dialogue?' *The Journal of Philosophy*, 86 (1), 5–22.
Benhabib, Seyla (1996), 'Toward a Deliberative Model of Democratic Legitimacy', in Seyla Benhabib (ed.), *Democracy and Difference. Contesting the Boundaries of the Political* (Princeton, NJ: Princeton University Press), 67–94.
Blamey, R.K., McCarthy, P., and Smith, R. (2000), *Citizens' Juries and Small Group Decision-Making* (Canberra: Australian National University).
Bohman, James (1995), 'Public Reason and Cultural Pluralism: Political Liberalism and the Problem of Moral Conflict', *Political Theory*, 23 (2), 253–279.
Bohman, James (1996), *Public Deliberation: Pluralism, Complexity and Democracy* (Cambridge, MA: MIT Press).
Bohman, James (1998), 'The Coming of Age of Deliberative Democracy', *The Journal of Political Philosophy*, 6 (4), 400–425.
Bovenkerk, Bernice (2009), 'Evaluating Participatory Methods', in Kate Millar, Pru Hobson West, and Brigitte Nerlich (eds.), *Ethical Futures: Bioscience and Food Horizons. 8th Congress of the European Society for Agricultural and Food Ethics* (Wageningen: Wageningen Academic Publishers), 305–309.
Chambers, Simone (1995), 'Discourse and Democratic Practices', in S.K. White (ed.), *The Cambridge Companion to Habermas* (Cambridge: Cambridge University Press), 233–259.
Chambers, Simone (2005), 'Measuring Publicity's Effect: Reconciling Empirical Research and Normative Theory', *Acta Politica*, 40, 255–266.
Christiano, Thomas (1997), 'The Significance of Public Deliberation', in James Bohman and William Rehg (eds.), *Deliberative Democracy. Essays on Reason and Politics* (Cambridge, MA/London: MIT Press), 243–277.
Cohen, Joshua (1989), 'The Economic Basis of Deliberative Democracy', *Social Philosophy & Policy*, 6, 25–50.
Cohen, Joshua (1996), 'Procedure and Substance in Deliberative Democracy', in Seyla Benhabib (ed.), *Democracy and Difference. Contesting the Boundaries of the Political* (Princeton, NJ: Princeton University Press), 95–119.
Cohen, Joshua (1997), 'Deliberation and Democratic Legitimacy', in James Bohman and William Regh (eds.), *Deliberative Democracy. Essays on Reason and Politics* (Cambridge, MA: MIT Press), 67–91.
Cohen, Joshua (1999), 'Reflections on Habermas on Democracy', *Ratio Juris*, 12 (4), 385–416.
Cooke, Maeve (2000), 'Five Arguments for Deliberative Democracy', *Political Studies*, 48, 947–969.

Cooke, Bill and Kothari, Uma (2001), *Participation: The New Tyranny?* (London: Zed Books).

Cunningham, Frank (2002), *Theories of Democracy* (London: Routledge).

Dryzek, John S. (1987), *Rational Ecology: Environment and Political Economy* (Oxford/New York: B. Blackwell).

Dryzek, John S. (1996), *Democracy in Capitalist Times: Ideals, Limits, and Struggles* (New York: Oxford University Press).

Dryzek, John S. (2000), *Deliberative Democracy and Beyond: Liberals, Critics, Contestations* (New York: Oxford University Press).

Dryzek, John S. and Niemeyer, Simon J. (2003), 'Pluralism and Consensus in Political Deliberation', paper given at Annual Meeting of the American Political Science Association, August 28–31.

Dryzek, John S. and Tucker, Aviezer (2008), 'Deliberative Innovation to Different Effect: Consensus Conferences in Denmark, France, and the United States', *Public Administration Review*, 68, 5–6, 864.

Eckersley, Robyn (2000), 'Deliberative Democracy, Ecological Representation and Risk: Towards a Democracy of the Affected', in Michael Saward (ed.), *Democratic Innovation: Deliberation, Representation and Association* (New York: Routledge), 117–132.

Eckersley, Robyn (2006), 'From the Liberal to the Green Democratic State: Upholding Autonomy and Sustainability', *International Journal of Innovation and Sustainable Development*, 1 (4), 266–283.

Elgesem, Dag (2005), 'Deliberative Technology?' in May Thorseth and Charles Ess (eds.), *Technology in a Multicultural and Global Society* (Trondheim: NTNU University Press), 61–76.

Elster, Jon (ed.) (1998), *Deliberative Democracy* (Cambridge: Cambridge University Press).

Fearon, James D. (1998), 'Deliberation as Discussion', in Jon Elster (ed.), *Deliberative Democracy* (Cambridge: Cambridge University Press), 44–68.

Ferrara, Alessandro (2001), 'Of Boats and Principles, Reflections on Habermas's "Constitutional Democracy"', *Political Theory*, 29 (6), 782–791.

Fishkin, James S. and Luskin, Robert C. (2005), 'Experimenting with a Democratic Ideal: Deliberative Polling and Public Opinion', *Acta Politica*, 40, 284–298.

Fossen, Thomas (2008), 'Agonistic Critiques of Liberalism: Perfection and Emancipation', *Contemporary Political Theory*, 7, 376–394.

Fung, Archon (2003), 'Survey Article: Recipes for Public Spheres: Eight Institutional Design Choices and Their Consequences', *The Journal of Political Philosophy*, 11 (3), 338–367.

Goi, Simona (2005), 'Agonism, Deliberation, and the Politics of Abortion', *Polity*, 37 (1), 54–81.

Gould, Carol (1996), 'Diversity and Democracy: Representing Differences', in Seyla Benhabib (ed.), *Democracy and Difference. Contesting the Boundaries of the Political* (Princeton, NJ: Princeton University Press), 171–186.

Gutmann, Amy and Thompson, Dennis (1990), 'Moral Conflict and Political Consensus', *Ethics*, 101, 64–88.

Gutmann, Amy and Thompson, Dennis (1996), *Democracy and Disagreement* (Cambridge, MA: Harvard University Press).

Gutmann, Amy and Thompson, Dennis (2004), *Why Deliberative Democracy?* (Princeton, NJ/Oxford: Princeton University Press).

Gutteling, Jan M. (2002), 'Biotechnology in the Netherlands: Controversy or Consensus?' *Public Understanding of Science*, 11, 131–142.

Habermas, Jürgen (1989), *Structural Transformation of the Public Sphere: An Inquiry into a Category of Bourgeois Society*, trans. Thomas Burger and Frederick Lawrence (Cambridge, MA: MIT Press).

Habermas, Jürgen (1990), *Moral Consciousness and Communicative Action* (Cambridge, MA: MIT Press).

Habermas, Jürgen (1996), *Between Facts and Norms. Contributions to a Discourse Theory of Law and Democracy*, trans. William Regh (Cambridge: Polity Press).

References 145

Habermas, Jürgen (2001), 'Constitutional Democracy: A Paradoxical Union of Contradictory Principles?' *Political Theory*, 29 (6), 766–781.

Hall, Stuart (1992), 'New Ethnicities', in J. Donald and A. Rattansi (eds.), *Race, Culture and Difference* (London: Sage and Open University), 252–259.

Hennen, Leonhard (1999), 'Participatory Technology Assessment: A Response to Technical Modernity?' *Science and Public Policy*, 26 (5), 303–312.

Hickerson, Andrea and Gastil, John (2008), 'Assessing the Difference Critique of Deliberation: Gender, Emotion and the Jury Experience', *Communication Theory*, 18, 281–303.

Hisschemöller, Matthijs and Hoppe, Rob (1995), 'Coping with Intractable Controversies: The Case for Problem Structuring in Policy Design and Analysis', *The International Journal of Knowledge Transfer and Utilization*, 8 (4), 40–60.

Honig, Bonnie (1993), *Political Theory and the Displacement of Politics* (Ithaca, NY: Cornell University Press).

Janis, I.L. (1982), *Groupthink: Psychological Studies of Policy Decisions and Fiascoes* (Second edn.; Boston, MA: Houghton).

Kirejczyk, Marta (2005), 'Het Belang van Publieke Betrokkenheid bij Medisch-ethische Onderwerpen (The Importance of Public Involvement with Medical-Ethical Subjects)', in Koos Van der Bruggen and André Krom (eds.), *Debat ter Discussie. Wie mag meepraten over medische technologie? (Debate Up for Discussion. Who Should Join in the Debate about Medical Technology?)* (The Hague: Rathenau Instituut), 115–120.

Kymlicka, Will (1989), 'Liberal Individualism and Liberal Neutrality', *Ethics*, 99, 883–905.

Manin, Bernard (1987), 'On Legitimacy and Political Deliberation', *Political Theory*, 15 (3), 338–368.

Mansbridge, Jane (1999), 'Should Blacks Represent Blacks and Women Represent Women? A Contingent "Yes"', *The Journal of Politics*, 61 (3), 628–657.

Meijboom, Franck, Visak, Tatjana, and Brom, Frans W.A. (2006), 'From Trust to Trustworthiness: Why Information Is Not Enough in the Food Sector', *Journal of Agricultural and Environmental Ethics*, 19, 427–442.

Miller, David (1993), 'Deliberative Democracy and Social Choice', in David Held (ed.), *Prospects for Democracy. North, South, East, West* (Cambridge: Polity Press), 74–92.

Mørkeberg, Annette and Porter, John R. (2001), 'Organic Movement Reveals a Shift in the Social Position of Science', *Nature*, 412, 677.

Mouffe, Chantal (2000), *The Democratic Paradox* (London/New York: Verso).

Mutz, Diana C. (2006), *Hearing the Other Side. Deliberative Versus Participatory Democracy* (Cambridge: Cambridge University Press).

Norton, Bryan (2003), *Searching for Sustainability: Interdisciplinary Essays in the Philosophy of Conservation Biology* (Cambridge: Cambridge University Press).

Olson, Kevin (2011), 'Legitimate Speech and Hegemonic Idiom. The Limits of Deliberative Democracy in the Diversity of Its Voices', *Political Studies*, doi: 10.1111/j.1467-9248.2010.00875.x.

O'Neill, John (2006), 'Who Speaks for Nature?' in J. Haila and C. Dyke (eds.), *How Nature Speaks. The Dynamics of the Human Ecological Condition* (Durham, NC: Duke University Press).

Paula, Lino and Birrer, Frans (2006), 'Including Public Perspectives in Industrial Biotechnology and the Biobased Economy', *Journal of Agricultural and Environmental Ethics*, 19, 253–267.

Pettit, Philip (2004), 'Depoliticizing Democracy', *Ratio Juris*, 17 (1), 52–65.

Phillips, Anne (1996), 'Dealing with Difference: A Politics of Ideas, or a Politics of Presence?' in Seyla Benhabib (ed.), *Democracy and Difference. Contesting the Boundaries of the Political* (Princeton, NJ: Princeton University Press), 139–152.

Plumwood, Val (1998), 'Inequality, Ecojustice, and Ecological Rationality', in John Dryzek and David Schlosberg (eds.), *Debating the Earth. The Environmental Politics Reader* (Oxford: Oxford University Press), 559–584.

Porsborg Nielsen, Annika, Lassen, Jesper, and Sandøe, Peter (2007), 'Democracy at Its Best? The Consensus Conference in a Cross-national Perspective', *Journal of Agricultural and Environmental Ethics*, 20 (1), 13–35.

Postema, Gerald J. (1995), 'Public Practical Reason: Political Practice', in Ian Shapiro and Judith Wagner DeCrew (eds.), *Nomos XXXVII: Theory and Practice* (New York: New York University Press), 345–385.

Purdue, Derrick (1995), 'Whose Knowledge Counts? "Experts", "Counter-experts", and the "Lay" Public', *The Ecologist*, 25 (5), 170–173.

Rawls, John (1987), 'The Idea of Overlapping Consensus', *Oxford Journal of Legal Studies*, 7 (1), 1–25.

Rescher, Nicholas (1993), *Pluralism. Against the Demand for Consensus* (Oxford: Oxford University Press).

Rosenberg, Shawn (2007), 'Types of Discourse and the Democracy of Deliberation', in Shawn Rosenberg (ed.), *Deliberation, Participation, and Democracy: Can the People Govern?* (New York: Palgrave), 130–158.

Sanders, Lynn Moss (1997), 'Against Deliberation', *Political Theory*, 25 (3), 347–376.

Shapiro, Ian (1999), 'Enough of Deliberation. Politics Is About Interests and Power', in Stephen Macedo (ed.), *Deliberative Politics. Essays on Democracy and Disagreement* (Oxford: Oxford University Press), 28–38.

Slovic, Paul (1999), 'Trust, Emotion, Sex, Politics, and Science: Surveying the Risk-Assessment Battlefield', *Risk Analysis*, 19 (4), 689–701.

Smith, Graham (2000), 'Toward Deliberative Institutions', in Michael Saward (ed.), *Democratic Innovation* (New York: Routledge), 29–39.

Smith, Graham (2008), 'Deliberative Democracy and Mini-publics', paper given at Political Studies Association (PSA) Annual Conference, Swansea University.

Sunstein, Cass R. (1995), 'Commentary: Incompletely Theorized Agreements', *Harvard Law Review*, 108, 1733–1772.

Sunstein, Cass R. (1997), 'Deliberation, Democracy and Disagreement', in R. Bontekoe and M. Stepaniants (eds.), *Justice and Democracy: Cross-Cultural Perspectives* (Honolulu: University of Hawaii Press), 93–117.

Sunstein, Cass R. (2002), 'The Law of Group Polarization', *The Journal of Political Philosophy*, 10 (2), 175–195.

Van den Brink, Bert (2005), 'Liberalism Without Agreement. Political Autonomy and Agonistic Citizenship', in John Christman and Joel Anderson (eds.), *Autonomy and the Challenges to Liberalism. New Essays* (Cambridge: Cambridge University Press), 245–271.

Van der Burg, Wibren (2003), 'Dynamic Ethics', *The Journal of Value Inquiry*, 37, 13–34.

Van Gunsteren, Herman (2006), *Vertrouwen in Democratie (Trust in Democracy)* (Amsterdam: Van Gennep).

Warren, Mark E. (2002), 'Deliberative Democracy', in A. Carter and G. Stokes (eds.), *Democratic Theory Today* (Cambridge: Polity Press), 173–202.

Wickson, Fern (2006), 'From Risk to Uncertainty: Australia's Environmental Regulation of Genetically Modified Crops', Doctoral Thesis (University of Wollongong).

Wynne, Brian (2001), 'Public Lack of Confidence in Science? Have We Understood Its Causes Correctly?' in Matias Pasquali (ed.), *Third Congress of the European Society for Agricultural and Food Ethics. Food Safety, Food Quality and Food Ethics* (Florence, Italy: University of Milan, 2001), 103–106.

Wynne, Brian (2003), 'Seasick on the Third Wave? Subverting the Hegemony of Propositionalism: Response to Collins & Evans (2002)', *Social Studies of Science*, 33, 401–417.

Young, Iris Marion (1990), *Justice and the Politics of Difference* (Princeton, NJ: Princeton University Press).

Young, Iris Marion (1996), 'Communication and the Other: Beyond Deliberative Democracy', in Seyla Benhabib (ed.), *Democracy and Difference: Contesting the Boundaries of the Political* (Princeton, NJ: Princeton University Press), 120–135.

Young, Iris Marion (1999), 'Justice, Inclusion, and Deliberative Democracy', in Stephen Macedo (ed.), *Deliberative Politics. Essays on Democracy and Disagreement* (Oxford: Oxford University Press), 151–158.

Young, Iris Marion (2003), 'Activist Challenges to Deliberative Democracy', in James S. Fishkin and Peter Laslett (eds.), *Debating Deliberative Democracy* (Malden, MA: Blackwell), 102–120.

Part III
Deliberative Fora: Deliberative Democracy Put to the Test

Chapter 5
Committees: The Politics of Containment

> Humphrey: "I am fully seized of your aims and of course I will do my utmost to see that they are put into practice".
> Minister: "Good….."
> Humphrey: "And to that end, I recommend that we set up an interdepartmental committee with very broad terms of reference so that at the end of the day we will be in the position to think through the various implications and arrive at a decision on the basis of long term considerations rather than rush prematurely into precipitated and possibly ill-conceived action which might well have unforeseen repercussions".
> Minister: "You mean, 'no'".
> Yes Minister (BBC): Doing the Honours

5.1 Introduction

In cases of intractable disagreement about complex issues such as biotechnology, governments are faced with the need to make regulatory decisions.[1] One way in which governments have sought to make expedient decisions in the face of public disquiet and persistent moral disagreement is to delegate the decision-making process to a politically independent committee. Committees have been established as a way to respond to the increasing call for more public deliberation. This move can also be regarded as an attempt to depoliticise the problem. As I argued in Chapter 2, when we delegate decision-making to an expert committee we assume that we are dealing particularly with scientific uncertainty. In the words of Hisschemöller, we are treating the policy problem as moderately structured (goal). It was my hypothesis that committees that acknowledge that normative uncertainty exists as well, and therefore allow for public input, are moving in the direction of treating the policy problem as the unstructured one it really is, and that these committees should be able to deal better with intractable disagreement. In this chapter I want to test this

---

[1] After all, even a decision not to regulate a certain controversial issue requires a choice by government.

B. Bovenkerk, *The Biotechnology Debate*, Library of Ethics and Applied Philosophy 29, DOI 10.1007/978-94-007-2691-8_5, © Springer Science+Business Media B.V. 2012

152   5 Committees: The Politics of Containment

hypothesis by making a comparative analysis between two biotechnology ethics committees in two countries, the Netherlands and Australia.

At the end of the previous chapter, I formulated three core conditions that, ideally, a deliberative democracy should fulfil in order to deal with intractable disagreement appropriately. These conditions are operationalised in a list of guiding criteria that could be used in the empirical analysis of deliberative fora, such as committees. In this chapter and the next I will apply these criteria in my analysis of deliberative fora. My aim with this analysis is not only to test to what extent the specific fora under investigation conform to my ideal view of deliberative democracy, but also to test the normative claims of deliberative democracy. Liberal democrats, deliberative democrats, and their critics each make certain normative assumptions about reality and about the character of persons. It is one of my aims here to test whether the normative assumptions implicit in deliberative democracy are warranted by examining to what extent underlying claims are realised in practice. In other words, the outcomes of my empirical analysis should have consequences for political theory. My purpose in the empirical part of this book is, furthermore, to identify what obstacles there are to genuine public deliberation in practice, in order to find out what assumptions of the theory of deliberative democracy need to be modified. The central question of this chapter is: To what extent can biotechnology ethics committees in the Netherlands and Australia deal with intractable disagreement and what implications does this have for the theory of deliberative democracy?

The three core conditions that I formulated in the previous chapter will also structure my analysis in this chapter. But first I will make a comparative analysis between the political cultures of the Netherlands and Australia, including the question how each deals with biotechnology, and I will introduce the biotechnology ethics committees of these countries in more detail.

5.2 Comparative Analysis Netherlands – Australia

I have chosen to compare the Netherlands and Australia because these countries have certain important similarities, but also enough institutional differences to be able to determine to what extent broader structures and contexts matter.

5.2.1 Political Cultures

Both the Netherlands and Australia are (Western) liberal democracies, but Australia's election system is representational – with a winner takes all district system in the lower house – whereas the Dutch system is proportional. In practice, the first leads to control of one dominant party, with a strong opposition party (and often a minority party that holds the balance of power in the upper house), whereas the second tends to lead to the existence of a wide range of parties and the establishment of a coalition government with an oversized cabinet. Australia has a Westminster style adversarial political culture and its style of democracy could be

5.2 Comparative Analysis Netherlands – Australia

termed majoritarian, as decisions are made on the basis of majority rule, whereas the Netherlands is regarded as having a consensual type of democracy, in which decisions are made on the basis of consensus between different minority groups.[2] This latter style of democracy has also been termed consociational, which refers to 'the role of elite accommodation in mitigating conflict in divided societies'.[3]

The emergence of this style of democracy and the Dutch practice of forming coalitions can be linked to the Netherlands' history of religious tolerance and pillarisation. In the past, the Netherlands provided a safe haven for religious minorities that were oppressed in their own countries.[4] The context of freedom of thought provided a fertile ground for public debate about theological matters on the basis of respect for other people's viewpoints. Different groups learned to co-exist through the system of pillarisation:[5] Denominational segregation led to the so-called 'school-struggle' in 1920, with Catholics and Protestants demanding separate schools because they did not agree with each other's teachings.[6] This struggle was resolved by creating separate schools and granting schools of each denomination the same rights and subsidies and leaving them free, within boundaries, to give content to their own teaching methods. Through pillarisation conflict was avoided, but the actual content of the disagreement was never discussed publicly. The decision-making process had the character of compromise and tolerance of other people's viewpoints, rather than the resolution of substantive arguments. The Dutch consensus style democracy is linked to a corporatist decision-making model,[7] which appeals to the same ideals of compromise and tolerance and the avoidance of (moral) conflict, and which is also termed the 'poldermodel', after the reclaimed land the Netherlands is famous for; the reclaimed land can only stay dry if all the parties involved co-operate on the basis of consensus.[8] Corporatism relies on co-operation between different interest groups, which are usually divided on socio-economic grounds; the most common corporatist model is that of co-operation between trade unions, industry, and government representatives who make decisions based on consensus formation. According to Jan Gutteling, 'the consensus that has driven social-economic developments is

---

[2] It should be noted that consensus in this context does not refer to the Habermasian notion of consensus, but refers more to a process of accommodation and compromise; rather than a decision that is 'wholehearted' and in the common interest, it refers to a decision that is supported by all important players. I would like to thank Bert van den Brink for bringing this to my attention.

[3] Herman Bakvis (1984), page 315.

[4] For example, the French Huguenots, Seardic Jews from Portugal, and Ashkenazi Jews from Eastern Europe found refuge and free thinkers such as Spinoza and Descartes were also welcome in the Netherlands. Johannes J.M. Van Delden, Jaap J.F. Visser, and Els Borst-Eilers (2004).

[5] Lijphart argues that this pillarisation has two dimensions, a primary religious, and a secondary class dimension, which has created fragmentation of Dutch society into four main blocks: a liberal block, which is secular and consists of middle and upper-middle class groups, a socialist block, which is also secular, but mainly working-class, and a Catholic and a Protestant block, which cut across class differences. Arend Lijphart (1975).

[6] Hans Daalder (1995).

[7] Arend Lijphart and Markus Crepaz (1991).

[8] Koos Van der Bruggen (1999).

likely also active in the development of modern biotechnology, resulting in a more participatory, rather than controversial, role for NGOs in some stages of policy making'.[9] At the same time, the absence of a strong opposition party reduces the need to justify policy decisions publicly and the Dutch government has been criticised for 'back-room politics' and bureaucracy.

Majoritarian democracies, like Australia, on the other hand, tend to go together with the pluralist interest group system. The pluralist model assumes that political and economic power are separated and that a variety of interest groups exist with different ideological, cultural, social or ethnic backgrounds and class interests, none of which is dominant in the arena of policy making.[10] In its policy formation the state reacts to lobbying from these interest groups, often leading to compromises in which the interests of opposing groups are reflected. This model has been criticised for overlooking the fact that power is not distributed equally in politics and that there is a bias towards powerful groups.[11]

Finally, the Netherlands employ a more active concept of neutrality than Anglo-Saxon countries, like Australia and the United States, where neutrality primarily refers to non-interference.[12] Presumably, each model places different demands on the way in which decisions are made, such as the need for making compromises and the need for defending one's points of view in public. Because collective decision-making through compromise or consensus, and co-operation between minority groups are central to consensus democracies, this style of governance has more affinity with deliberative democracy – in which consensus, and co-operation are also a major focus – than majoritarian styles of governance. It therefore seems reasonable to expect that the Netherlands is better able to deal with intractable disagreement than Australia with its more adversarial or antagonistic political culture. My hypothesis in this chapter is, then, that the Dutch committee can deal better with intractable disagreement than the Australian one, because even though it still focuses primarily on scientific uncertainty, in a consensus democracy there is more room to supplement this with a discussion on value uncertainty. This hypothesis will be tested in this chapter and the next.

5.2.2 Biotechnology in Australia and the Netherlands

Regarding biotechnology the two countries also have important similarities and differences. Agriculture plays an important role in the culture and economy of both countries. Moreover, both the Netherlands and Australia have set up biotechnology ethics committees as a response to public disquiet about genetic modification and

---

[9] Gutteling (2002), page 137.

[10] Hindmarsh (1994), page 106.

[11] Ibid. Hindmarsh quotes Ralph Miliband (1989). See also Kenneth Newton (1969). See also Hugh V. Emy and Owen E. Hughes (1991).

[12] Wibren Van der Burg (1998).

both have run consensus conferences on biotechnology-related topics. The committees of both countries appear to play a role in stimulating public debate, which is of particular interest given the central question of this book. An important dissimilarity, however, is that the focus of the biotechnology debate in Australia tends to be on genetically modified crops and food, whereas in the Netherlands the debate has centred on genetically modified animals. The debate concerning genetically engineered animals in the Netherlands has focused primarily on medical applications, whereas the Australian debate seems to emphasise agricultural applications, giving rise to different moral concerns. Some argue that the difference in focus can be explained by the fact that unlike European countries, Australia has not experienced great outbreaks of animal disease, such as BSE and foot and mouth disease.[13] The idea of using animals for human purposes may also be regarded more as 'common sense' in Australia, as the use of animals in agricultural production figures largely in the primary industries that Australia is dependent on. This is reflected in the historical view that 'Australia rides on the sheep's back'. Even though Australia is the most highly urbanised country in the world – around 80 per cent of all Australians live in the cities of over 50,000 inhabitants, primarily on the coast – its citizens still identify to a large extent with life in rural areas and the outback.[14] The role of farms differs between the Netherlands and Australia as well. In the Netherlands – as in the rest of Europe – farms are smaller than in Australia and it is therefore harder to segregate GM from non-GM crops. Moreover, in the Netherlands farms also play a role in recreation and nature preservation, as there is less 'pristine' nature left than in Australia.[15] I have chosen not to compare committees in the United States or the United Kingdom, because the Australian and Dutch publics seem more aware of and concerned about biotechnology than that of the United States, whereas the British public seems more averse to it. Finally, I have chosen to analyse Australian committees because Australia is (to an extent) independent from the European situation.

If we compare the regulatory systems regarding biotechnology in the Netherlands and Australia it appears that the Netherlands have a stronger and more active system. In fact, according to the biotechnology critic Richard Hindmarsh, Australia has one of the weakest regulatory systems of all OECD countries.[16] He argues that the regulatory system in Australia has been strongly influenced by the biotechnology industry, which, for example, pressed for a product instead of a process approach in the legislation of genetically engineered products.[17] In his estimation,

---

[13] Interview with member of interest group on 11/2/2005.

[14] See I.H. Burnley (ed.) (1974).

[15] Paula and Birrer (2006).

[16] Hindmarsh (1994).

[17] As I already briefly mentioned in Chapter 2, advocates of the 'product approach' argue that an estimation of health risks can only be based on examination of the components of the food consumed, whereas advocates of the 'process' approach argue that all products created using GMOs in the production-process should be labelled, since it is often the process that the consumer is concerned about and not merely the end-product.

creating public acceptance and not debate or awareness has been the driving force of Australian government agencies and government sponsored organizations, such as the Australian Science & Technology Council.[18] He characterises the Australian culture as one 'where political power is held by select groups of dominant individuals, a minority of the population, a culture where public participation is not actively sought or encouraged'.[19] When public participation is welcomed, according to Hindmarsh, it is as part of a co-optation strategy. In 1990, a parliamentary inquiry into genetic engineering was initiated in Australia; this included public hearings and submissions and resulted in the report 'Genetic Manipulation: The Threat or the Glory'. Critics have labelled this inquiry an 'absorption of protest strategy'.[20] Even though some public comments were taken seriously – for example, the recommendation that the biotech industry should be liable for foreseeable damage – critics argue that one important issue emanating from the submissions was overlooked, namely, the question of whether the Australian public wanted genetic modification in the first place.[21] Hindmarsh's views are to a certain extent underscored by other researchers. For example, Hans Löfgren and Mats Benner argue that the Australian government is 'neither neutral nor detached with respect to the growth of biotechnology but an interventionist and purposeful actor that seeks to shape public attitudes... government efforts to influence consumer attitudes are more conspicuous in biotechnology than in any other domain of techno-science.'[22] In contrast to Hindmarsh, they do not see this as a sign of weak regulation; in fact, they argue that contrary to its tradition of economic liberalism – in which the government was careful not to favour certain economic sectors over others – the Australian government has been unusually active in promoting the development and public acceptance of biotechnology and organising its implementation, including the provision of great financial incentives to the biotechnology industry and the adoption of a strong regulatory framework (albeit one that tends to favour the biotech industry).[23]

Another difference between the Dutch and Australian regulatory contexts is that the latter operates in more of a vacuum, whereas the former cannot operate independently from European regulation. Being an island continent, Australia has no contiguous border with other countries, whereas the Netherlands borders on Germany and Belgium and is close to other European countries. This means that even though the biotechnology industry lobby has also influenced the Dutch regulatory system, they have other options. In particular, when the Dutch government

---

[18] Hindmarsh (1994), pages 267–280.

[19] Ibid., page 271.

[20] Ibid., page 401.

[21] Richard Hindmarsh and Kees Hulsman (1992).

[22] Hans Löfgren and Mats Benner (2003), pages 34–35.

[23] Löfgren and Benner argue that 'state intervention premised on partnerships with business, for the purpose of facilitating technology development and innovation, is now a central theme in Australian policy deliberations'. Ibid., page 30.

forbids certain biotechnological procedures they can relatively easily carry them out in a neighbouring country with a more relaxed regulatory system, such as Belgium.

Regarding public acceptance of genetic engineering, attitudes in the Netherlands and Australia appear to be broadly similar. While early surveys concluded that the Australian public was supportive of genetic engineering, later studies have found that Australian citizens do have a number of concerns about particularly GM foods and their effects on human health and the environment.[24] One reason why many Australians might be hesitant about releasing GMOs into the environment is that this could be compared to releasing exotic species, a practice that has led to many problems throughout (white) Australia's history. Surveys of Dutch attitudes towards biotechnology also showed that citizens were generally accepting of applications that benefit human health or the environment, but that they have concerns about products without an obvious benefit or that raise moral issues.[25] The majority of both Australian and Dutch consumers support labelling of GM foods, but while Australian governments first rejected labelling, the unique corporatist tradition in the Netherlands has led to a dialogue between biotechnology companies, retailers, consumer and environmental organizations regarding the desirability of labelling and of general information exchange. This led to the labelling of GM foods when this was not yet demanded on the European level. Interestingly, labelling has not led to a decline in sales or to consumer protests in the Netherlands.[26] Finally, it should be noted that Dutch citizens tend to have a more trusting attitude towards government agencies and regulations than Australian citizens.[27]

5.3 A Tale of Two Committees

First, I describe the context, terms of reference, and procedures of the Dutch Committee for Animal Biotechnology (CAB) and after that of the Australian Gene Technology Ethics Committee (GTEC). Comparing specifically these two committees is not a self-evident choice, as the two committees differ significantly both in their procedures and in their subject matter. Nevertheless, I chose to compare these committees, because they are both the main national committees centring on the ethical aspects of biotechnology, and because they are both meant to fulfil a function in stimulating public debate.

---

[24] Respondents displayed more moral qualms about cross-species gene transfer and the genetic engineering of humans or the transfer of human genes into other organisms than with genetic engineering of animals, and they had even less moral problems with plant biotechnology. Norton, Lawrence, and Wood (1998).

[25] J.J. Beun et al., Ibid., pages 164–167.

[26] Ibid.

[27] http://ec.europa.eu/public_opinion/cf/subquestion_en.cfm (accessed on March 24, 2009). This observation is supported by the Eurobarometer.

5.3.1 Committee for Animal Biotechnology

Compared to other countries, the debate about animal biotechnology has started relatively early in the Netherlands, as a response to the creation of a transgenic bull, Herman, in 1990.[28] Herman's genome was modified by insertion of a human-identical gene, in order for his female offspring to produce milk with human lactoferrin, which could serve as the basis of medicines in the treatment of intestinal infections.[29] Following widespread public concern about this transgenic experiment, a parliamentary debate about the moral acceptability of animal biotechnology ensued, resulting in the 'Decree of Animal Biotechnology'. According to CAB chairman Egbert Schroten, this policy should be regarded as the result of the aforementioned Dutch 'poldermodel', which he describes as follows: 'if there is, somewhere in society, a danger of polarisation or deadlock someone takes the initiative to get the various interest groups together in order to talk and to create a compromise'.[30] The decree is part of the Animal Health and Welfare Law and states that as of April 1, 1997, carrying out biotechnological procedures with animals is prohibited without a permit.[31] With this decree, the Parliament has expressed its position as not favouring a complete rejection of animal biotechnology, but only allowing it under extraordinary circumstances and under strict control. The decree, therefore, has been given shape in a so-called 'no, unless-policy'. Applications for a permit are lodged with the Minister of Agriculture, Nature conservation and Fisheries[32] and will be granted when 'a) the procedures have no unacceptable consequences for the health and welfare of the animals, and b) there are no moral objections to the procedures'.[33]

In order to determine whether these conditions are met, the Minister seeks advice from the CAB, which consists of nine experts with a background in human medicine, animal experiments, veterinary science or zoology, social sciences, medical or animal biotechnology (2 representatives, one of which is nominated by the Committee of Genetic Modification, which deals with plant biotechnology), ethics, and ethology. The ninth member is the chairman, currently from a theological background.[34] The members have been appointed on the basis of independent scientific expertise and not as representatives of interest groups or of the opinion of particular segments of society or society as a whole. There is no requirement regarding male to female ratio of the members. The formal task of the CAB is to advise the Minister

---

[28] Brom (1997).

[29] Huub Schellekens (1993).

[30] Egbert Schroten (1999), page 260.

[31] Committee for Animal Biotechnology (1998).

[32] Now the Minister for Economic Affairs, Agriculture & Innovation.

[33] Animal Health and Welfare Law (1997). Ethicists will be quick to point out that b) should state 'no *other* moral objections'. However, this is the text as laid down in the Act.

[34] Committee for Animal Biotechnology, 'Annual Report', Staatsblad van het Koninkrijk der Nederlanden, 1997 # 5, Besluit van 9 December 1996.

on a case-by-case basis about the acceptability of proposed biotechnological procedures, and thus about the issuing of licenses. The Decree describes three functions which the committee should serve:

a) the clarification of the moral position of animals in view of biotechnology,
b) the strengthening of the moral position of animals in view of biotechnology, and
c) the identification, formulation, and assessment of problematic developments at an early stage in order to assist and stimulate public discussion about animal biotechnology.[35]

After receiving the Committee's recommendation, the Minister draws up a draft decision stating whether or not a license will be issued, and invites the general public to comment on the CAB opinion and the draft decision through a public consultation procedure. The CAB takes the comments of the public into consideration and advises the Minister on a response. Finally, the Minister decides whether or not to issue the license.[36] The CAB does not stand on its own, but is imbedded in a wider network of committees; it has links to the Committee of Genetic Modification (COGEM) and to animal experimentation committees (AEC's). The latter review all experiments carried out with animals and when it concerns genetically modified animals, they receive the CAB's recommendation on the specific license.

Animal biotechnology in the Netherlands is currently directed towards human medicine and scientific knowledge and not towards agricultural goals. In 1996, in preparation for the instalment of CAB, a committee of external experts created an assessment model for the ethical review of animal biotechnology.[37] According to this model, the ethical assessment should take place in five steps. A license application should be rejected in case 1) the goal of the project for which a license is requested is not of fundamental importance, 2) the harm to the health and well-being of the animals as a result of the biotechnological procedures is unacceptable, 3) the violation of the animals' integrity is unacceptable, or 4) if a realistic alternative to the (goal of the) proposed research is present. If none of the above steps have led to a rejection of the license, in step five an integral assessment is made, in which the degree of harm to the health, well-being, and integrity of the animals is balanced against the importance of the goal of the project. This assessment model is schematically displayed below (Table 5.1).

During its monthly meetings, the CAB tries to reach consensus on each of these steps. When consensus is not reached, a split recommendation is presented to the Minister, setting out the arguments of both the majority and the minority. Split recommendations usually reflect a conflict within the committee about how to understand the norms in the decision model. For example, a minority of the

---

[35] Lino E. Paula (2001). See also the Explanatory Memorandum of the Decree of Animal Biotechnology.

[36] Committee for Animal Biotechnology (1998).

[37] Frans W.A. Brom et al. (1996).

Table 5.1 Assessment model CAB

Step 1:
Goal of the project $\Rightarrow$ fundamentally important? $\Rightarrow$ no : **reject permit**
$\Downarrow$
Step 2:
Harm to health and welfare of the animals $\Rightarrow$ nature and degree of harm? $\Rightarrow$ unacceptable: **reject permit**
$\Downarrow$
Step 3:
Violation of integrity of the animals $\Rightarrow$ nature and degree of violation? $\Rightarrow$ unacceptable: **reject permit**
$\Downarrow$
Step 4:
Alternatives $\Rightarrow$ realistic alternative available? $\Rightarrow$ realistic alternative: **reject permit**
$\Downarrow$
Step 5:

Acceptability of the biotech- nological procedures goal: **permit**	$\Rightarrow$	degree of harm to health and well-being and violation	$\Rightarrow$	not proportional to the
		of integrity = proportional to		**reject**
		the importance of the goal		
		$\Downarrow$		
		Procedure acceptable: **issue permit**		

Committee members takes the existence of an alternative to the *ultimate goal* of the proposed project to constitute a realistic alternative, whereas the majority considers only an alternative to the *specific project* to be realistic. Also, a majority of the Committee regards the attainment of knowledge on its own ('fundamental research') to constitute an important societal goal, whereas a minority does not consider the mere increase of knowledge to weigh up against moral objections.[38]

It should be mentioned that at the time of writing the Dutch government is planning to abrogate the advising procedure for all biotechnological procedures

---

[38] A full list of divided issues within the CAB can be found in Paula (2008), page 81. At the time of writing, only two applications had been rejected and the rest had received a positive recommendation, albeit often subject to conditions, such as a time limit for the research. The reason for rejecting one application was that the research was regarded as premature at that point in time in the context of the overall research program of the applicant. The other was rejected because the researchers failed to provide additional information as requested by the committee. A few applications were withdrawn during the review process.

with animals within a biomedical context and thereby cease the work of the CAB, because it feels that after more than ten years the review procedure does not lead to any more new insights. The CAB can be called together on an ad hoc basis, if extraordinary applications are lodged. Needless to say, this decision is disputed by stakeholder groups such as the Animal Protection Agency.[39]

5.3.2 Gene Technology Ethics Committee

Australia occupies a special position within the biotechnology debate because it is the only developed country in the world that is considered to be 'megabiodiverse': it possesses 10 per cent of the world's biodiversity.[40] Australia is also one of the main biotechnology producer countries in the world.[41] Even though the federal government and many state governments have a positive stance on biotechnology, all the states have used their right to diverge from federal regulation and have declared GM-free zones.[42] While the federal government has approved the commercial release of GM crops, until recently – when Victoria and New South Wales lifted their bans – all states had moratoria on commercial crops.[43] In July 2000, the Australian government established the National Biotechnology Strategy, which gives shape to the development of the Australian biotechnology sector. A government agency was set up in order to facilitate this Strategy. This agency, named Biotechnology Australia, comprises five different government departments[44] and besides implementing the Strategy has the task of creating public awareness by providing the public with information, creating and disseminating educational materials, organizing rural forums and checking public attitudes relating to biotechnology. The aim of Biotechnology Australia is to ensure that Australia reaps the benefits of biotechnology, while at the same time ensuring that human and environmental health are protected.[45] When we take a look at its public awareness campaign it becomes clear that Biotechnology

---

[39] Personal communication with a member of an animal protection association.

[40] Monica Seini (2004), page 194.

[41] See Edna F. Einsiedel, Erling Jelsøe, and Thomas Breck (2001).

[42] 'Under the mirror legislation of States, it is possible for state governments to call GM-free zones based on marketing grounds'. See http://www.non-gm-farmers.com/news (accessed on April 20, 2005).

[43] At the time of writing, South Australia, Western Australia and Tasmania all have bans in place and the latter also has a ban on GM animals. The conditions in Queensland and the Northern Territory are unsuitable to growing GM canola. See http://www.non-gm-farmers.com/news (accessed on April 20, 2005) and http://www.greenleft.org.au/2008 (accessed on February 27, 2009). One reason for this divergence might be that the premiers of all state governments are from the Australian Labour Party, whereas the federal government is a Liberal/National Party coalition.

[44] Agriculture, Fisheries and Forestry; Environment and Heritage; Health and Ageing; Industry, Tourism and Resources; and Science Education and Training. See www.biotechnology.gov.au (accessed on April 22, 2005).

[45] As stated on their website: www.biotechnology.gov.au.

162   5 Committees: The Politics of Containment

Australia places most emphasis on the former. This view is supported by critics of the Australian government's role in negotiations over the Biosafety Protocol, who argue that Australia has pushed trade interests rather than health and environmental ones.[46] Many of the arguments that critics of the biotech industry have identified as rhetoric can be found in their information and education materials. A glossy brochure handed out to Australian schoolchildren provides a good example. It contains statements reminiscent of technological determinism: 'it seems inevitable that we will come to rely on gene technology to feed the world in the same way that, today, we could not manage without canning, refrigerated transport and storage..'. Also, the idea that biotechnology is no different from how humans have always interfered with nature figures highly in the brochure: 'In the distant past, people manipulated the process of fermentation to create alcoholic drinks and cheese and yoghurt. These are further examples of products made from biotechnology'. The potential benefits of genetic modification are presented quite optimistically in the brochure and criticism is brushed aside:

> Gene modification may be able to improve a crop's nutritional value, make it look better, last longer and even taste nicer! But the idea doesn't appeal to everyone, Why should we change the genes of the plants and animals we use for food? Well. . . .why not?[47]

In Australia, most procedures involving gene technology are regulated under the Commonwealth Gene Technology Act 2000 (from now on: the Act), which took effect in June 2001. The purpose of this Act is to create Australia-wide uniform regulation on gene technology, involving animal, plant, and human applications. This regulation aims to protect human and environmental health and safety by identifying and managing risks resulting from gene technology. The Act's objectives are carried out by the Gene Technology Regulator (GTR), who is supported by the Office of the Gene Technology Regulator (OGTR) and is a division of the Therapeutic Goods Administration within the Australian Government Department of Health and Ageing.[48] Three committees were established to be available to consult with and advise the GTR with his decisions – at the request of either the GTR himself or of the Gene Technology Ministerial Council – namely:

- the Gene Technology Technical Advisory Committee (GTTAC), advising on technical matters,

---

[46] Chin (2000), note 41.

[47] All quotes are from the brochure 'Juggling Genes'.

[48] Prior to the establishment of the OGTR biotechnology research in Australia was monitored by the Genetic Manipulation Advisory Committee (GMAC), established in 1987. However, compliance to GMAC guidelines was not compulsory. While the OGTR regulates gene technology research and development, regulation of the final use of genetically modified organisms is the responsibility of product-specific regulating agencies, such as the Australia New Zealand Food Authority (ANZFA) in the case of food and the Therapeutic Goods Administration (TGA) in the field of medicine. This regulatory framework has been criticised on the grounds that the end-use agencies such as ANZFA do not possess sufficient expertise to adequately assess the risks of releasing GMOs into the environment. See M. Hain, C. Cocklin, and D. Gibbs (June 2002).

5.3 A Tale of Two Committees

- the Gene Technology Community Consultative Committee (GTCCC), advising on matters of 'community consultation' or public debate, and
- the Gene Technology Ethics Committee (GTEC), advising on ethical matters.

During the writing of this book, as of 1 January 2008, as a consequence of the statutory review of the Act in 2005, – which found that there had been considerable overlap between the roles of GTEC and GTCCC – the functions of the two non-technical committees were combined into one advisory committee, the GTECCC.[49] As this book is about moral disagreement, I have concentrated on the functioning of the GTEC. I will also briefly examine the GTCCC, as its name suggests that this committee gives shape to the public debate function of the Act, and the GTTAC, as it is the committee with most influence.

The function of GTEC is:

> to provide advice to the Regulator and GTMC on the following: (a) ethical issues relating to gene technology;
> (b) the need for, and content of, codes of practice in relation to ethics in respect of conducting dealings with GMOs;
> (c) the need for, and content of, policy principles in relation to dealings with GMOs that should not be conducted for ethical reasons.[50]

According to Löfgren and Benner, 'the development of this national system has been accompanied by a great deal of lobbying and public debate. Groups such as the Australian Gene Ethics Network and the Australian Conservation Foundation argue that economic interests have been unduly privileged'.[51] This privileged treatment may be classed as a manifestation of the antagonistic political culture in Australia that is characterised by a pluralist interest group model.

The GTEC consists of 12 members and also has two advisors who are bioethics experts. The expertise of the members is wide ranging and includes ethics, public health, (environmental) law, theology, environmental science, social ecology, reproductive technology, animal health and welfare and plant pathology. The male to female ratio of the members is about 50:50. The three aforementioned committees have some cross-membership as well, but there are no cross-sessional meetings.[52] Both GTEC and GTCCC are recent committees that have experienced a steep learning curve. Unlike the GTTAC (and unlike the Dutch CAB), they are not dealing with specific license applications, but reflect on general questions. Issues that GTEC has reported on are, for example, risk analysis, globalisation, and transkingdom gene transfer. According to my interviewees, the GTEC or GTCCC were never asked for recommendations and were left to create their own projects and working document. After every committee meeting a communiqué is released to the public on the

---

[49] http://www.ogtr.gov.au/internet/ogtr/publishing.nsf/Content/gtecgtccc-1 (accessed on August 6, 2009).

[50] From: www.biotechnology.gov.au (accessed on April 22, 2005).

[51] Löfgren and Benner (2003), page 34.

[52] Interview with a committee member on 10/04/2005.

OGTR website. The following three sections provide a critical review of the CAB and GTEC, each focussing on one of my three core conditions.

5.4 Goals of Deliberation

In Chapter 4, I distinguished three main goals of public deliberation: striving for consensus, broadening the debate (inclusiveness) and deepening the debate (quality) and I pointed out tensions between these goals. A similar tension can be found when we look at the goals of committee deliberations. The Dutch CAB has two main functions, firstly advising on specific license applications and secondly, stimulating public debate. The goal of the first function is to achieve consensus, whereas the second function focuses more on broadening, but with a more tentative aim of consensus in the future. We could say that in both functions the Committee strives for deepening by aiming to offer a well considered opinion as a basis of the license advice and of public debate. I will argue that a tension exists within the CAB between all three goals. The main function of the GTEC is not reaching consensus, but rather deepening the debate by providing advice in the form of high quality expert reports. A tension exists here between deepening and broadening the debate. Moreover, another problem emanates in the Australian context – its outcomes do not have much, if any, influence in the decision-making process.

5.4.1 CAB: Tension Between Different Roles

The guidance on the Decree of Animal Biotechnology explicitly states that the goal of the case-by-case approach is ultimately to formulate general rules that reflect social consensus about the acceptability of animal biotechnology, the assumption being that it is feasible that such a consensus is reached. However, due to its role in the issuing of permits, the CAB is bound to a time limit and this has repercussions for the possibility of achieving consensus. From the moment the CAB receives an application, it is given six months to make its recommendation. Even though this is a longer time frame than, for example, animal experimentation committees have, the time limit naturally places constraints on the decision-making process. As Paula has argued, the CAB has met its time limits, but at the cost of delaying choices regarding the (normative) content of recommendations by way of 'evading argumentation styles'.[53] This was necessary because there is no agreement within the Committee about certain values and interpretations of core notions and there has been an ever returning, and so far unresolved, discussion on certain issues, such as what constitutes an alternative and to what extent a violation of integrity is acceptable. On the other hand, it can be argued that setting a time limit forces a committee to address their differences and to not put these off indefinitely. Moreover, the Committee

---

[53] Paula (2001), page 42.

5.4 Goals of Deliberation

members themselves do not experience the time frame as problematic.[54] This is in part due to the fact that the Committee organises annual 'retreats' in which ideas are exchanged independently from specific applications.

Against Paula it could further be argued that it is the fundamental nature of the moral disagreement that makes it intractable and impossible to find consensus, rather than the time limit. Nevertheless, a salient point in the evaluation of the CAB is that over 90 per cent of its recommendations were in fact reached unanimously, and are therefore considered to be based on consensus.[55] Between 1997 and 2004 only 14 divided recommendations were made; in all of these a minority of the Committee argued against issuing a license.[56] Meijer et al., who evaluated the functioning of the Decree of Animal Biotechnology, conclude from the high level of unanimity of Committee recommendations that the assessment model the Committee uses has proven to be effective.[57] One can wonder what is meant by this notion of effectiveness, however. Would the assessment model not have been effective if more divided recommendations had been drafted? Does the fact that so many unanimous decisions were reached mean that no moral problems persist? Or does it mean that the more divisive moral problems were excluded from Committee discussions?

I will argue in the next main section that the terms of reference of the CAB in fact do exclude certain issues from the discussion, which appears to demonstrate my earlier point that the risk of consensus is that it can often only be reached at the cost of the exclusion of viewpoints. I argued in Chapter 4 that when we are dealing with an unstructured problem, such as biotechnology policy, we should not in the first place strive for consensus, as in unstructured problems disagreement exists about the facts and values involved on different levels; for example, the problem definition and interpretation of core values are already disputed. In such a circumstance we benefit from fostering a diversity of opinions, while a focus on consensus can function to prematurely narrow down the debate. As Hisschemöller and Hoppe argued, when unstructured problems are treated as moderately structured or even structured problems they tend to become intractable. Recall that a structured problem could be dealt with by leaving decisions up to experts or bureaucrats who use a predetermined decision model in a depoliticised context. Because in the Netherlands licensing decisions are largely left up to experts – for the Minister nearly always follows the CAB's advice – and the Committee members follow a standard decision model, it appears that animal biotechnology is being treated as a moderately structured problem. When certain core notions have not crystallized yet superficial agreements can hide more fundamental disagreements and in such a context it is problematic to give such an important role to one committee made up of only experts. An example of an uncrystallised concept in the Dutch context is the notion

---

[54] Committee for Animal Biotechnology (2001).

[55] Albert Meijer et al. (2005).

[56] Albert Meijer et al. (2006).

[57] Meijer et al. (2005), page 14.

of 'animal integrity'.[58] Some people use this as a 'container concept' that refers to all moral objections besides those to do with animal health and welfare issues, whilst others reject the idea that there are moral objections besides health and welfare in the first place. Moreover, some people interpret integrity as a gradual notion and others regard it as more absolute.[59] Incorporating such a notion into a decision model and basing decisions solely on the advice of one group of experts that apply this model suggests more convergence on this notion than is present.

The CAB has been criticised for combining too many different roles, from problem-solver to promoter of interests (of the animals involved) to the stimulation of public debate. Paula, for example, argues that in order to promote public debate, one has to take as neutral a stance as possible and allow all different views and arguments to be explored. When promoting the interests of animals, on the other hand, this is not necessarily the best attitude. After all, how neutral is advice when one of the outcomes of the deliberations is already given, namely the strengthening of animal interests? Also, problem solving by drafting recommendations calls for expediency and this might obstruct the exploration of different viewpoints.[60] The interviewed committee members, however, did not see problems with the combination of these different roles. I concur with these members that the relevant question to ask in this context is what should be understood by stimulating public debate. It is not inherently contrary to public debate to take a stance on the issue that is being debated. The CAB does not have to be regarded as primary facilitator of public debate, but could also be regarded as a participant who brings in relevant information and arguments. Nevertheless, in practice the CAB's role in stimulating public debate has been ambiguous. Even among Committee members there is no clear picture about what exactly its role should be. In the early years the CAB did not take a proactive role in public debate – but merely a reactive one.[61] In later years, the Committee has played a more active, although still limited, role in public debate, through participation in six-monthly discussion meetings organized by the Ministry of Agriculture and through co-authorship of a Trend Analysis of developments in biotechnology.

Despite these recent efforts, a major obstacle has been the overly legalistic context that has evolved around the Committee's decisions, as it is experienced by everyone involved – Committee members, license applicants, and other stakeholders. Recommendations from the CAB can be contested in court and while disclosure of conflicting argumentations and views within the Committee could support public debate – by assisting people to make up their own minds about animal biotechnology – too much transparency about conflicts within the Committee could cause legal

---

[58] See Bernice Bovenkerk and Lonneke M. Poort (2008).

[59] This difference between the gradual and absolute versions of integrity is explored in more detail in Bovenkerk, Brom, and Van Den Bergh (2002).

[60] Paula (2001).

[61] Committee for Animal Biotechnology (2001).

complications.[62] When the Committee drafts a recommendation and interest groups formulate objections both parties anticipate the legal feasibility of their arguments. The discourse therefore has the character of a legal document, rather than one of moral argumentation. At the same time, both parties have to engross themselves in the arguments of the opposite party and this has at least led to a legal learning process by the different stakeholders.[63] Unfortunately, this learning process has now reached a 'saturation point' and has not led to the opinion transformation desired by deliberative democrats.[64] The fact that so much depends on the advice of the CAB, in terms of research projects being granted and possible medical breakthroughs made, might obstruct the CAB's role in public debate as well. Perhaps for the same reason, in its response to public submissions of comments, the Committee often takes a rather defensive stance. Some Committee members themselves oppose this attitude and argue that public submissions should be taken more seriously;[65] when the Committee cannot find an adequate response, it would do well to openly admit this rather than use evasive argumentation styles. It can be concluded that due to the legalistic context within which the CAB operates, the outcomes of its deliberations are not easily revisable.

The CAB has mainly focused on its primary role of adviser regarding license applications and as we have seen this included a focus on consensus, albeit in a narrow sense as consensus between Committee members. As will be explained in more detail later, the goal of deepening the debate has to a certain extent been reached, because the Committee has thoroughly discussed all the relevant viewpoints that are involved, but only within the limits set by its terms of reference. A clear tension exists between the goals of consensus and deepening on the one hand and inclusiveness on the other hand, due to the legalistic context in which the CAB operates and for other reasons that will become clearer in the next sections.

5.4.2 GTEC: A Toothless Tiger

The main role of the GTEC is to provide in depth advice to the Gene Technology Regulator (GTR). It remains unclear what specific goal this advice fulfils for the GTR, or for government in general. As the GTEC recommendations are of a high standard from an academic point of view, and many different perspectives are included in its discussions, one could conclude that the main goal is that of deepening. However, this goal does not appear to be aimed at furthering public debate. Rather, I think that the GTEC, and also the GTCCC, could be understood as institutions designed to substitute public debate rather than stimulate it. There are no mechanisms in place to seek direct public comment on GTEC publications. The

---

[62] Paula (2001); Meijer et al. (2005).

[63] Meijer et al. (2005).

[64] Albert Jacob Meijer and Frans W.A. Brom (2008).

[65] Interviews with several committee members on 3/4/2005, 5/4/2005 and 7/4/2005.

public does have the chance to make submissions regarding applications for release of GMOs into the environment and the accompanying risk assessment and management plan (RARMP).[66] However, according to Renato Schibeci et al., 'there is little indication that this involvement actually affects policy outcomes'.[67] Of the three GTR committees, only the technical committee (GTTAC) plays a role in drafting the risk assessment and management plan. Public input is, therefore, confined to the question of risk and concerns that do not fit the risk paradigm are overlooked. Critics also point out that licensing is untransparent, as quite significant changes to initial license conditions can be made without public consultation.[68] However, public submissions were sought and public forums were held by the government in the initial drafting of the Gene Technology Bill and also during the review of the Gene Technology Act.[69]

Even though the third OGTR committee, the Gene Technology Community Consultative Committee, does have a consultation function, it only consults stakeholder groups and it 'lacks any form of citizen power'.[70] As Geraldine Chin notes, this committee should not be considered as a substitute for input from a broader public.[71] Similarly, Schibeci et al. argue that the regulatory framework does not leave any room for public involvement and is structured in a top-down fashion. They noted that after two years in operation the GTCCC had not made any significant contribution to public involvement and when they approached committee members to find out why this was the case, they were told by the chair of the committee that the members were not allowed to discuss what went on in committee meetings.[72] Some members on this Committee felt they had the responsibility to organise public forums, but this was not supported by the OGTR.[73] Members of interest groups feel that government consultations are merely an exercise of going through the motions, and are actually meant to avoid divisive debate.[74] They would like the government to play a more active role in stimulating public debate, for example, by assisting with the organization of public forums and by distributing more information. Some members of the GTEC argue that this committee does play a role in stimulating public debate by publishing its reports on the OGTR website; unfortunately, there is no indication that these reports actually promote public debate. Furthermore, the chairman did speak at various forums on behalf of GTEC. However, as Renato Schibeci and Jeffrey Harwood point out, the OGTR approaches community involvement in a top-down manner and rejects public concerns because these would be based on

---

[66] Chin (2000).

[67] Renato Schibeci, Jeff Harwood, and Heather Dietrich (March 2006), page 442.

[68] Interview with committee member on 10/03/2005.

[69] Chin (2000).

[70] Ibid.

[71] Ibid.

[72] Schibeci, Harwood, and Dietrich (2006).

[73] Drawn from interviews with committee members on 10/03/2005 and 10/04/2005.

[74] Interview with member of interest group on 11/2/2005.

5.4 Goals of Deliberation 169

ignorance. Moreover, the GTCCC website does not invite dialogue with the public either.[75]

Some members of GTEC, as well as the consultative committee GTCCC, regret not having more influence on licensing decisions. In fact, none of the interviewees seemed to have a clear idea about their specific tasks as a committee and its role in the whole regulatory process. Still, others consider the functioning of the GTEC as very successful and productive. The members met twice a year for two days each time and therefore had an intensive deliberation. They also installed smaller working groups that would look into specific issues and report back to the committee. Many high quality reports were drafted – considering its infrequent meetings. In theory, the GTEC could veto an application if it held that there were strong ethical concerns. However, this has never happened. It is suggested that the reason why only the technical committee GTTAC has input in licensing decisions, is that the GTR's underlying view of science is positivistic, carrying the assumption of value-neutrality.[76] The move to separate a technical from an ethics committee is open to criticism. Levidow and Carr, for example, suggest that separating technical issues from ethical issues and from public concerns is a strategy employed to 'manage away' public disquiet, while allowing the important decisions to be solely based on 'sound scientific evidence', rather than 'emotions'.[77] By separating the technical from the ethical aspects of biotechnology, and only considering the former in the licensing procedure, the latter are in effect rendered less important.[78] This has led some to feel that 'the ethics gets tacked on at the end'.[79] As the GTEC has no decision-making influence and does not seem to provide much input for public debate either, this Committee seems to be toothless and one can wonder, therefore, whether even the goal of deepening is reached.

Fern Wickson, who has made an in-depth study of the regulation of biotechnology in Australia, argues that not only is there a sole focus on risk within the OGTR, but on a specific, positivistic, picture of risk; in the original risk analysis framework of the OGTR, risk assessment is described as 'a scientific process that does not take political or other non-scientific aspects of an application to use a GMO into account'... 'risk assessment will be transparent, objective, and scientifically based'.[80] However, as we saw before, particularly regarding the potential effects of the release of GMOs on the environment, a lot of scientific disagreement exists and there are a lot of 'unknown unknowns', so the assumption that decisions regarding biotechnology can be solely made on a scientific basis (and a narrow one at that) is

---

[75] Renato Schibeci and Jeffrey Harwood (2007).

[76] Interview with committee member on 10/3/2005.

[77] Levidow and Carr (1997).

[78] Ibid. Fern Wickson gives other examples to substantiate the claim that GTEC is awarded less importance and influence the GTTAC. For example, while GTEC was in the process of writing a report on the ethics of transkingdom crosses, the Regulator had already approved licences for such procedures. Wickson (2006).

[79] Interviews with committee members on 10/3/2005, 10/4/2005 and 8/4/2005.

[80] Wickson (2006), page 151.

170   5   Committees: The Politics of Containment

a controversial one.[81] The dominance of a positivistic risk discourse illustrates the technocratic character of biotechnology policy in Australia. Like in the Netherlands, this means that in Australia biotechnology is treated as a structured or moderately structured problem, rather than an unstructured one.

5.5 What Types of Argument Are Valid?

As became clear in the previous chapter, I hold that no views should be excluded from public debate at the outset, unless these views include threats of violence or counteract the possibility of holding a debate in the first place. I argued that participants should be allowed to draw on their comprehensive views and that alternative lay person perspectives should be taken into account. Moreover, in order not to exclude marginalised groups' communication styles not only rational argumentation should be allowed. How do the committees perform on these points?

5.5.1 CAB: Narrow Terms of Reference

Through parliamentary debate in the Netherlands, it was decided that while animal biotechnology is morally problematic (hence the 'no'), it is sometimes necessary and acceptable (hence the 'unless'). The discussion between those that squarely favour biotechnology – such as biotechnology companies, many researchers, and some patient organizations – and those that squarely oppose it – such as proponents of animal rights and certain religious groups – is in a certain respect settled by the instalment of the Committee. When this discussion resurfaces in the media or parliament, the government can point out that a solution has been found by appointing an independent committee of experts. The instalment of the Committee does not only lead to the exclusion of more extreme viewpoints, but also more general and fundamental moral questions, such as 'what do we want to achieve with this new technology?' and 'what constitutes a good life for humans and animals?', are bypassed. In effect, the only exercise that remains to be done is the shaping of the practical conditions under which biotechnology with animals is permitted, whereas the 'meta questions' go unaddressed.[82] When participants to public hearings raise objections of a more general nature than those concerning the specific concept license under discussion, they are told they are not in the right forum to do so.

The terms of reference of the CAB also serve to exclude certain views. Through the case-by-case and step-by-step approach the CAB aims to achieve shared values and directives on the basis of casuistry. This entails that the Committee's

---

[81] Ibid.

[82] This point has also been raised by animal protection associations. See Paula (2001), page 72.

5.5 What Types of Argument Are Valid? 171

recommendations concern only individual applications for research, and within these applications only the genetic modification of the animals involved, and not the animal experiments carried out afterwards.[83] Within this framework only direct consequences are taken into account, but the indirect, wider consequences of the research are not addressed.[84] In a case-by-case approach only the moral concerns arising from that specific research can be considered, whereas many moral concerns arise on a more general level. For example, the fact that nearly all the research focuses on 'Western diseases' affecting relatively few people, rather than 'Third World diseases' like malaria, could not be discussed as part of the Committee's review.[85] Moreover, the cumulative consequences of several individual research projects are overlooked; an effect that is not very problematic after one procedure may become problematic after a whole range of procedures.[86]

Another criticism of the case-by-case approach is that only the moral concerns invoked by the specific applications presented for review are addressed. So, for instance, if there are no applications for transgenesis in order to improve the meat to fat ratio of pigs, the moral aspects of this possible use of biotechnology will not be discussed. Even though one of the formal tasks of the CAB is to fulfil exactly this kind of signalling function and the chairman does actively search for information on possible future applications of biotechnology, there has been little room to address future applications either within the Committee or with the wider public, because the emphasis has always been on the Committee's role in licensing.[87] Furthermore, the step-by-step approach, delineating that the Committee must use the assessment model depicted above, does not ensure that all the moral concerns that pertain to an application are addressed. Objections against biotechnology because it amounts to 'playing for God' or because it is a sign of human *hubris*, for example, cannot be taken into consideration within the assessment model. Strictly speaking, these considerations should not be part of the review either, because according to its formal task, the CAB should only give its opinion about the consequences of the genetic modification *for animals*, raising the question what platform there is to discuss these wider considerations.

---

[83] The follow up research is dealt with by AEC's. Moreover, the CAB's recommendations exclude imported genetically modified animals.

[84] Paula (2001), page 75.

[85] Interview with committee member on 3/4/2003.

[86] For example, some argue that the use of animals in medical experiments leads to an instrumentalisation of the animals involved. In their eyes, instrumentalisation is a matter of degrees; each further use of animals brings them closer to being instruments solely created for our use. See Brom (1997). Separating the review of animal experiments through animal experimentation committees (AEC's) and that of the actual creation of transgenic animals in the CAB does not seem to do justice to this cumulative effect. Taken on its own, a specific case might not seem to be quite as morally objectionable as when it is regarded as part of an overall trend. Paula (2001), page 79.

[87] Personal communication with the chairman of the CAB.

172   5 Committees: The Politics of Containment

In practice, there appears to be a self-censuring role within the CAB regarding broader moral considerations as well. For example, whenever concerns of a religious or spiritual nature were raised, either by Committee members or public submissions, these tended to be brushed aside as unscientific or too subjective.[88] As most CAB members have a scientific background, it is not surprising that the majority holds that biotechnological procedures for the sole purpose of gaining knowledge are acceptable. Moreover, an analysis of CAB recommendations over its first period reveals that the emphasis in Committee argumentations was on the aim of research and the assessment of alternatives; both more technical issues. The analysis reveals a lack of emphasis on the more explicitly 'ethical' criteria of the assessment, particularly animal integrity. They point out that a smaller amount of text is devoted to integrity than to other steps in the assessment and that the wording under the sub-heading 'integrity' does not vary much from application to application. Moreover, 'no application is considered to cause a major violation of an animal's integrity, and there is no elaborated communication on the reasoning why this is so'.[89] On the other hand, the Committee members participate in annual retreats where they discuss ethical issues separately from specific applications. During these retreats issues such as integrity and other more ethical issues are focussed on explicitly.[90] It can be concluded then, that the Committee does discuss ethical issues, but that it has difficulty firstly, in reaching consensus on these issues, and secondly, in incorporating the results of their discussions in their official recommendations. On the positive side, my interviews reveal that the different committee members have learned from interaction with each other and have broadened their outlook, especially concerning ethical issues. For example, one committee member with a technical background has conceded that even though from the outset he had no high opinion of the notion of animal integrity, he now at least understands and accepts the distinction between animal welfare and animal integrity.[91]

5.5.2 GTEC: Narrow Problem Framing

The understanding of risk within the Australian Gene Technology Act and the risk assessment methodology applied by the GTR have received criticism, not only by outsiders, but also by some of the GTEC's own members, for their unqualified assumption of objectivity.[92] The Act works within the framework of substantial equivalence, meaning that the safety of a genetically modified crop is judged in

---

[88] Interviews with committee members on 2/4/2003 and 6/4/2003.

[89] Lino Paula and Tjard De Cock Buning (2000), page 1517. In later years, violation of integrity has been considered serious in some applications, but this has not been sufficient reason to reject them.

[90] Drawn from personal experience.

[91] Interview with committee member on 6/4/2003.

[92] Interviews with committee members on 10/3/2005 and 8/4/2005.

5.5 What Types of Argument Are Valid?

comparison to the safety of the original crop. When conventional agriculture is taken as the standard the underlying assumption is that this is an accepted and acceptable practice. Even though these assumptions are controversial, they are understood in the Act as the objective and scientific basis of risk assessment.[93] While the precautionary principle is explicitly adopted in the Act, in practice the GTR has failed to use it; as has been argued by Lawson, the principle is interpreted solely as a method of managing risks and the interpretation does not allow for a risk avoidance strategy.[94] The positivistic image of risk as adopted by the Act precludes contestation on the interpretation of risk or of the very dominance given to risk in dealing with biotechnology. As has been argued convincingly by Langdon Winner, the discourse of risk functions to minimize discussion on other ethical or socio-political concerns.[95] Lawson, furthermore, argues that certain evaluative judgements are inherent in the Act that lead the GTR to 'waver on the side of releasing GMOs', rather than to err on the side of caution in the absence of hard and certain evidence.[96] This is because of the unquestioned assumption in the Act that biotechnology will provide great benefits that Australia cannot afford to lose out on.[97]

Another criticism directed at the GTR is that it does not deal with the social or economic consequences of planting GM crops for GE-free farms, because it focuses too narrowly on health and environmental risks.[98] In fact, not only does the OGTR explicitly exclude economic and social issues, but it also frames health and environmental risks narrowly.[99] While the Act uses a much broader definition of the environment, the OGTR interprets it as pristine nature only and therefore does not regard risks to, for example, neighbouring farm environments as problematic.[100] Concerns voiced in the community have also been excluded by the narrow problem framing, which centres on 'scientifically quantifiable dangers'. Objections to gene technology per se, concerns about the social-economic context in which it will operate, doubts about the gene technology industry, or criticism of the scientific research that decisions are based on, go unaddressed within the OGTR process.[101] In practice

---

[93] C. Lawson (2002).

[94] In his eyes, this shows that the pretension of scientific objectivity in risk assessment and management is incorrect, because the decision to allow certain low probability risks is value-laden and can be disputed. Ibid.

[95] Langdon Winner (1986).

[96] Lawson (2002), pages 203–204.

[97] The Act states in note 15, pages 18104–18105, that 'there is no doubt that biotechnology holds great potential for this country. In terms of health, agriculture, industry, primary production and the environmental benefits we have seen only the prelude to the possibilities... For Australia to lose the benefits of this technology when we are able to manage those risks would be an irresponsible and unsupportable step for government to take'.

[98] For example Hindmarsh (2001); Nicole Rogers (2002).

[99] Interview with committee member on 10/3/2005 and see Schibeci, Harwood, and Dietrich (2006).

[100] Kerry Ross (2007).

[101] Ibid., Wickson (2006).

174   5 Committees: The Politics of Containment

the GTEC excludes certain views as well. A member of GTEC, for example, criticised the narrow focus of the discourse within the Committee, as the underlying
assumptions of the Act and many of the Committee members were anthropocentric – including an overly instrumental view of nonhuman animals and nature – and
ethnocentric, while broader perspectives, such as those informed by environmental
ethics or feminism, were rejected.

Moreover, apart from license applicants and holders, and other accredited organizations, nobody has a right of review once a decision to grant a license for the release
of GMOs has been taken.[102] This means that affected parties, such as non-GM
farmers neighbouring on GM fields and consumers cannot raise objections once a
decision has been made. The Act, then, favours the biotechnology industry, because
in its deliberations the GTR has to take into consideration possible appeals when
they refuse to grant a license, whereas no appeals will be made when they grant
a license. This power imbalance is also present during the OGTR process, because
industry actors play a role in the early stages of the licensing process – when they are
asked to provide information – as well as the later stages, while citizens and other
stakeholders only get a say in later stages, when problem definitions and solution
options have already been established.[103]

5.6 Who Should Participate?

In Chapter 4, I formulated six criteria pertaining to the core condition of inclusive
participation in the decision-making process. Within the context of the committee-
system the most relevant of these are that experts should not be the sole actors in
the decision-making process, that there should be open expert contestation, that the
lay public and stakeholders should be able to influence the decision-making process, and that opinion transformation of all participants should occur. The latter has
also been termed 'two-way transformative learning' to set it apart from a cognitive deficit model where public participation is simply a different expression for
top-down 'education' of the public.[104] How do our committees perform on these
counts?

5.6.1 CAB: Predominance of Experts

Both the CAB and the GTEC are made up of experts from different relevant fields.
Appointing only experts inevitably leads to an emphasis on technical considerations,
and those concerns that are viewed as 'unscientific' are disregarded. For example,
the CAB rejects the objection against the perceived unnaturality of crossing the

---

[102] Hain, Cocklin, and Gibbs (2002).

[103] Ross (2007).

[104] Wickson (2006).

species barrier, because people who raise this objection are thought to commit the naturalistic fallacy. However, as I argued before, reference to the unnaturality of gene technology should be regarded as merely the starting point of an objection that needs to be made more explicit; it is too easy to put this objection aside simply because, in words of the Nuffield Council, 'the "natural/unnatural" distinction is one of which few practicing scientists can make much sense'.[105]

The CAB's exclusive experts composition might unintentionally result in the exclusion of valuable experiential knowledge from the lay public. Even though studies into the positions of scientists on GM indicate that scientists by no means represent a homogeneous group, it can be expected that most adhere to a scientific worldview at least in a minimal sense.[106] According to such a worldview, only those opinions that are based on mainstream scientific findings are acceptable and the acquisition of scientific knowledge is valued highly and intrinsically.[107] Scientists tend to have a strong belief in the integrity of other scientists and tend to place a high level of trust in them, dismissing the possibility that some scientists might be driven solely by self-interested or economic motives.[108] Simply supplementing committee membership with lay members may not lead to the desired results of increasing public participation and gaining experiential lay knowledge, however. It can be expected that lay members of committees that deal with technical issues are too intimidated by the use of difficult terminology to be able to adequately express their views. Most interviewees did not regard the appointment of lay members as a viable option either, because firstly, they would not be lay members for long, but would soon become some kind of expert, and secondly, because it would be difficult to find lay members who would possess experiential knowledge, but who would not have a stake in the decisions. After all, what kind of lay people should these be? Patients, farmers, animal lovers? Thirdly, if there are inexperienced members on a committee, the discussion would be uneven and there would be a high chance that they would get trumped each time.

Besides the mechanism of lay people challenging the expert paradigm, another way of exposing the values inherent in scientific knowledge, is to make room for open expert controversy.[109] As stated succinctly by a biotechnology critic,

> science has been a poor source of information about how to regulate biotechnology because regulatory systems have favoured one viewpoint of a complex problem. The favoured viewpoint has been that of molecular biologists, who have based their advice on a positivist paradigm and gene theory, both of which are seriously flawed.[110]

---

[105] Nuffield Council on Bioethics (1999), page 15. See also Chapter 3 of this thesis about rhetoric.

[106] See Scott and Carr (2003).

[107] See Deckers (2005).

[108] Ibid.

[109] Ulrich Beck (1992).

[110] Kees Hulsman (2002), np.

However, even when 'dissenting' scientists have been part of a committee, this has often been explained by critics as token representation or co-optation of opposing viewpoints.[111] Some members of ethics committees argue that the degree of disagreement between scientists tends to be exaggerated. Research about scientists' views on genetic engineering, on the other hand, has shown that these are 'by no means uniform, yet that "policy makers and regulators. . . .tend to discount diversity among the views of scientists"'.[112] It has also been suggested, however, that the Committee members defend their own sectoral interests. According to Paula,

> personal interests, both scientific and financial, play an essential role in the discussion within the CAB. From the interviews with CAB-members and bureaucrats it becomes clear that the communication is open and respectful, but that the background of a CAB-member is indistinguishable from the point of view that he defends.[113]

While the decision-making structure does involve a public input process, and the experts are, therefore, seen to respond to the concerns of a broad public, research amongst interest groups and lay persons shows that the recommendations of the Committee are often too technical for non-experts to understand and do not always connect to the concerns of the public.[114] The main interaction between Committee members, researchers, and the public takes place during consultative hearings that are organised by the Ministry about a few proposed licenses each time. Although there is some opportunity for asking technical questions, these consultations are explicitly not meant to take the form of a discussion, but only to give the public the chance to submit verbal or written comments. The Committee members then discuss these comments behind closed doors and if they deem it necessary, they adjust their recommendation. Citizen submissions do not appear to have an influence on the outcome of the recommendations, although they have sometimes led to a change in a recommendation's formulation. The fact that the CAB usually already weighs the anticipated arguments of social organizations and interest groups in its first recommendation could explain why the comments do not substantially influence the decision-making process.[115] Persons submitting comments, however, do not feel that their comments are given adequate attention and, therefore, they resubmit the same comments over and over again. Some of those involved have characterised the public consultations as a 'ritual dance'.[116]

---

[111] Hindmarsh (1994). Co-optation means rendering opposition harmless by making dissenting voices take part in the decision-making process and thereby making them responsible for the collectively reached decisions.

[112] Deckers (2005), page 458, note 6.

[113] Paula (2001), page 46.

[114] Ibid., page 48.

[115] Ibid., page 47. This is supported in my own interviews with committee members on 5/4/2003 and 6/4/2003.

[116] Personal communication with the chairman of the CAB. See also Committee for Animal Biotechnology (2001) and Meijer et al. (2005).

5.6 Who Should Participate? 177

Animal protection associations do indicate that their point of view often closely matches that of the minority of the Committee in divided recommendations; so even though their views are represented, they do not influence the final decisions reached. The fact that the public does not feel its views are deemed of importance, may lead them to either lose interest or take more extreme positions than before.[117] In that case, a mutual exchange of ideas between the CAB and the public will not have been achieved and it is questionable whether the Committee has reached its task of promoting public debate. On the other hand, it has to be noted that the people who submit objections cannot be counted as a cross-section of society. It concerns a small group of people motivated to submit objections precisely because they hold strong views.[118]

From the point of few of the educational aspect of deliberation, and the criteria of opinion transformation a forum where researchers and concerned citizens can engage and learn about each other's viewpoints is desirable. Attempts to create such a forum were made after 2001 when the Ministry organised a series of four public debates about issues central to the CAB's weighing process, such as 'animal integrity' and 'the social importance of the goals of research'.[119] An attempt was made to increase public participation by inviting social organizations that were not yet directly involved in this discussion, and by publishing invitations for participation on the internet and in newspapers.[120] Another explicit focus of the meetings was to specify arguments that underlie the different standpoints in the discussion. There was a higher attendance than at the consultative hearings; around 60 participants attended each time – this time not only critics, but also biotechnology proponents.[121] Nevertheless, only a quarter of the participants were not affiliated with some organization or interest group. Participating CAB members note that they did not hear any new arguments and that the meetings did not influence their assessment.[122] The discussions primarily led to a repetition of earlier arguments and points of view. The discussion on animal biotechnology therefore seems to be satiated. On the one hand this can be explained by the high attendance of participants that were already involved in the procedures around animal biotechnology. On the other hand, Meijer et al. conclude that the Committee must have considered all the relevant arguments regarding animal biotechnology in its internal discussions.[123] However, some fundamental arguments were raised during these meetings that have not been discussed

---

[117] Paula (2001).

[118] At first between 10 and 20 people would show up to consultations, later usually less than 10, and sometimes consultations were cancelled because of lack of interest. Meijer et al. (2005), page 20.

[119] Other issues were 'alternatives to animal biotechnology' and 'the value of a power free dialogue'.

[120] Tatjana Visak and Franck L.B. Meijboom (2002).

[121] Ibid., Tatjana Visak and Franck Meijboom (2003a, b, 2004).

[122] Interviews with committee members on 3/4/2003 and 8/4/2003. See also Meijer et al. (2005).

[123] Ibid.

satisfactorily within the CAB, such as the objection to playing God and the emphasis in our health care system on curing rather than prevention.[124] The participants did feel that the meetings contributed to a positive 'discussion climate'; opponents and proponents appreciated meeting with each other and the tone of the discussion was respectful. Moreover, several participants think that the support for the Decree and for the CAB has been increased through these meetings.[125]

One of the outcomes of the meetings supports my earlier argument about core notions that are not yet crystallised. Participants had different interpretations of the notion of integrity. The underlying values and arguments behind these opinions remained unclear, however. The debate organisers conclude that taking people seriously entails that the discussion should be held on the deeper levels of the underlying values as well as on a more pragmatic level; a conclusion that underlines the points I made in Chapter 2.[126] In each of the meetings a different style of debate was used and it was concluded that the discussion was most fruitful when the group was broken up into smaller groups – the small size of the discussion groups contributed to more involvement of the participants, real deliberation on an equal footing, and an actual deepening of the arguments. Most people used arguments based on extrinsic considerations.[127] Intrinsic considerations were present in the background, but remained implicit. The organisers consider that perhaps the preference for extrinsic arguments was based on the wish to convince others and that to that end arguments were sought that other people could more easily accept.[128]

While the public participation procedure, therefore, creates the semblance of interaction between experts and the wider public, the discourse about animal biotechnology remains expert-oriented and the hearings are structured in such a way that only extreme opponents are involved, but that they at the same time are excluded from the dialogue. Moreover, the procedure seems to have no meaning for society at large, as an analysis of the CAB's contribution to public debate via the media, shows that only in one case – namely the application concerning cloning which was rejected – the views of the Committee have been reported.[129] The CAB's public debate function has, therefore, not been effective in reaching a wider audience.

5.6.2 GTEC: Tokenism

As we saw before, only the GTTAC has direct influence on decisions regarding gene technology policy and licensing in Australia. The exclusive focus on risks in

---

[124] Interview with committee member on 2/4/2003.

[125] Meijer et al. (2005), page 26.

[126] Visak and Meijboom (2002).

[127] Examples of cases were the creation of shrimps that did not cause allergic reactions and the use of biotechnology to bring back extinct species.

[128] Visak and Meijboom (2003b).

[129] Paula (2001), page 66.

5.6 Who Should Participate? 179

the GTTAC entails that some people – scientists – are labelled experts while others are excluded.[130] Even among scientists there appears to be a narrow representation; the GTTAC is dominated by experts from a molecular and cellular biology background rather than an ecology background. As Wickson argues, this leaves a gap in the expertise on the Committee about the environmental effects of the release of GMOs, and the risk assessment can therefore only be partial: 'the GM crop is essentially assessed in isolation rather than contextually in relation to how and where it will be grown in practice'.[131] Open expert contestation does not appear to take place in GTTAC. The dominance of molecular and cellular biologists also entails that the members have a direct interest in the promotion of gene technology or at least that they are likely to view gene technology favourably.[132] Some interviewees also thought conflicts of interest existed regarding some members of the GTEC.[133] Surprisingly, only the GTTAC has lay membership. The Community Consultative Committee, despite what its name suggests, does not contain any members from the lay public, but only from 'stakeholders'.[134] On the one hand, it could be argued that all of its members are lay members, as they are not scientific experts. On the other hand, however, no members of the public that are not in some way involved in gene technology sit on the Committee; there do not appear to be truly independent members. Its membership is composed of representatives of groups and associations, such as the 'Australian Women in Agriculture Network of Concerned Farmers' and the 'Plant Breeders Rights Advisory Committee'.[135] While this composition is typical of interest group politics, the lack of lay members is surprising as in AEC's in Australia it is common to have lay membership. This is in contrast to the Netherlands, 'where the legal requirements for the composition of committees are strongly based on expertise'.[136]

In order to be appointed as a member of the GTEC or GTCCC one needs to have demonstrable skills or experience relevant to gene technology. Interestingly, however, the lay member of the GTTAC can be drawn from cross-membership with either the GTEC or GTCCC.[137] This member is, therefore, considered a lay person in one context but not in the other. It should be noted that considering the controversy over biopiracy it is surprising that there is no representation from indigenous groups in any of the GTR committees. Also worth noting is that on the GTEC there is representation from members who object to biotechnology across

---

[130] Wickson (2006).

[131] Ibid., pages 165 and 182.

[132] Ibid., page 166.

[133] Interviews with committee members on 10/3/2005 and 10/4/2005.

[134] Of course, many would argue we are all stakeholders when it comes to genetic modification, as the whole (human and non-human) community is affected by developments in this field. By stakeholders here I mean organised groups who have a direct interest in genetic modification.

[135] See www.ogtr.gov.au/pdf/committee/gtccmembers.pdf (accessed on September 27, 2006).

[136] Jan Vorstenbosch (2000), page 1485.

[137] Wickson (2006), page 176.

the board, which seems to be possible because of GTEC's limited influence on the actual licensing process. In other words, the GTEC is under less pressure to reach decisions than the CAB and can afford to have more fundamental and open ended discussions.

The public could participate in the decision-making process of the GTR at two stages. Firstly, citizens and stakeholder groups could comment on the draft Gene Technology Bill before the OGTR was installed and they could take part in public forums that were held around Australia. While some critics argue that these were attended by only a low number of participants, the almost thousand citizens involved is a much higher number than the participants in CAB meetings.[138] As a result of public submissions it seems that the Act was amended only on one point, namely a requirement of insurance by license holders.[139] Secondly, citizens can write submissions on a draft Risk Assessment and Risk Management Plan (RARMP) and these will be considered by the Regulator before releasing a final RARMP. Besides these, several states have organised community forums about issues relating to gene technology, such as the precautionary principle, and specific gene technology applications, such as GM grape vines. The stated aim of these forums was to allow a community to hear a balanced discussion about the potential benefits and risks of gene technology, tailored to the community's specific interests and to make citizens acquainted with the regulatory framework. These forums were attended by twenty people on average and from my interviews some interesting points about them emerge. Firstly, the discussions seemed to work best when they had short introductions by speakers from different backgrounds and then broke up into smaller groups, which gave people the opportunity to have a more targeted discussion with the speakers and took away fears of speaking in front of larger groups. Most participants had a fixed view on gene technology before they attended, which did not change as a result of the discussion. Some attendees were very suspicious about the whole process and the discussion at times became quite adversarial. Participants did not receive middle of the road positions by speakers well, because it appeared they needed to have more polar positions as a reference point.[140] These circumstances confirm the antagonistic context in which gene technology is discussed in Australia.

The process of making submissions to a RARMP has been criticised for being a tokenistic public consultation rather than a true public participation.[141] Kerry Ross, who studied the level of public participation in a specific application of GM canola in detail, argues that within the space of the Act the OGTR has considerable room to involve the public, but has done so only in the most limited way.[142] It has

---

[138] Ibid.

[139] Hain, Cocklin, and Gibbs (2002).

[140] Interview with committee member on 10/4/2005.

[141] Public consultation is characterised by a one-way top-down information flow from government to the public, whereas public participation entails a 'two-way exchange of information between decision-makers and the public'. See Ross (2007).

[142] Ibid.

chosen to only call for public submissions in a late stage of the procedure, when problem framing had already been done and no new concerns raised by the public could be included. The OGTR has clearly adopted the cognitive deficit model of the public understanding of science, meaning that the public merely needed to be educated by the OGTR and could not offer its own experiential or contextual knowledge. Contrary to the custom regarding public input processes in Australia, the 256 submissions that were received were not made public by the OGTR. None of the people who sent a submission received a substantive reply to their concerns and their submissions do not seem to have influenced the decision-making process. Broader concerns from the public, such as those of a social, economic, moral, or even environmental nature, were considered 'OSA', outside the scope of the assessment. Most concerns were in fact redirected to other agencies, who do not have a public input procedure in place.[143] Ross argues that in this way the fragmented regulatory regime that characterises biotechnology policy in Australia leads to the containment of public participation. This is exacerbated by the fact that in order to make written submissions lay people have to read long documents filled with technical terminology. Moreover, they could often not check all the scientific publications quoted in these documents, because they were not always publicly available and were unobtainable from the OGTR – some of the quoted research was unpublished and a great amount was industry funded.[144] Research shows that those who made submissions felt that their input had not been seriously considered and that they felt discouraged from involvement in future. They experienced the public consultation process as a way to channel public opinion while at the same time ignoring it.[145] In light of these observations, Wickson's conclusion seems warranted: 'while the public has been granted an avenue for participating in decision-making, the avenue of written submissions has been framed in such a narrow way as to exclude the types of concerns that predominate in the community'.[146]

5.7 Committees of Containment: Discussion and Conclusion

From the foregoing review, it can be concluded that both committees experience persistent moral disagreement about biotechnology, within their own ranks and vis-à-vis the public, but that they have had difficulty dealing with this. Both have excluded viewpoints that are present in the wider community. The establishment of the CAB has precluded the more extreme viewpoints, while its terms of reference have excluded consideration of more fundamental or general 'meta' questions, indirect or cumulative consequences of animal biotechnology, moral concerns not

---

[143] These include issues that one would expect to be the OGTR's responsibility, such as 'safety and labelling of GM foods' and 'herbicide use and resistance management'. Ibid., page 216.

[144] Wickson (2006), page 183; Ross (2007).

[145] Ross (2007).

[146] Wickson (2006), page 182.

included in the assessment model, broader socio-economic concerns, and spiritual or religious concerns. In practice, due to the expert composition of the CAB, concerns that were not considered wholly scientific were excluded. Meta questions are not formally excluded from GTEC discussions, and its terms of reference therefore leave more room for broader concerns than that of the CAB, but in practice many of these issues appear to go unaddressed as well. The majority of this committee was not open to alternative viewpoints, such as those drawn from environmental ethics.

Licensing of GMOs in Australia only addresses questions surrounding risk and this risk discourse functions to minimise discussion on ethical or socio-political issues. The separation of ethical and technical committees shows that the wrong policy problem is addressed; biotechnology is treated as a moderately structured (goal) instead of an unstructured one. This separation works to disable the influence of ethics on decision making. Ethics is 'tacked on at the end' rather than given a place within all stages of decision making. The separation of the regulatory framework into three (or currently two) committees does not acknowledge that these domains are actually entangled and that value judgments are present in all of these domains; in other words, it fails to recognize that ethical concerns do not form a separate sphere at all. A positive aspect of the Dutch CAB is that technical and ethical issues are not in this way separated. On the other hand, in the actual weighing process the more technical considerations do tend to prevail over the ethical issues. Moreover, the set-up of the CAB has led to an evasion of the more fundamental value conflicts. This is to an extent compensated by the annual retreats in which more general issues are discussed separately from specific license applications. These annual retreats have not led to agreement about shared values, but they have in some instances led to a better understanding of each other's moral viewpoints. It can be concluded that both committees have dealt with intractable disagreement by evading the broader questions and more fundamental value differences. They both have the capacity to address such questions, but these do not influence their official standpoints and more fundamental discussions are not related to the general public. As one important aspect of dealing with intractable disagreement, in my view, consists in taking all sides seriously and also showing them that they have been taken seriously, I conclude that the committee system has failed in this respect. Relegating this unstructured policy problem to a committee could be regarded as an attempt to depoliticise the problem by letting a group of experts deal with it. However, as has become clear in this chapter, when unstructured problems are involved, not only values, but also facts or science should be discussed in the public sphere. This means that in fact politicisation – or repoliticisation – needs to take place in two directions; not only values should be a topic of discussion, but also scientific evidence and the status of scientific knowledge.

The influence of the consultation process on the CAB's decision-making process is minimal, although the points of view of the public are often apparent in divided recommendations. The avenue for public submissions to GTTAC decisions is framed narrowly and the GTR regards most citizen concerns as 'outside the scope of the assessment'. With regards to both committees the materials on the basis of which the public could make submissions were often too technical or difficult to

5.7 Committees of Containment: Discussion and Conclusion

come by. In the perception of the public in neither committee did public submissions have much influence on decisions. The media have not shown a great deal of interest in the advices of the CAB or in reports of the GTEC. In defence of the functioning of the CAB and GTEC it has to be acknowledged that they are relatively new committees, with no precedents, that have experienced a steep learning curve. Many of these criticisms actually appear to be inherent to the committee-system, rather than to the functioning of the specific committees under review.

The tensions that I noted in the previous chapter between the different goals of deliberative democracy can be discerned in the committee system as well. The CAB aimed to reach consensus, but this could only be achieved by evading argumentation styles and narrow terms of reference. Also, due its legalistic context, the CAB was made to uphold its recommendations and was therefore less open to public input or opinion transformation; two-way transformative learning only took place on a superficial level. The goal of consensus therefore worked counter to the goal of inclusiveness. The ideal of quality of deliberation was approached fairly well, both in the Netherlands and in Australia. However, this quality was also reached to the detriment of inclusiveness. The CAB seems to have been better able to gauge all the different arguments that are present in society, as is witnessed by the fact that it already anticipates submissions and that no new issues were raised in the public meetings. However, this is partly because stakeholder groups have chosen to fight the CAB on its own terms and have stopped raising broader concerns than the ones allowed through its terms of reference. The public consultation procedure of the CAB has made this committee's functioning more transparent than the functioning of the GTEC, creating more opportunity to expose arguments based on fallacies or self-interest. However, the absence of power imbalances and partisan interests has not been fully achieved in either of these committees. Indeed, one can wonder whether these can ever really be avoided. As will become clearer in the next chapter, this is one more reason to think that a process of depoliticisation does not work in practice.

The comparison of the two committees appears to conform quite well to the comparison of the institutional context of these committees as explained at the start of this chapter. The Dutch consociational style of democracy in which there is an important role for an elite to accommodate conflict can be discerned in the way the CAB functioned. The committee members are all experts and are in this sense elite, but they do not represent a specific segment in society or an interest group. The CAB strives for consensus (as can be expected in consensual styles of democracy) and is quite open to input from society, but this input comes mainly from interest groups. The public procedure, therefore, has the character of a dialogue between stakeholders, which conforms to the Dutch corporatist model. In public debates there is tolerance of each other's viewpoints, but the experts to a large extent make the decisions independently from public input on the basis of a predefined assessment model. The Australian GTEC has less public input, but has more representation of different viewpoints within its own ranks, conforming to the adversarial style of the Australian political culture. Moreover, as could be expected in an antagonistic political culture, critics cast more doubt on the independence of key players in the

184   5 Committees: The Politics of Containment

regulatory system. The pluralist interest group model is reflected in the instalment of the GTCCC. The separation of technical aspects, and ethical aspects and community views, as well as the power imbalance in problem framing within the OGTR process, seems reflective of the Australian government's bias in favour of biotechnology. This conforms to the criticism of the pluralist model that it reinforces the unequal distribution of power and a bias towards powerful groups, in this case the biotechnology industry.

On the one hand, it becomes clear from this comparison that views within society are taken more seriously by the CAB than by the GTEC, but on the other hand, this also makes it easier for the Dutch government to contain debate about animal biotechnology by pointing out that a committee has been installed to deal with these issues, while the untransparency and lack of public input into the GTEC appears to cause more opposition within Australian society. Interest groups in Australia do not shy away from strong language and protest and there seems to be less willingness on both sides of the debate to listen to each other's viewpoints. Still, both committees could be classified as committees of containment; they both leave meta-questions unaddressed, while at the same time providing a justification for governmental decisions, as formally ethical questions have been addressed. This is supported by more general literature about public involvement, in which hearings are criticised for throwing up barriers for the participation of ordinary citizens (such as location, time of the hearing, and the accessibility of information), for their one-way direction (presentations and testimonies rather than discussion), and their limited scope. In the words of Gene Rowe, 'public hearings often seem designed to contain and control participation...by allowing only limited choices on narrow, short-term questions at a late stage of the policy process'.[147]

The central question of this chapter was to what extent biotechnology ethics committees in the Netherlands and Australia can deal with intractable disagreement. My hypothesis was that the Netherlands could deal with it better than Australia. This is true to a certain extent; the debate both within the CAB and between the CAB and the public did appear to be more open and respectful. While in both countries citizens were disillusioned with their lack of influence and this could lead to more intractability, in the Netherlands at least the dialogue was kept open between regulators, opponents and proponents of animal biotechnology. In Australia there appeared to be more lobbying and influence from the biotechnology industry and more suspicion of vested interests by the key players. It can be concluded that the political culture of each country was highly influential in shaping regulation, the decision-making process itself, and responses to it. The decision to delegate decision-making to an 'independent' committee of experts, therefore, may be an attempt to depoliticise the conflict in the 'Pettit-meaning' of the word, but in reality fails to do so. Both committees – but particularly the Australian one – did not function to counter lobbying and vested interests. Despite the mentioned differences between the committees, both appear to be used as a way to contain public involvement, rather than stimulate

---

[147] Gene Rowe and Lynn J. Frewer (2000), page 18.

5.7 Committees of Containment: Discussion and Conclusion

it. Even though concerns of the general public might have been eased by pointing to an independent committee that has 'thoroughly examined' the moral aspects of biotechnology, the more vocal critics in society – such as animal protection associations or organizations critical of genetic engineering – were left disgruntled by the limited terms of reference or influence of the committees, by the lack of transparency and the lack of public involvement. This has made them even more vocal in their criticism, increased their protests and lobbying efforts, and in the end will lead to more politicisation rather than less. It appears, then, that depoliticisation in the 'Hisschemöller-sense' of the word was attempted, and containment of the debate was achieved, but this strategy may backfire in the end.

In both countries the policy problem was dealt with within an expert-discourse, while from my analysis of intractable disagreement it followed that imposing 'the morally right decision' via an expert committee is contrary to both liberal and democratic democracy and is treating the wrong policy problem. In my eyes, depoliticising an unstructured conflict by attempting to treat it as a structured problem is problematic. When unstructured problems are involved public debate needs to be broadened and deepened and avenues for debate should not be prematurely excluded. Especially when norms are involved that are the subject of different interpretations and that have not been crystallized it is premature to contain debate by installing a committee that has a decisive role. On the other hand, a committee with a high status can ensure that moral considerations influence decision-making from the early stages.[148]

What, then, is the proper role of an ethics committee? Committees should have a role in ensuring that ethics is incorporated in the decision-making process. However, they should not be the 'be all and end all' of the regulatory process. Two circumstances are important in this context: Firstly, the goals of broadening and deepening are both important, but can conflict with each other, and secondly, involving a broader public in decision-making can run counter to decision-making expedience. Therefore, an ethics committee should play a role in deepening, but should not take the place of other social actors. Moreover, the primacy of decisions should ultimately lie in the political arena. Committees, then, should not have the final say or even a leading role in decisions, but should primarily provide expert accounts for the benefit of both the public and political debate. However, their expertise should not be taken at face value, but rather treated as only one perspective on the issue. Committees should acknowledge that no consensus exists on both the facts and the values. In this view, the committees do not need to aim for consensus and it is not even relevant whether their decisions are based on majority/minority views; they merely need to provide information and analysis and not make a decision. The buck of making decisions should not be passed to committees; decisions should be made in the political arena, on the basis of a public debate that involves a broader public and in which the committee provides input. In light of these criticisms it can be concluded that the committee system is not the most appropriate way of giving shape

---

[148] See Bovenkerk and Poort (2008).

186   5 Committees: The Politics of Containment

to the call for more public deliberation in situations of intractable disagreement. In the next chapter, I will examine whether organised public debates conform better to deliberative democrats' view of public deliberation.

References

Animal Health and Welfare Law (1997), 'Animal Health and Welfare Law, article 66, subsection 1'.

Bakvis, Herman (1984), 'Toward a Political Economy of Consociationalism. A Commentary on Marxist Views of Pillarization in the Netherlands', *Comparative Politics*, 16 (3), 315–334.

Beck, Ulrich (1992), *Risk Society. Towards a New Modernity* (London: Sage).

Beun, J.J. et al. (1998), 'Biotechnology in The Netherlands', *Australasian Biotechnology*, 8 (3), 164–167.

Bovenkerk, Bernice and Poort, Lonneke M. (2008), 'The Role of Ethics Committees in Public Debate', *International Journal of Applied Philosophy*, 22 (1), 19–35.

Bovenkerk, Bernice, Brom, Frans W.A., and Van den Bergh, Babs J. (2002), 'Brave New Birds. The Use of 'Animal Integrity' in Animal Ethics', *The Hastings Center Report*, 32 (1), 16–22.

Brom, Frans W.A. (1997), *Onherstelbaar verbeterd. Biotechnologie bij dieren als een moreel probleem (Irrepairibly Improved. Animal Biotechnology as a Moral Problem)* (Utrecht: University of Utrecht).

Brom, Frans W.A., Hilhorst, M.T., Meulen, Ter, Vorstenbosch, R.H.J., and Jan, M.G. (1996), *Het Toetsen van biotechnologische handelingen bij dieren. Rapport van een commissie van externe deskundigen ten behoeve van de Commissie Biotechnologie bij Dieren (Assessing Biotechnological Procedures with Animals. Report of a committee of external experts for the Committee for Animal Biotechnology)* (Utrecht: Raad voor Dierenaangelegenheden). http:// www.rda.nl/home/31?highlight=toetsen+van+biotechnologische+handelingen+bij+dieren# [180]

Burnley, I.H. (ed.) (1974), *Urbanization in Australia. The Post-war Experience* (London: Cambridge University Press).

Chin, Geraldine (2000), 'The Role of Public Participation in the Genetically Modified Organisms Debate', *Environmental and Planning Law Journal*, 17 (6), 519.

Committee for Animal Biotechnology (1998), *Annual Report* (Utrecht: Ministry of Agriculture, Nature Conservation and Fisheries).

Committee for Animal Biotechnology (2001), *Self-Evaluation* (Utrecht: Ministry of Agriculture, Nature Conservation and Fisheries).

Daalder, Hans (1995), *Van Oude en Nieuwe Regenten. Politiek in Nederland (Of Old and New Regents. Politics in the Netherlands)* (Amsterdam: Bert Bakker).

Deckers, Jan (2005), 'Are Scientists Right and Non-scientists Wrong? Reflections on Discussions of GM', *Journal of Agricultural and Environmental Ethics*, 5 (18), 451–478.

Einsiedel, Edna F., Jelsøe, Erling, and Breck, Thomas (2001), 'Publics at the Technology Table: The Consensus Conference in Denmark, Canada and Australia', *Public Understanding of Science*, 10, 83–98.

Emy, Hugh V. and Hughes, Owen E. (1991), *Australian Politics: Realities in Conflict* (Second edn.; South Melbourne: McMillan Education Australia Pty Ltd).

Gutteling, Jan M. (2002), 'Biotechnology in the Netherlands: Controversy or Consensus?' *Public Understanding of Science*, 11, 131–142.

Hain, M., Cocklin, C., and Gibbs, D. (2002), 'Regulating Biosciences: The Gene Technology Act 2000', *Environmental and Planning Law Journal*, 19 (3), 163–179.

Hindmarsh, Richard (1994), 'Power Relations, Social Ecocentrism, and Genetic Engineering: Agro-Biotechnology in the Australian Context', Doctoral Thesis (Griffith University).

Hindmarsh, Richard (2001), 'Constructing Bio-Utopia: Laying Foundations Amidst Dissent', in Richard Hindmarsh and Geoffrey Lawrence (eds.), *Altered Genes II. The Future?* (Melbourne: Scribe Publications), 36–53.

References 187

Hindmarsh, Richard and Hulsman, Kees (1992), 'Gene Technology: The Threat or the Glory?' *New Scientist* (Australian Supplement), 25 April, 4.

Hulsman, Kees (2002), 'The Role of Science in the Regulation of Genetically Modified Organisms', *Environment, Culture, and Community* (Brisbane).

Lawson, C. (2002), 'Risk Assessment in the Regulation of Gene Technology Under the Gene Technology Act 2000 (Cth) and the Gene Technology Regulations 2001 (Cth)', *Environmental and Planning Law Journal*, 19 (3), 195–216.

Levidow, Les and Carr, Susan (1997), 'How Biotechnology Regulation Sets a Risk/Ethics Boundary', *Agriculture and Human Values*, 14 (1), 29–43.

Lijphart, Arend (1975), *The Politics of Accommodation: Pluralism and Democracy in the Netherlands* (Berkeley, CA: University of California Press).

Lijphart, Arend and Crepaz, Markus (1991), 'Corporatism and Consensus Democracy in Eighteen Countries', *British Journal of Political Science*, 21, 235–246.

Löfgren, Hans and Benner, Mats (2003), 'Biotechnology and Governance in Australia and Sweden: Path Dependency or Institutional Convergence?' *Australian Journal of Political Science*, 38 (1), 25–43.

Meijer, Albert Jacob and Brom, Frans W.A. (2008), 'Biotechnology and Social Learning: An Empirical Analysis of the Dutch Animal Biotechnology Act', *Technology in Society*, 31 (1), 117–124.

Meijer, Albert et al. (2005), *Evaluatie van het Besluit Biotechnologie bij Dieren (Evaluation of the Decree of Animal Biotechnology)* (Utrecht: Utrechtse School voor Bestuurs-en Organisatiewetenschap, Ethiek Instituut).

Meijer, Albert et al. (2006), *Besluit Biotechnologie bij Dieren: maatschappelijk leerproces? (Decree of Animal Biotechnology: Social Learning Process?)* (Utrecht: Utrechtse School voor Bestuurs-en Organisatiewetenschap, Ethiek Instituut).

Miliband, Ralph (1989), *Divided Societies: Class Struggle in Contemporary Capitalism* (Oxford: Clarendon Press).

Newton, Kenneth (1969), *The Sociology of British Communism* (London: Allen Lane).

Norton, Janet, Lawrence, Geoffrey, and Wood, Graham (1998), 'Australian Public's Perception of Genetically-Engineered Foods', *Australasian Biotechnology*, 8 (3), 172–181.

Nuffield Council on Bioethics (1999), *Genetically Modified Crops: The Ethical and Social Issues* (London: Nuffield Council on Bioethics).

Paula, Lino E. (2001), *Biotechnologie bij dieren ethisch getoetst? Een onderzoek naar het functioneren van het Besluit Biotechnologie bij Dieren (Animal Biotechnology Ethically Evaluated? An Analysis of the Functioning of the Decree for Animal Biotechnology)* (The Hague: Rathenau Instituut).

Paula, Lino (2008), *Ethics Committees, Public Debate and Regulation: An Evaluation of Policy Instruments in Bioethics Governance* (Amsterdam: Free University).

Paula, Lino and Birrer, Frans (2006), 'Including Public Perspectives in Industrial Biotechnology and the Biobased Economy', *Journal of Agricultural and Environmental Ethics*, 19, 253–267.

Paula, Lino and De Cock Buning, Tjard (2000), 'Governmental Review and Decision Making in Ethics: A Study of the Working Practices of the Dutch National Committee on Animal Biotechnology', in Michael Balls, A.M. Van Zeller, and M.E. Halder (eds.), *Progress in the Reduction, Refinement and Replacement of Animal Experimentation* (Amsterdam: Elsevier Science), 1505–1518.

Rogers, Nicole (2002), 'Seeds, Weeds and Greed: An Analysis of the Gene Technology Act 2000 (Cth), Its Effect on Property Rights, and the Legal and Policy Dimensions of a Constitutional Challenge', *Macquarie Law Journal*, 2, 1–30.

Ross, Kerry (2007), 'Providing "Thoughtful Feedback": Public Participation in the Regulation of Australia's First Genetically Modified Food Crop', *Science and Public Policy*, 34 (3), 213–225.

Rowe, Gene and Frewer, Lynn J. (2000), 'Public Participation Methods: A Framework for Evaluation', *Science, Technology, & Human Values*, 25 (1), 3–29.

Schellekens, Huub (1993), *De DNA-Makers (The DNA-Makers)* (Maastricht/Brussel: Natuur & Techniek).

Schibeci, Renato and Harwood, Jeffrey (2007), 'Stimulating Authentic Community Involvement in Biotechnology Policy in Australia', *Public Understanding of Science*, 16, 245–255.

Schibeci, Renato, Harwood, Jeff, and Dietrich, Heather (2006), 'Community Involvement in Biotechnology Policy? The Australian Experience', *Science Communication*, 27 (3), 429–445.

Schroten, Egbert (1999), 'Consensus Formation in Bioethics', *Jahrbuch für Wissenschaft und Ethik*, 4, 259–266.

Scott, Maggie and Carr, Susan (2003), 'Cultural Theory and Plural Rationalities: Perspectives on GM Among UK Scientists', *Innovation*, 16 (4), 349–368.

Seini, Monica (2004), 'Commodification and Access. Biotechnology and Australia's Indigenous Flora', in Richard Hindmarsh and Geoffrey Lawrence (eds.), *Recoding Nature. Critical Perspectives on Genetic Engineering* (Sydney: UNSW Press), 192–205.

Van Delden, Johannes J.M., Visser, Jaap J.F., and Borst-Eilers, Els (2004), 'Thirty Years of Experience with Euthanasia in the Netherlands: Focussing on the Patient as a Person', in T. Quill and M. Battin (eds.), *Physician-Assisted Dying: The Case for Palliative Care and Patient Choice* (Baltimore, MD: John Hopkins University Press), 202–216.

Van der Bruggen, Koos (1999), 'Dolly and Polly in the Polder: Debating the Dutch Debate on Cloning', *The Public Debate on Cloning: International Experiences* (The Hague: Rathenau Institute), 16–17.

Van der Burg, Wibren (1998), 'Beliefs, Persons and Practices: Beyond Tolerance', *Ethical Theory and Moral Practice*, 1, 227–254.

Visak, Tatjana and Meijboom, Franck L.B. (2002), *Integriteit van dieren: Bouwsteen of Struikelblok? Verslag van de 1ste discussiebijeenkomst over biotechnologie bij dieren (Animal Integrity: Building Block or Stumbling Block? Report of the First Discussion Meeting About Animal Biotechnology)* (Utrecht: Centre for Bioethics and Health Law).

Visak, Tatjana and Meijboom, Franck (2003a), *Op Zoek naar Alternatieven – maar voor Welke Doelstelling? (Searching for Alternatives – but for Which Goal?)* (Utrecht: Ethiek Instituut).

Visak, Tatjana and Meijboom, Franck (2003b), *Het Maatschappelijk Belang van een Doelstelling: Zoeken naar goede argumenten. Verslag van de derde discussiebijeenkomst over biotechnologie bij dieren (The Social Importance of an Aim: Seeking Good Arguments. Report of the Third Discussion Meeting About Animal Biotechnology)* (Utrecht: Ethics Institute).

Visak, Tatjana and Meijboom, Franck (2004), 'De Waarde van een Machtsvrije Dialoog voor Verdere Verdieping. Verslag van de vierde discussiebijeenkomst over biotechnologie bij dieren (The Value of a Powerfree Dialogue for Further Depth. Notes of the Fourth Discussion Meeting about Animal Biotechnology' (Utrecht: Ethiek Instituut).

Vorstenbosch, Jan (2000), 'Session Summary: The Role of Ethical Committees', in Michael Balls, A.M. Van Zeller, and M.E. Halder (eds.), *Progress in the Reduction, Refinement and Replacement of Animal Experimentation* (Amsterdam: Elsevier Science), 1485.

Wickson, Fern (2006), 'From Risk to Uncertainty: Australia's Environmental Regulation of Genetically Modified Crops', Doctoral Thesis (University of Wollongong).

Winner, Langdon (1986), *The Whale and the Reactor: A Search for Limits in an Age of High Technology* (Chicago, IL: University of Chicago Press).

Chapter 6
Consensus Conferences: The Influence of Contexts

> *The peculiar evil of silencing the expression of an opinion is that it is robbing the human race; posterity as well as the existing generation; those who dissent from the opinion, still more than those who hold it. If the opinion is right, they are deprived of the opportunity of exchanging error for truth; if wrong, they lose, what is almost as great a benefit, the clearer perception and livelier impression of truth, produced by its collision with error.*
> *John Stuart Mill, On Liberty*

6.1 Introduction

Will the cloning of animals like Dolly lead to human cloning? What constitutes an acceptable risk of introducing GMOs into the food chain? Should employers be allowed to test the DNA of their prospective employees? These and similar questions have been the focus of recent exercises in public debate. What these questions have in common, and what makes them particularly fit for public debate, is that they involve complex issues, often of a technological nature, that presuppose – often contested – expert knowledge and about which citizens disagree, often deeply. In the previous chapter, I concluded that even though committees do have a role to play, the committee system is not the best, and certainly not the only, way of giving shape to the call for more public deliberation about such complex cases. In previous chapters I also called into doubt the predominance of official expert knowledge when unstructured problems are involved. Organised public debates, also termed deliberative fora or deliberative mini-publics,[1] at first sight appear to offer a more self-evident method of increasing public deliberation and dealing with issues concerning values and worldviews. In this chapter, I will examine whether this appearance is correct. The central question of this chapter is as follows: To what extent can deliberative mini-publics deal with intractable disagreement regarding novel technologies?

---

[1] Dryzek and Tucker (2008).

B. Bovenkerk, *The Biotechnology Debate*, Library of Ethics and Applied Philosophy 29, DOI 10.1007/978-94-007-2691-8_6, © Springer Science+Business Media B.V. 2012

In Chapter 4, I argued for three core conditions and some more specific guiding criteria that should be met when we want to make 'dealing with intractable disagreement' operational. These regard the character of the goals, the scope of valid arguments, and the range of participants. As I explained before, my final aim is to seek the implications of these conditions when they are applied to real world deliberative fora. In other words, I want to both examine to what extent deliberative fora conform to my criteria, and identify the obstacles to genuine public deliberation in practice, in order to find out what aspects of the theory of deliberative democracy need to be strengthened. My assumption was that while committees would perform better if they acknowledged uncertainty regarding values, consensus conferences would perform better if they allowed for more discussion on scientific uncertainty, besides discussion on values. This includes calling into question the status of expert knowledge. In order to determine what institutional conditions are favourable to an open and inclusive debate I make a comparative analysis between the Netherlands and Australia. I hypothesised that the Netherlands would be better able to deal with intractable disagreement than Australia and this was true to a certain extent for committees. In this chapter I ask whether this is also the case for consensus conferences. I furthermore want to test the suggestion made in the Introduction, that when intractable value disagreement is involved, the depoliticisation of debate might be helpful, because it would allow values to be brought into the debate outside of the context of sectional interests.

In this chapter, I first provide a general background to my analysis by shortly describing three different types of deliberative fora that have been employed so far and their strengths and weaknesses and by comparing the cultural climates of Australia and the Netherlands in the context of public debates. I introduce the two consensus conferences in Australia (on gene technology in the food chain) and the Netherlands (on cloning) and then critically review these on the basis of my three core conditions regarding the goals of deliberation, the arguments that are valid, and the range of participation. I discuss my findings with particular attention to what we can learn from my comparative analysis and what my findings mean for the potential of depoliticisation.

6.2 Background

Exercises in public debate have become increasingly popular since the 1990s and – besides in the Netherlands and Australia – have been carried out in many different countries, such as Denmark, the UK, Canada, Norway, France, Germany, Argentina, and South-Korea.[2] Genetic engineering is one of the most debated topics at these meetings. Three main reasons why this particular topic has been popular are the

---

[2] The American LOKA Institute lists 77 consensus conferences or citizen panels held worldwide until now. 32 of these focussed on biotechnological issues. See http://www.loka.org/pages/panel. htm (accessed on 30 May 2007 and 19 July 2011).

6.2 Background

loss of public confidence in government and the scientific experts governments rely on, the resulting need to 'de-monopolize expertise',[3] and the need for procedural efficiency in decision-making.[4] According to Reinier Keller and Angelika Poferl, who have analysed German deliberative forums, or so-called Alternative Dispute Resolutions, such 'procedures attempt to channel citizen protests into new organized institutional forms'.[5] If this were true, deliberative forums would just be another way to contain public concern and dissent. Another criticism of public involvement is that it can be used by politicians to bypass lobbyists and press their own preferences.[6] Whether these qualifications of deliberative forums are correct, amongst other things, depends on the question as to what extent the outcomes of these forums can influence policy making. This in turn, as becomes clear in what follows, depends on views regarding the political legitimacy of deliberative forums.

6.2.1 Deliberative Mini-publics

Many different experiments have been carried out with the aim of increasing public participation, including consensus conferences, citizens' juries, deliberative polls, alternative dispute resolution, and planning cells.[7] Even though in the next sections I will only compare lay panel deliberations that can best be characterised as *consensus conferences* in Australia and the Netherlands, it is helpful to give an overview of the arguments for and against citizens' juries and deliberative polls as well. This will enable us to get a clearer picture of the obstacles to genuine public debate, because we can then examine whether possible shortcomings are typical of consensus conferences in particular or of deliberative mini-publics in general.

The model of the citizen consensus conference was developed first in Denmark, by the Danish Board of Technology.[8] The set-up of consensus conferences varies in a few respects from country to country, but the following characteristics remain constant. A panel of around 15 lay citizens is chosen to deliberate about a certain well-defined and morally controversial topic, usually regarding a novel technology, during an intensive period, such as two weekends. Beforehand, the panel goes through a learning process, including self-study of materials selected for them by the conference facilitators and lectures given by a wide variety of experts. The panel

---

[3] Einsiedel, Jelsøe, and Breck (2001), page 95. The authors adopt this notion from Ulrich Beck.

[4] Reinier Keller and Angelika Poferl (2000).

[5] Ibid. It should be mentioned that Alternative Dispute Resolutions differ from consensus conferences in an important aspect, namely that they do not use forms of random selection.

[6] John Parkinson (2004), quoted by Dryzek and Tucker (2008).

[7] Other methods of achieving public input are referenda, public hearings or consultations, opinion surveys, and negotiated rulemaking. See Rowe and Frewer (2001). I do not focus on these here, because they are less characteristic of the deliberative notion of preference transformation.

[8] Porsborg Nielsen, Lassen, and Sandøe (2007) In Denmark 22 consensus conferences have been carried out since 1987.

drafts a list of questions that in their eyes need to be answered in order for them to make a final recommendation. This part of the process usually takes place behind closed doors in order to allow the participants to become acquainted with the topic in as unbiased a way as possible and in order to lower possible personal participation barriers of the panellists. During the public part of the conference – which is open to the public and the media – the experts (which will often also have been selected by the panel from a list prepared by the conference facilitators) address these questions. Often, on the second day of the public conference, the audience is also allowed to address the experts. On the third day, the panel withdraws to deliberate about the topic in private. The aim of the conference is to reach consensus – although this is not obligatory – which is laid down in a lay panel report.[9] This report is presented to the public, the media, and policy makers. A consensus conference is best held at a time when no policy decisions have yet been made, so that the outcome can still influence and inform the decision-making process, and the topic of the consensus conference should be chosen such that even though it is complex, it is possible to delimit. One question that can be raised in the evaluation of consensus conferences is who should be considered a lay person. As already became clear, the lay/expert distinction is in a sense artificial as lay people can contribute experiential knowledge and experts are often familiar only with their narrow field of expertise. Moreover, during a process such as a consensus conference, the participants become more knowledgeable on the topic and soon cease to be real lay people. The main aspect that distinguishes them as lay appears to be that they do not have a personal (material or power) interest in the topic.

The model of citizens' juries is based on courtroom juries in the criminal justice system. Like consensus conferences, the jury consists of lay persons. However, unlike consensus conferences – where the panel is usually self-selected, as it is recruited on the basis of interest expressed by the participants – citizens' juries are composed of randomly selected citizens with the use of a quota system, so as to reflect the composition of the community in question.[10] Citizens' juries can call expert witnesses and witnesses that represent affected interests and, like consensus conferences, a neutral moderator facilitates the discussion. The hearings can last for up to a week, at the end of which the jury presents its recommendations. A response is elicited from the organising body, such as a government department. Unlike consensus conferences, usually no part of the citizens' jury process is open to the public.[11]

James Fishkin and Robert Luskin focus on the problem of scale in deliberative democracies. They propose deliberative polls in order to let as many people as possible come to informed opinions before voting: 'In contrast to ordinary polls, showing public opinion as it *is*, these deliberative fora attempt to show public opinion as it

---

[9] Smith and Wales (2000).

[10] Ned Crosby (1995).

[11] Rowe and Frewer (2000).

6.2 Background

would be if its members learned, thought, and talked more about the issues'.[12] Or, in the words of Dag Elgesem, 'deliberative polling is a kind of counterfactual polling, where it is measured what the people would have thought if they had the time to consider the issues more closely'.[13] A random sample of the relevant population is interviewed and invited to a deliberative weekend. They engage in small group discussions and plenary sessions where they can question a panel of experts and politicians. The debate may be broadcast on television. Participants are sent briefing materials to study beforehand with the main policy options and arguments for and against. After the period of deliberation they are polled on the same questions as before the deliberation. In Fishkin and Luskin's eyes, the advantages of deliberative polls are that they form a better cross-section of the general public – and thus hopefully offer a broader representation of the public's views – than alternative public fora and thereby contribute more to political equality. Because education and deliberation precede the vote, deliberative polling also ensures better quality of decisions than, for example, traditional opinion polls or referenda. The aim is not to reach consensus, but to help the participants clarify their own position before polling. The voting afterwards is secret, which maximizes equality, because people are not so easily influenced by others in the group to conform. Participants in deliberative polls are not self-selected, which increases the representativeness of the group. From their studies of the model of deliberative polling, Fishkin and Luskin draw several conclusions, some of which are important for our purposes.[14] Opinions and voting intentions change under the influence of a deliberative poll and this change is related to the increase of knowledge about the topic under discussion. This result is also found in evaluations of citizens' juries and consensus conferences.[15] Even if participants do not reach more agreement, they do agree more about what it is exactly that they disagree about.

Gene Rowe and Lynn Frewer have proposed a quite extensive list of criteria to compare the relative merits of different deliberative fora, some of which are useful for our purposes, because they focus on ways of dealing with disagreement.[16] One of their criteria is that of transparency, which means that the public is made aware of the way in which decisions are made, because otherwise the public might suspect bias. Organisers should also take care to avoid the *appearance* of bias in the way the panel is formed. The consensus conference scores better on this point than the citizens' jury, because part of the conference takes place in public. Deliberative polls are completely public, but this criterion is also less relevant here, as decisions are in the end made individually. Consensus conferences and citizens' juries strive

---

[12] James S. Fishkin and R.C. Luskin (2000).

[13] Elgesem (2005), page 65.

[14] Fishkin and Luskin (2005).

[15] For citizens' juries see, for example, Smith and Wales (2000); Robert E. Goodin and Simon J. Niemeyer (2003). For consensus conferences see, for example, Igor Mayer, Jolanda De Vries, and Jac Geurts (1995).

[16] Rowe and Frewer (2000).

194   6 Consensus Conferences: The Influence of Contexts

for consensus and polls do not. One problem with the use of voting in deliberative polls is that the same problems with traditional voting, such as voting cycles, occur, because the choice is limited to only a few alternatives. Moreover, there is less room for finding alternative solutions or reaching compromises. Another possible drawback of debate on such a large scale is that it might be hard to organise and for some the barrier to speaking in front of such a large audience might be too big.[17] For this reason deliberative polls often supplement large group discussions with intensive discussion in small groups. The benefit of small group discussion is that it 'reduces the scope for demagogy and allows all speakers to be heard' and that there is less likelihood that the debate is 'dominated by a small number of skilled participants and charismatic speakers'.[18] Participants in a public deliberation that are part of a small group are more likely to identify with this group.[19] This could lead to group conformity. However, in deliberative polls this problem is addressed to a certain extent as the participants move from discussions in smaller groups to discussion in larger groups continuously and in the larger groups there are opinion leaders that supplement the range of arguments.[20]

A final criterion of Rowe and Frewer's that is relevant here refers to the quality of the mechanisms used for structuring the decision-making process. If this criterion is not met very well, the outcome of the deliberation will be compromised due to problems with group dynamics. Preferably, these mechanisms ensure that the underlying arguments and not only the outcome of the deliberation are documented. This in turn will help fulfil the criterion of transparency. The structuredness of the decision-making is highly dependent on the quality of the facilitator. The decision-making process is more structured in citizen's juries and consensus conferences than in deliberative polls, although as I will argue later, the focus on a consensus document often entails that underlying arguments are not reported.

One final word about deliberative mini-publics, before we move on to the comparative analysis between the Netherlands and Australia. When we are thinking about ways to increase public participation it makes sense in this technological age to search for technological ways of doing so. One could think of communication technologies such as internet fora or political mobilisation sites on the internet. Benefits of using such technologies could be greater inclusiveness, access to balanced information, and lower participation barriers, as it may be less time consuming and people would not have to travel anywhere to be able to participate. Some have treated communication technologies as simply another way of carrying out public deliberation, but one can wonder if it is simply a matter of just transferring deliberative fora models to ict. Elgesem warns against too simplistic a picture of internet deliberation: 'it would clearly be naïve to believe that one could just introduce tools for open discussion of political issues and hope that deliberative

---

[17] Archon Fung points out some further problems of deliberative polls. Fung (2003).

[18] Smith and Wales (2000), page 59.

[19] Blamey, McCarthy, and Smith (2000).

[20] Elgesem (2005), page 74.

democracy will emerge'.[21] Some experiments have been carried out with deliberation on the internet and these tended to yield disappointing results. Either nobody participated or the forum was just used for the purpose of holding monologues, without people really communicating or listening to one another. Moreover, the problem of groupthink was present; the opinions on the forum tended to be quite homogenous.[22] One reason for this tendency is likely that internet fora – at least up until now – have been used quite exclusively by groups with a particular profile. For example, 'in the Digital City of Amsterdam... 91% of the inhabitants were men, and 58% of the active users were between 18 and 30 years of age'.[23] These results cast doubt on the aforementioned benefit of inclusiveness of internet fora. Elgesem's scepticism about using ict to create deliberative democracy is also based on the fact that social concepts tend not to be easily transferred into technological applications in general; the technology brings in its own characteristics that can change the social concepts themselves. A simple example is email; this has turned out to not just be regular mail in electronic form; email has introduced different norms and usages that nobody could envisage beforehand. Norms of privacy and communication are different and features can be added that go well beyond the possibilities of regular mail. This causes Elgesem to suggest that 'there is no such thing as a neutral implementation of a social institution into digital form. It is more reasonable... to assume that the implementation would have side-effects on the democratic properties of the model'.[24] While I do not mean to suggest that there is no future for new technology in deliberative democracy, these findings do prompt us to move forward with caution. More research needs to be done on the potential of ict for public deliberation.

6.2.2 Comparative Analysis

As explained in the previous chapter, Australia's political culture is more adversarial than that of the Netherlands. People appear to advance more conspiracy theories, there appears to be more industry influence over the regulation of gene technology, and there is less dialogue between stakeholders. The public awareness or education campaigns that Australian government bodies have organised have been heavily criticised for being biased in favour of gene technology and for seeking to foster acceptance rather than providing citizens with neutral information so that they can make up their own minds.[25] Some characterise public consultation exercises in Australia as risk management strategies rather than as part of informed or active

---

[21] Elgesem (2005), page 62.

[22] Ibid.

[23] Ibid., page 70.

[24] Ibid., page 68.

[25] Richard Hindmarsh (August 1992), Löfgren and Benner (2003).

democracy.[26] Australians are known for speaking their mind and for not holding back strong opinions. It is not unusual to see banners or hear loud protests at public consultations or debates, for people to read out long statements rather than enter a discussion, or for participants to accuse their opponents of acting purely out of self-interest. Australians tend to be suspicious of people in positions of authority. Amongst other things, this is likely to be the result of the antagonistic political climate. During elections, political parties in Australia often denigrate their opponents rather than focus on presenting their own political program. In contrast, even though the Netherlands has witnessed such scenes in recent years as well, this country has a tradition of respectful dialogue and in comparison to Australia there is more public debate in the opinion pages of newspapers and in talk shows on television; often scientific experts and politicians are involved in these. Moreover, in countries with a proportional election system, such as the Netherlands, a greater diversity of political parties tends to be represented in parliament and this creates a need to debate issues more thoroughly, but also more respectfully, because these parties still need to be able to work together in a potential future coalition. It is not unimaginable that this style of parliamentary debate rubs off on debate in society. In contrast, parliamentary debates in Australia tend to be characterised by strong attacks on one's opponents and a polarisation and simplification of issues under debate, as the debate primarily takes place between two large parties, who have to clearly distinguish themselves from each other, but who in practice tend to move to the centre in order to increase the size of their support base.

Annika Porsborg Nielsen et al. warn that when making a comparative analysis of consensus conferences in different countries it is easy to overlook the fact that the aims of the consensus conferences in the different countries might be different. They argue that the concept of participation varies between different countries and that this has consequences for perceptions of the political legitimacy of consensus conferences. The interpretation of the concept of participation in turn depends on the general understanding of democracy in the background political culture. According to these authors, the legitimacy question is important because it appears to underlie people's evaluation of the goals and the success of consensus conferences, and their opinion about what the role of experts and lay people should be.[27] They follow Habermas' distinction between three different ideals of democracy: a liberal one, a republican one, and a proceduralist-deliberative one. Of course, these three models are used as ideal types for comparative purposes, and do not neatly fit real life democracies, but the authors do point out some exemplar countries. In the liberal ideal, legitimacy is based on the equal opportunity of all citizens to influence political decisions by voting for representatives. Deliberation primarily takes place between elected representatives and the purpose of consensus conferences is primarily to inform citizens and to give them an inside look into the world of experts. The authors mention France as an example of this model. In the republican ideal the

---

[26] Interview with committee member on 10/3/2005 and interest group member on 11/2/2005.

[27] Porsborg Nielsen, Lassen, and Sandøe (2007).

political process is legitimate in as far as political decisions correspond to the shared values and culture of a particular community. The aim of consensus conferences is then to discover these shared values and the role of lay people is important, because their views and values hold the ultimate moral authority. Norway provides an example of this model. In the proceduralist-deliberative ideal legitimacy is conferred on political decisions through the transparency and inclusiveness of deliberative procedures. The aim of consensus conferences is not so much to find shared values, but to give everyone the opportunity to voice their views and to better inform those involved, including elected representatives, on the topic at hand. The value of lay perspectives is taken to lie in the fact that lay people do not deliberate with self-interested motives. The authors argue that Denmark approximates this model. Against the background of this political culture, it is not surprising that the citizen consensus conference model stems from Denmark. Moreover, in Denmark there is a strong tradition of public debate, based on the so-called 'high-school' tradition that provides adult education and discussion.[28]

Porsborg Nielsen et al. argue that evaluations of consensus conferences should take into account the background political culture and cannot generalise across cultures, but they do not claim that one political background culture is better at achieving the aims of consensus conferences than another. In their eyes, this cannot necessarily be the case, because the aims of consensus conferences in each culture may be different.[29] However, in my view this does not mean that we cannot say that these aims themselves cannot be evaluated as better or worse for their potential to deal with intractable disagreement through deliberation. A similar point is argued by John Dryzek and Aviezer Tucker, who also compared consensus conferences about GM food and their impact in three different countries, in this case France, Denmark, and the United States. They conclude that 'mini-publics such as consensus conferences can play a role in deliberative democratisation, though their potential is radically different in different sorts of political system'.[30] They distinguish three different kinds of system on the basis of whether states are inclusive or exclusive – inclusive states welcome input of public interests in public policy and exclusive states restrict this input – and whether they are active or passive – active states aid the organization and representation of interests, and passive states do not interfere in civil society. They classify Denmark as active inclusive, because its government actively seeks input from interest groups, France as passive exclusive, 'with policy making normally the preserve of an elite political class' and 'closed input structures', and the United States as passive inclusive, a pluralist state with lobby groups trying to influence a government that neither supports or hinders them.[31] In this classification scheme, the Netherlands with its corporatist tradition would tend

---

[28] For an explanation of this tradition, see Tarja Cronberg (1995). See also Ida-Elisabeth Andersen and Birgit Jaeger (1999).

[29] Porsborg Nielsen, Lassen, and Sandøe (2007).

[30] Dryzek and Tucker (2008), page 34.

[31] Ibid., page 16.

Table 6.1 Passive/active and inclusive/exclusive states

	Active	Passive
Inclusive	Denmark (and other Nordic countries) and Netherlands: actively seeks input from interest groups/corporatism	United States and Australia: pluralist tradition
Exclusive	Have existed prior to the 1980s, but not currently[32]	France (and some conso-ciational regimes); elitists/input from organizations in civil society not encouraged

towards actively inclusive – although consociational democracies have exclusive tendencies as well, due to the power of elites and the practice of backroom politics. Australia with its pluralist tradition would be more likely to be passively inclusive. In order to give a clear overview of this distinction, I have schematically depicted it in Table 6.1 below.

The differences noted between the consensus conferences in the different countries were not so much the result of the differences in set-up of the consensus conferences, but rather these differences in set-up were the result of the political cultures. According to the authors, the impact of consensus conferences 'is likely to be restricted without supportive structures and processes in the public sphere and state, the character of which will again vary by system type'.[33] For example, in passive inclusive states, like the US and Australia, the role of the media needs to be much greater if the lay panel has to have any hope of influencing policy decisions.

How can the difference in the political cultures of Australia and the Netherlands that I noted before be related to these categories? I pointed out that the Netherlands is well-practiced in the tradition of stakeholder dialogue, while Australia has a pluralist interest group tradition. In the interest group system, a main way for organizations to exert influence is through lobbying of government officials. This system has been criticised because it gives more powerful groups a better chance of influencing policy decisions. In such a system, where not all parties are given equal opportunities to influence decision-making and disputes are not settled through dialogue or compromise, one can expect the different actors to become more strongly opposed to each other. This is likely to explain the antagonistic political culture of Australia, and as we shall see later, this is illustrated well in my case study.

---

[32] These were states that 'under the sway of market liberal ideology have tried to destroy the basis for the organization of particular interests – notably labour unions'. Ibid., page 9.

[33] Ibid., page 34.

6.3 Cloning and GM Food: Two Consensus Conferences

In this section, I will elaborate the set-up and organization of the Australian consensus conference on gene technology in the food chain and in the next section, I will describe the Dutch lay panel deliberation on cloning.

6.3.1 Australian Consensus Conference

The first Australian consensus conference about gene technology in the food chain was held in March 1999. Prior to this conference Australia knew little public awareness or media coverage about GM food. After protests in the UK and New Zealand, public concern increased in Australia as well and this was the immediate cause for organising this conference.[34] The conference was initiated by the Australian Consumers' Association (ACA), whose primary aim was to stimulate public discussion about GM food. The Australian Museum, an organization that was viewed by the general public and stakeholders as independent, was asked to host the consensus conference. Even though the conference was not directly connected to any particular policy initiative, it did take place just before a federal regulatory framework for GM foods was established and the Australian New Zealand Food Authority updated its standards regarding GMOs, which particularly touched on the issue of labelling.[35]

A steering committee consisting of 17 members from government, academia, industry, and two NGO's selected the conference facilitator, the expert panel, and decided on communication strategies.[36] Three members of the steering committee wrote the briefing paper that would form the basis of the lay panel members' knowledge of the topic and their starting point for deciding which experts to interrogate. In the briefing paper they aimed to present an unbiased and up to date picture of gene technology in the food chain, and a broad range of views on the topic.[37] The steering committee also compiled a collection of newspaper clippings and magazine articles about gene technology for the lay panel, which some panel members decided not to read for fear of becoming biased.[38] Some lay panel members also searched for information materials on their own and distributed these materials amongst only some of the other panellists. Alison Mohr, who made a detailed study of the Australian consensus conference, argues that while the briefing paper aimed to demarcate the range of issues that were up for discussion, and in effect frame the debate, the panellists' own initiatives to look for information ran counter to this effort.[39]

---

[34] Einsiedel, Jelsøe, and Breck (2001).

[35] Alison Mohr (2002).

[36] Einsiedel, Jelsøe, and Breck (2001), page 88.

[37] See Mohr (2002).

[38] This fear appears to be well-founded, as studies have shown that the media base their reports mainly on statements from private corporations and public organizations. Ibid.

[39] Ibid.

The lay panel was selected with the help of an independent market research consultancy through advertising, which resulted in 200 applications. The advertisement did not mention the topic of the consensus conference, but asked for people who were interested in issues that will affect Australia's future to be involved in a forum.[40] Through an interview process, 14 panellists (eight women and six men from all states of Australia and with diverse educational and professional backgrounds) were selected by the steering committee. Through the interviews those with strong or predetermined views on gene technology were excluded.[41] One of the panellists had an aboriginal background, but no one born overseas was selected and there was disproportionate representation of people from rural areas.[42] The organisers aimed to have a 'slice of Australia' on the panel, without having the pretension that all possible perspectives were represented.[43] During two preparatory weekends, the facilitator taught the lay members discussion, interpreting, and question preparing skills and worked on their team spirit and sense of trust.[44] They also became familiar with the topic and drafted the key questions to pose to the experts during the actual conference. The list of eight questions corresponded to the topics that were emphasised in the briefing paper, and focused on risks and potential negative impacts on human health and the environment, the effects of a possible Australian moratorium, ethical issues, the possible dangers of control by multinational agencies, the context of international treaties and trade agreements, and on consumer information.[45]

On the first day of the conference the various experts responded to questions in 18 presentations over the course of 10 hours, and on the night of the first day the lay panel drafted clarificatory and additional questions, which were posed to the expert panel on the second day, when the audience could also interrogate the experts. The panel members not only determined the questions to be asked and chose the 13 experts answering them, but they also decided that the experts had to declare their position on gene technology beforehand. Moreover, the panel determined demanded inclusion of information about the role of lay persons, ethicists, and stakeholders in the regulatory process regarding GM foods. They also sought experts that were not on the steering committee's list, such as a nutritionist and an expert from a religious background. The panel chose not to consult any government officials, save one from the science and technology research ministry, which is illustrative of the general

---

[40] Interview with organiser of consensus conference on 10/4/2005.

[41] Aidan Davison and Renato Schibeci (2000).

[42] This was due to the fact that a stratified selection was aimed for and, therefore, from each state both a metropolitan and a rural person were selected, whereas Australia is a highly urbanised country. Elaine McKay and Peter Dawson (1999).

[43] Interview with organiser of consensus conference on 10/4/2005. See also Alastair Crombie and Colin Ducker (2000).

[44] Jan McDonald (June 1999).

[45] For the precise wording of the questions see http://www.austmus.gov.au/consensus (accessed on 30 May 2007).

6.3 Cloning and GM Food: Two Consensus Conferences

distrust of Australians towards regulatory authorities.[46] Moreover, partly due to the distribution of extra information materials by some of the panellists – also dubbed 'the Monsanto File' – the panel appeared to be suspicious of the Monsanto representative who was invited to provide expertise.[47] This suspicion was exacerbated by the attitude of the representative who initially treated the consensus conference as a public relations exercise and evaded answering some of the panel's questions.[48]

After the cross-examination of experts, the report writing phase began, which lasted from 2 pm until 7 am the next morning. The lay panel members felt they were rushed through the questioning and report writing stages and some felt they lacked the opportunity to quietly reflect on the current of information that washed over them.[49] In an evaluation of the consensus conference some members suggested that the final consensus was reached under so much time pressure and exhaustion that they just conceded to viewpoints they disagreed with and that a number of members were in bed when some decisions were made.[50] Some members indicate that those members with the most stamina and strongest personalities in the end had most influence over the lay panel report. As one of those involved characterised it, 'it was survival of the fittest'.[51] The lay panel strived for consensus and choose not to include minority viewpoints in its report.[52] The final consensus report contained compromises and concessions were made in order to achieve other preferences.

In the lay panel report, the panel takes a very positive stance on the consensus conference method; it even recommends that the government provide for public participation more structurally using a similar mechanism. It adopts quite a critical position on GM foods; the panel is particularly wary that developments in gene technology primarily benefit private biotech companies and is sceptical of the argument that biotechnology will solve world hunger.[53] The panel is quite concerned with the possibility of negative impacts on human health and the environment. The panel is in favour of the precautionary approach to risks and rejects the standard of substantial equivalence. Moreover, it is opposed to the narrow focus on risk in decision-making regarding GMOs and recommends the consideration of cultural, moral and religious beliefs in the decision-making process. This was primarily translated into the recommendation that an ethicist have a say in GMO policy decisions. The panel recommends the establishment of a new authority, financed with license fees from

---

[46] Einsiedel, Jelsøe, and Breck (2001), page 92.

[47] McKay and Dawson (1999), page 24.

[48] Carolyn M. Hendriks (2004). The representative essentially had one answer to all questions, which was that Australia has a good regulatory system for GMOs and that Monsanto complies with this. W.M. Blowes (1999).

[49] Mohr (2002).

[50] Ibid. Mohr adds that even though extending the process with a few days might be desirable, the Danish experience is that with more available time the final finetuning is often postponed to the last minute and sessions still last through the night.

[51] Interview with organiser of consensus conference on 10/4/2005.

[52] Mohr (2002), page 153.

[53] The Australian Museum (1999).

companies that commercialise GM products. This fee should also be used as insurance against unforseen harms. The panel calls for a moratorium on the commercial release or unlabelled import of new GM foods until a new authority is established, a comprehensive labelling system is adopted, and risks are better evaluated. The panel emphasises the inclusion of citizen perspectives and argues that 'the vision citizens hold for the future of their country must be taken into account'.[54] Amongst other things, this entails an examination of impacts on and the viability of alternatives to gene technology, such as organic agriculture, and better processes to inform the public about GMOs.

Both the effectiveness of the consensus conference as a tool for public participation (phase 1) and the outcomes and impact of the consensus conference (phase 2) were officially evaluated.[55] When considering these evaluations, it has to be kept in mind, however, that they were not independent evaluations, as they were commissioned by the conference organisers. This would explain why their results are more positive about the success of the conference than other evaluations. The pre- and post-conference attitudes of the lay members were tested by four different evaluators.[56]

6.3.2 Dutch Lay Panel

After Denmark, the Netherlands was the second country to stage a consensus conference, in 1993 about the genetic modification of animals.[57] The public debate on cloning, which I analyse here, was held over the period of March 1998 until June 1999 and was initiated by the Ministry of Health, Welfare and Sport and the parliament, who invited the Rathenau Institute to organise it.[58] This institute is an independent organization that supports the formation of social and political opinions regarding scientific and technological developments. The Dutch debate on cloning was somewhat different in character than the Australian consensus conference. First of all, the lay panel did not explicitly strive to reach consensus, even though the lay panel report did turn out to approach a consensus document. Moreover, the lay panel conference did not stand on its own, but was part of a larger group of initiatives to stimulate public debate on this issue. The start of the public debate was a public hearing in which experts were interrogated and in which citizens only attended in the audience. Of these experts, four had a background in philosophy or theology, three were scientists, one was a representative of a biotechnology company that uses

---

[54] Ibid., page 5.

[55] Phase 1 was evaluated by McKay and Dawson (1999). Phase 2 by Crombie and Ducker, 'The First Australian'.

[56] By the official evaluators, Mohr, the conference facilitator, and Market Attitude Research Services.

[57] Davison and Schibeci (2000).

[58] Van der Bruggen (1999).

6.3 Cloning and GM Food: Two Consensus Conferences

cloning, one was the coordinator of a patient organization, and one a representative of an animal protection organization.[59] The emphasis on ethics and philosophy shows that it was acknowledged from the start that cloning is a topic with important moral and metaphysical dimensions. The hearing was used as an agenda-setting device, in order to demarcate the topics for debate and to determine on which issues general consensus already existed.[60]

The hearing made clear that the topic of cloning did not stand on its own, but had overlap with other topics, most notably genetic engineering. During this initial phase it immediately became clear that the different participants of the hearing were talking at cross purposes, because they already started out with a different definition of the problems concerning cloning.[61] A book was published that compiled articles from the different experts that were heard.[62] Besides the hearing and the lay panel conference, five discussions were held, each covering a different aspect of the cloning debate. For example, representatives of different faiths – including Christianity, Islam, Judaism, and Humanism – discussed cloning and (religious) ideology. Other partial debates were about cloning cattle, cloning animals for the creation of medicines, therapeutical cloning in humans (for gene therapy), and about political parties and ideologies and cloning. Members of the lay panel did not take part in these discussions, but were encouraged to attend them. Furthermore, the Rathenau Institute and SWOKA – an institute for strategic consumer research – organised a series of regional debates, for example with student associations and a rural women's network. The reason for taking this multi-layered approach was that the Ministry of Health wanted as many people as possible to participate in the broad public debate on cloning.[63] Finally, the conclusions from the public debates were supplemented with an opinion poll among 847 Dutch citizens. At a final meeting to close off the official public debate, both the lay panel and the Rathenau Institute presented their recommendations. The latter was based on an analysis of all the arguments and viewpoints brought forward in the different stages of the debate. Also, a philosopher of technology presented his views on the debate.[64]

The whole process of the actual lay panel conference took around a year; the panel met ten times in total, on Saturdays every 5–6 weeks. The panel drafted its own agenda, and some of its activities were open to the public. The lay panel conference was organised before policy was made on cloning. The 17 citizens of the lay panel were recruited by placing an advertisement in a national newspaper and selected from the 200 responses to this advertisement. Unlike the Australian advertisement, the Dutch one did mention the topic of the conference. For this reason, it could be

---

[59] Frank Biesboer (1998).

[60] Consensus for instance existed on the desirability of a ban on making identical humans via cloning, and on placing embryo's that had been subject to experiments back into the uterus. Ibid.

[61] Van der Bruggen (1999).

[62] Bout (ed.) (1998).

[63] Interview with organiser of public debate on 2/4/2003.

[64] Tsjalling Swierstra (2000).

said that the process was even more self-selecting than that of the Australian conference and it is expected that mainly people that were already interested in the topic replied. Most responses were from people who enjoyed higher education, but the Institute did try to compose a panel as representative of the general population as possible, and therefore included some people with lower levels of education.[65] They also tried to compose a group with a variety of religious, ethnic, and professional backgrounds, different regional origins, and an even division of men and women.[66] The respondents to the advertisement had to explain in a few sentences why they wanted to participate, and the majority replied that they were concerned about or opposed to cloning as yet another development in a line of questionable technological developments. However, those with more extreme viewpoints were not chosen, because it was thought that they already had an entrenched point of view about which no debate would be possible.[67]

The lay panel was free to choose the experts they wanted to interrogate. They became acquainted with experts through the literature, through the initial public hearing and through suggestions by the Rathenau Institute. According to a representative of the Rathenau Institute, the organisers were never directive; the panel was never told that they should not consult certain experts.[68] The Institute hired an independent person to make a compilation of publications on cloning, which was given to the panel at the start. The panel members also found articles from newspapers and magazines themselves, which they distributed amongst the rest of the panel. In contrast to the Australian situation, this was encouraged by the Dutch organisers. In order to create a sense of variation, the panel chose different work-forms during their discussions, ranging from plenary discussions, subgroup discussions, and short exercises, to writing future scenarios. Part of the panel also visited a company involved with animal cloning.[69]

Even though the lay panel did not absolutely reject cloning the report was quite critical of this technique, particularly when used in conjunction with genetic modification. Animal cloning was regarded as less of a problem than human cloning.[70] The report states that medical benefits are too easily claimed and demands more emphasis on looking for alternative solutions. The majority of the members support a 'no-unless' policy, because they feel that the burden of proof about the acceptability of cloning should lie with researchers. Also, they reject the comment of one of the ethical experts that emotional arguments are merely 'gut feelings' with no moral standing. Rather, they see emotions as markers of intuitions that are important to

---

[65] One of the lay panellists had even studied ethics in the past. Interview with organiser of public debate on 2/4/2003.

[66] Interview with organiser of public debate on 2/4/2003.

[67] Interview with organiser of public debate on 2/4/2003.

[68] Interview with organiser of public debate on 2/4/2003. The lay panel members also state in their report that they were never directed in their choices.

[69] Emine Bozkurt (ed.) (1999).

[70] Van der Bruggen (1999).

people and should be taken seriously. Intuitions can be a motivation to find more rational arguments. One such argument that is often labelled as emotional, but that the panel feels should play a role, is the objection to crossing natural species barriers. Their advice to politicians is that cloning should be made an unattractive option, which requires new regulation, such as the obligation to carry out a health impact assessment for each proposed cloning procedure. Also, they argue that the costs of unintended negative consequences of cloning should be borne by those who carry it out. Finally, they suggest to politicians that they could install moratoria and they call for a more international focus of regulation.[71]

The process of the Dutch lay panel conference appears to have been less pre-structured than the Australian one. Also, the panel members were allowed to have contact with the media; some contacted their regional newspapers to let them know they were in the lay panel and this resulted in interviews. No effort was undertaken to avoid contact between the lay panel and the experts or audience. As stated before, there were no signs that the organisers tried to influence the information sources or the selection of experts. These observations suggest that there was less focus on the avoidance of bias than in the Australian experience, possibly because the climate was less antagonistic. Framing of the lay panel debate took place through the agenda setting public hearing.

6.4 Goals of Deliberation

The first condition for a deliberative democracy to be able to deal with intractable disagreement that I proposed in Chapter 4 was that the aims of the deliberation should match the type of policy problem. What goals did the organisers of the Dutch and Australian consensus conferences have and did these match the policy problem? In Chapter 4, I distinguished three main goals of public deliberation: striving for consensus, broadening the debate (inclusiveness) and deepening the debate (quality) and I pointed out a tension between these goals. I also argued that this tension is less problematic when we let go of the aim of consensus. In unstructured policy problems such as the biotechnology debate aiming for consensus is premature. It is more important at this stage to include as many different viewpoints and groups as possible, so that everyone has a say already from the first stage of problem demarcation. The next step is to deepen the debate by making sure that all relevant arguments surface and are discussed thoroughly. In Chapter 4, I also formulated three more specific criteria regarding goals of deliberation, namely that opinion transformation should take place, that the deliberation should influence the decision-making process, and that outcomes should be revisable. To what extent were these criteria met in practice?

---

[71] Bozkurt (ed.) (1999).

6.4.1 Australia: Focus on Consensus

The Australian museum states five aims of holding the consensus conference: (1) to facilitate broad public debate, (2) to allow lay members to become informed and involved in decision-making, (3) to show that a plurality of views exists on GM food, (4) to 'bridge the gap between experts and lay people', and (5) to 'integrate the consensus conference model into decision-making on important matters of public policy in Australia'.[72] These aims seem to incorporate all three goals of consensus, inclusiveness, and deepening and reflect a view of two-way transformative learning, in which a dialogue is held with mutual influence. Some expert participants, however, seemed to view the aim of the conference as educating the public, so as to take away their irrational fear of biotechnology.[73] While the organisers did not seem to expect a tension between their different goals, and in their statement the goal of consensus appears to be the weakest one, I think in practice there was an emphasis on reaching consensus to the detriment of inclusiveness. There was a strong wish to present a consensus report; the report was written under extreme time pressure and some member's views were excluded, as they had already retired to sleep or just caved in, wishing to finish the report. Neither did the report reflect minority arguments. Because of the focus on consensus it was not taken into account sufficiently that there were different dimensions of more or less fundamental disagreement about facts and values, starting already at the level of problem definition. Particularly in the context of an adversarial political culture in my view we should not expect much from a consensus report. This view is supported by the fact that its outcomes only influenced policy to a very limited extent.

The panel acknowledged the adversarial context in which legislation had to be drafted, due to the many vested interests lobbying government, and argued that government should refrain from taking a stance, but rather facilitate stakeholder dialogue. As we saw in the previous chapter, the government partly complied with this request by installing the Gene Technology Community Consultative Committee. It deviated from the recommendation by not appointing lay members on this committee, however, and as we saw, this committee does not exert much influence over decisions. Some of the details of the lay panel recommendation, such as a regular review of the regulations, were also adopted. Moreover, some Ministers have referred to the consensus conference as influencing their budget allocations.[74] The panel also recommended more comprehensive labelling of GM foods, but this

---

[72] McDonald (1999).

[73] For example, in his speech, one expert stated that 'regardless of the benign nature of gene technology, its use in the production of food has become topical. As a result there is a need to educate consumers about the technology such that they are more confident about its application'. Geoffrey Annison (1999), np.

[74] Crombie and Ducker (2000). Crombie and Ducker, who were the official Phase II evaluators, argue that a large part of the lay panel recommendations were followed up. However, in hindsight this does not appear to the case.

6.4 Goals of Deliberation

was not complied with. It appears that important decisions regarding labelling had already been taken before the consensus conference was held.[75]

A copy of the lay panel report was forwarded to all Australian senators and put in the parliamentary library. Many lay panellists and experts were disappointed that the report was not actively brought under the attention of more policy-makers, did not get a wider reception, and that no politicians expressed a commitment to it.[76] On the contrary, since the conference, several politicians have publicly demonstrated their support for gene technology, not acknowledging the critical comments of the lay panel.[77] Before the conference, no explicit agreements were made with political decision-makers, and since there is no direct tie to the Australian parliament – as there is for example in Denmark – the status of the lay panel report remained vague. In some cases a consensus conference can influence policy decisions in an indirect way, as the result of media reports and subsequent discussions in response to the lay panel report.[78] There is no evidence for such indirect influence in the Australian consensus conference. Regarding the criterion of revisability it is telling that to date no second conference has been organised, creating the appearance of closure.

More than half of the conference participants (including lay panellists, experts, audience, and steering committee) changed their views compared to prior to the conference.[79] The majority of the lay panel became more critical of gene technology, while most steering committee members became more positive. As could be expected, members of the expert panel hardly changed their viewpoints. Particularly the preparation before the conference seems to have caused the panel members to change their views. Before the conference 54% of the panellists appeared to have a positive attitude towards GM food, and 40% had no view on it. After reading the briefing paper, these attitudes shifted considerably to a more negative stance.[80] This could be interpreted either as a sign that the briefing paper was not as unbiased as the authors purported it to be, or as support of the view of many biotechnology critics that the more people become informed on the topic, the more critical they become. At least a third of the participants indicate that before the actual conference and the panel deliberations began they had formed more or less definite opinions. Not only did the participants become more aware of the complexities of the issue of GM food, but their confidence in the food they consumed also decreased, and they started to doubt the noble intentions of scientists. However, after hearing the expert presentations, some of those who were absolutely opposed to gene technology became more

---

[75] Mohr (2002).

[76] This was partly due to the inadequate handling of the closing stages of the conference by the organisers. See Ibid.

[77] For example, the Premier of Queensland claimed he was going to 'pursue biotech industry's opportunities vigorously' and the Treasurer's speech refers to 'reaping the benefits from this cutting edge area of science'. McDonald (1999).

[78] This was the case in Danish consensus conferences, for example. See Cronberg (1995).

[79] Interview with organiser of consensus conference on 10/4/2005. See also Mohr (2002).

[80] McKay and Dawson (1999), pages 23–24.

positive about some applications, such as solutions to blow fly problems in sheep. Generally, the conference served to move panel members away from more extreme viewpoints to more moderate and complex views.[81]

It can be concluded that the condition of matching the aim of the consensus conference to the policy problem was not fulfilled very well. The organisers stated too many aims, some of which were in tension with each other. Predominance was given to reaching consensus, which I argued is premature in the case of unstructured problems, and was detrimental to the success of other goals, such as deepening. Deepening was only achieved to a limited extent; while some new views came to light, experts did not seem to be open to these. Experts took their role to mean simply educating the lay members. Broadening took part only to a limited extent, because the consensus conference had little influence directly or indirectly – via the media – on the decision-making process.

6.4.2 *Netherlands: Deepening the Debate*

It appears that various, sometimes conflicting, interpretations of the goal of organising public debates lay behind the Dutch public debate on cloning. One goal of the Ministry was to make as many citizens as possible aware of the topic so that they could form their own opinion. The emphasis would then be on providing information that was as objective as possible, and this could be regarded as a top-down approach to educating the public. This aim conforms both to the deepening and inclusiveness goals and appears to reflect the liberal ideal of democracy described by Porsborg Nielsen et al. (see above). Another goal, in the words of the Minister of Health at the time, was to 'have a discussion with as many rational arguments as possible and to make clear with the help of moral, social, and political arguments what we as a society think about cloning and where we will draw the line'.[82] The Minister's description of this goal appears to favour the goal of consensus and is congruent with the republican ideal of democracy as set out above, because it views the aim of public debate as finding values that are widely shared in society. However, other aspects of the Dutch debate suggest that the proceduralist-deliberative ideal was followed, because the aim was to involve as many different views as possible and to inform those involved, particularly political representatives, and because consensus was not explicitly aimed for. Moreover, the Rathenau Institute wanted to follow a more bottom-up approach, in which all the participants to the debate would provide expertise and information and this would be analysed in order to get 'to the bottom of underlying considerations'.[83] In other words, the emphasis for them was on inclusiveness and deepening. The fact that the lay panel did not aim for consensus and that both minority and majority views were described in the report suggests

---

[81] Ibid.

[82] Minister E. Borst-Eilers as quoted in Biesboer (1998).

[83] Frank Biesboer (1999), page 13.

6.4 Goals of Deliberation

that deepening was in practice the main goal. As I will elaborate later in this chapter, I do not think that the goal of inclusiveness was reached or that it was regarded as the most important one.

What was the influence of the recommendations on policy decisions? In 1998, prior to the conference, parliament had accepted a motion by political opposition parties for a moratorium on cloning experiments until public debate had reached conclusions. According to the Rathenau Institute, this means that 'the societal debate was given political weight from the outset'.[84] The panel report was officially handed to the Minister of Agriculture at the final meeting and it was distributed to stakeholders and politicians. During the final meeting, the Minister of Agriculture stated that the outcomes of the public debate supported the 'no-unless' policy for animal biotechnology and that he had no intention of changing it into a 'yes-if' policy under the influence of lobbying by scientists and biotechnology companies. Other than that, the outcome of the lay conference does not appear to have been sufficiently conclusive to directly influence debates in parliament or policy decisions. Moreover, the lay panel report was only one document in a series of recommendations to government. Some cynically argue that public debates such as these merely function as delaying mechanisms for politicians who are not yet ready to make certain decisions.[85]

The facts that the lay panel debate was part of a whole series of public debates and that several books were published as a result of the debate suggest that the debate is open-ended and ongoing. On the other hand, after the lay panel report was handed to the Minister no more panel meetings have taken place and neither did the panel get the chance to clarify their views in public; something which could have been done in a workshop, for example.[86] Opinion transformation did occur to a certain extent; while initial attitudes towards cloning did not change, members did change their minds on certain issues and their views became more subtle as they grasped the complexity of the topic.[87]

It can be concluded that if we take the main aim of the lay panel deliberation to be deepening, this matched the policy problem fairly well. The Dutch situation does show the pitfall of letting go of the aim of consensus; while the possibilities were present to influence the decision-making process, this did not happen, because the results were not conclusive enough. Moreover, the lay panel debate was part of a series of debates, which also reflects the aims of deepening and broadening, but also functions to limit the lay panel's direct influence. On the other hand, if we recall the three-track approach that I proposed in Chapter 4, the stage in which citizens deliberate conforms to the second deliberative track, and does not play a direct role in the decision-making process, but only an indirect role of informing the institutional track. In my opinion this latter track is aided more by a majority/minority report and

---

[84] Ibid., page 10.

[85] Interview with organiser of public debate on 2/4/2003.

[86] This was suggested by one of the organisers. Interview on 2/4/2003.

[87] Bozkurt (ed.) (1999).

210　　6 Consensus Conferences: The Influence of Contexts

a reflection of underlying arguments than by a consensus report, because political actors will need to make up their own mind on the basis of arguments and views that are present in society.

6.5 Arguments

In Chapter 4 I argued that no relevant arguments or viewpoints should be excluded from public deliberation and I proposed three more specific criteria: Lay person input should be included, there should be open expert contestation, and self-interested motives or fallacies should be uncovered. How did our consensus conferences fare?

6.5.1 Australia: Trust on Trial

On the first day of the conference, the scene was set by the introductory remarks made by the Federal Minister for Agriculture Fisheries and Forestry, which emphasised the importance of gene technology to the Australian economy and to global food security, downplayed its potential risks, and contained rhetorical remarks, such as 'we can not turn back the clocks'.[88] His remarks were conceived by many of those involved as an indication that the government had a predetermined agenda, did not take the panellists' achieved level of understanding of the topic seriously, and did not accord high priority to the outcomes of the consensus conference.[89] Moreover, as could be expected within an antagonistic political culture, some panellists were suspicious of bias and agenda-setting by the steering committee.[90] As the evaluators of Phase I put it, 'trust was on trial during the consensus conference'.[91] Some panellists criticised the facilitator for pressuring them too much and for being too directive; for example, the facilitator would end disagreements and 'provide a definite answer'.[92] On the other hand, the role of the facilitator in creating a team spirit and ensuring equal and respectful discussion was deemed as very important by everyone; many commented on the facilitator's skilful handling of group dynamics.[93]

The expert speakers were required to give a politically and value-'neutral' answer to the key questions, but many were unable to leave out their own opinions, or the views of the organizations they represented. Besides the example of the Monsanto representative I already mentioned, other speakers' presentations also contained

---

[88] Hendriks (2004), page 111.

[89] For panellists' reactions to the speech, see Mohr (2002), page 153.

[90] Before the second preparatory weekend the conference facilitator arranged a meeting between the panellists and some of the steering committee members in order to counter this suspicion, but this appeared to have the opposite effect. See Ibid.

[91] McKay and Dawson (1999), page 28.

[92] Mohr (2002), page 134.

[93] Ibid.

6.5 Arguments

elements of a public relations-nature, such as the following statement: 'Gene technology as a source of innovation in products and processes will profoundly impact the quality of life of all Australians – either as consumers or beneficiaries of economic growth and of a sustainable environment'.[94] One opponent ended his speech with the doomsday words 'the future will curse us'.[95] On the one hand, this supports the idea that facts and values are not so easily separated. On the other hand, some speakers deliberately omitted certain information or evaded answering questions, suggesting that self-interested motives were present. This was especially the case with the Monsanto representative, who was subsequently met with suspicion. Several commentators argue that the expert speakers polemically used the conference to push their own interests, and that the lay panel had to 'sift fact from propaganda'.[96] In the written transcripts of their speeches several statements can be found that in Chapter 2 I earmarked as rhetorical devices.[97]

The way the conference was 'pre-structured' by the organisers was criticised by many of the participants, both experts and lay people. Several commentators argue that the set-up of the public part of the conference, particularly the interaction with the experts, created an adversarial atmosphere, leading to a stating of opposite positions rather than a dialogue. The only part of the conference that had the shape of a real debate was the discussion between the lay panel members during the report writing stage. Based on an evaluation of the participants' experience of the 'highly prescribed nature' of the formal discussions taking place during the consensus conference, Mohr concludes that 'the rules of interaction had in fact stifled any sense of an open debate'.[98] Dialogue between the experts and the lay panel, between the experts themselves, and between the audience and the formal participants was not possible, because it was structured like a court room examination, in which experts could only respond to a small defined set of questions and were not allowed to respond to each other or to different questions. Moreover, the members of the lay panel were forbidden to have contact with the expert panel outside of the formal consensus conference.[99]

Mohr likens the consensus conference to jury deliberations – in which jury members might come to a decision not so much on the basis of arguments, but on how persuasive the witnesses are: 'rather than facilitating broad public debate from a plurality of perspectives, the jury metaphor suggests that the consensus conference

---

[94] Annison (1999), np.

[95] Peter Wills (1999), np.

[96] McDonald (1999).

[97] For example, a director of the Australian Food and Grocery Council uses the argument that humans have been breeding and cultivating animals and plants for thousands of years. Also, he defines biotechnology as 'any technique that uses living organisms to make *beneficial* materials or products' (my italics) and claims that 'Australia has no choice but to embrace the new technology'. Annison (1999).

[98] Mohr (2002), page 191.

[99] However, some panellists did seek more informal 'unauthorised' contact with the expert speakers. Ibid.

model facilitates debate from a duality of views'.[100] In a jury system only two contradictory positions can be expressed, and not the plurality of views that are actually held about GM food. McDonald even concludes that this jury-like character was created deliberately and that this shows the adversarial nature of the conference.[101] Also, the expert and lay panels were placed on opposite sides of the Senate chamber, where the consensus conference was held, which reinforced the divide between these two groups. It has been suggested that this hindered the creation of mutual understanding of the topics under discussion and that the gap between experts and lay persons that the Australian Museum – in recognition of the value of lay experiential knowledge – wanted to bridge, was in fact augmented. As I argued in Chapter 4, one of the main purposes of organising public debates is exactly to benefit from lay experiential knowledge. Moreover, as we saw, in unstructured problems the lay/expert divide is often an artificial one and the official status of expert knowledge should be called into question. However, it can be concluded from my analysis of the Australian consensus conference that the strict expert/lay divide was reinforced. As Purdue also pointed out regarding the British consensus conference, when a strict 'division of labour' is created between lay people 'who ask questions' and experts, who 'provide answers', the fact that expert knowledge is contestable and often relies on subjective judgments is overlooked.[102]

Unlike other countries that stage consensus conferences, such as Denmark and the UK, no process was included in which experts could respond to each other's presentations, much to the frustration of the participants. Moreover, because experts were chosen to present two different sides of each question, the positions presented to the lay panel were quite polarized and an adversarial atmosphere was created. Some lay panel members would have liked to hear more neutral, or middle-of-the-road positions.[103] This polarization led to a duality of viewpoints, and this suggests that the Australian Museum's aims of bringing to light the plurality of views about biotechnology were not met. In other countries, a draft of the lay panel report is usually given to the expert presenters who can clarify issues or suggest certain refinements before the final report is produced. The organisers of the Australian conference decided against this option, because members of the steering committee worried that the experts would use this process to try and influence the outcomes, rather than just focus on the technical issues. These decisions again show that this consensus conference took place in an antagonistic climate.

According to the post-consensus conference evaluation carried out by Mohr, 'the majority of expert speakers believed that some members of the audience had been primed to ask certain questions of them in order to score "political" points and therefore presented comments dressed up as questions'.[104] Lay panel members also felt

---

[100] Ibid., page 216.

[101] McDonald (1999).

[102] Purdue (1995), page 171.

[103] Mohr (2002).

[104] Ibid., page 173.

6.5 Arguments

that some questions from the audience were statements backed by certain organizations rather than genuine questions. Moreover, the session that involved the audience was under great time constraints and this excluded many people wanting to ask questions. This illustrates that the conference organisers tried to contain the debate by setting strict boundaries. In fact, the general structure of the consensus conference appears to have been very rigid and formal; there were many constraints, because the steering committee feared that it would be difficult to keep the process in check and that experts (and audience members) would inappropriately try to influence the position of the lay panel. This would suggest that the steering committee was very aware of the politically charged character of the gene technology debate. Lay panel members, for their part, interpreted it as a lack of trust in their capacities to recognize an attempt to misinform or influence them.[105]

Was there room for a discussion of broader moral and metaphysical values and how were value disagreements dealt with? As became clear from the lay panel report, the consensus conference provided an opportunity to reflect on broader moral and even metaphysical issues than had dominated the GM debate until then. Even though the panel extensively discussed risks, they also criticised the narrow focus on risk in decision-making, pointed out that risk analysis and science in general were value laden, and wanted to include visions about the future. They also spoke about the moral aspects of interfering with nature and the philosophical aspects of what was happening on the cellular level.[106] Other explicitly moral issues they discussed were biopiracy and implications of consuming GM foods for those with religious or moral prohibitions to certain foods.

Fundamental value disagreements arose especially around the idea that 'humans should not tamper with nature'. For some, but not all, this had a religious background.[107] Not all panellists felt that moral and value disagreements had been dealt with satisfactorily. The value-related and religious questions surrounding gene technology were not dealt with during the preparatory weekends. Also, one panel member was frustrated after repeatedly bringing up a value-related subject that the facilitator kept on postponing until it was finally never addressed. Some panellists were also disappointed that no expert speaker with broad knowledge of ethics was interrogated.[108] The one person who was explicitly selected as an ethical expert, a Catholic priest, was the speaker most appreciated by the lay panel.[109] Speakers who were not ethical experts were also asked to comment on ethical issues, but their contributions remain very superficial. One confined himself to stating some areas of concern, followed by the rhetorical question 'is this unethical?', implying that

---

[105] Ibid.

[106] Interview with organiser of consensus conference on 10/4/2005.

[107] Interview with organiser of consensus conference on 10/4/2005.

[108] Mohr (2002).

[109] McKay and Dawson (1999).

biotechnology was not.[110] None of the speakers provided the lay panel with frameworks or a background in moral theories that could help them reflect on and discuss moral problems themselves. Neither did they get training in argumentation and the ability to recognize fallacies. On the other hand, the facilitator did work on their strategic questioning skills and team development.

Was a diversity of views from lay people achieved in this consensus conference? As was mentioned above, the panel was not supposed to represent all views that exist in Australian society, but to constitute 'a slice of Australia'. What exactly is meant by this remains unclear; this term gives the appearance of a random cut out of the whole population, but as we have seen, this is not the case. The panel was initially self-selected, which is likely to exclude certain groups, such as minorities, and less-educated people. Probably, then, a slice of Australia denotes that a demographically stratified selection was made, aiming at gender equality and a diversity of ethnic, religious, socio-economic, and educational backgrounds. This has been achieved fairly well. However, as we saw in Chapter 4, this presupposes false essentialisms, for example, as if one aboriginal person could speak for all aborigines. A diversity of views was not necessarily aimed for and neither was it achieved, because the steering committee held interviews expressly to exclude those with a special interest in the topic, or with an extreme view on it. In fact, according to Aidan Davison and Renato Schibeci, the selection method excludes all 'community members who have previously deliberated on the matter at issue'.[111] It appears, therefore, that the organisers were not aiming to bring to light all relevant arguments.

Opinion polls taken after the conference show that the recommendations made by the lay panel reflected the views of the majority of the Australian public, suggesting that the lay panel was at least an adequate representation of the majority of the wider public.[112] Still, the supposed representativeness of a lay panel consisting of only 14 persons will always remain problematic. This is aptly described by one participant of the internet forum put up by the ABC: 'the selection of 14 citizens out of 18 million has nothing to do with democracy either. These people weren't elected, so they have no mandate whatsoever to "represent" the Australian public. They certainly don't represent me'.[113]

To sum up, while there was room to discuss fundamental worldviews and value-related issues, there seemed to be less room to discuss scientific uncertainty. On the one hand, scientific uncertainty was reflected by open contestation between experts, but this was done in such a polarised way that more middle of the road positions were excluded. Also, while the normative aspects of science were the topic of discussion, particularly regarding risks, and scientific methods were therefore challenged, the strict division between experts and lay people was not challenged, but

---

[110] Jim Peacock (1999).

[111] Davison and Schibeci (2000), page 59.

[112] For example, research carried out by Ernst and Young and the Commonwealth Department of Industry, Science and Resources in 1999. Mohr (2002).

[113] Davison and Schibeci (2000), page 58.

even reinforced. Fallacies and self-interested arguments were uncovered by the lay panel. In fact, due to the adversarial atmosphere in which the conference took place, the fear of self-interested motives was always at the forefront. It appears that in order to keep this from getting out of hand, the organisers tried to contain the debate by rigidly pre-structuring the debate and by excluding potential participants with a special interest in the topic. Ironically, by these choices they may have contributed to the adversarial atmosphere. As I will elaborate in more detail in the discussion of this chapter, this finding raises questions for the potential of consensus conferences to provide depoliticised fora.

6.5.2 Netherlands: A Philosophical Framework

Quite fundamental moral and metaphysical questions were discussed by the Dutch lay panel, including questions such as 'should we use nature in this way?'; 'is there a genetic basis to human identity?'; and questions focussing on our proper relationship to animals and on justice between rich and poor countries.[114] Following the presentation by philosopher Tsjalling Swiersta, who argued that the basic attitudes that people have regarding science and technology determines to a large extent what their starting point is in the discussion about cloning, the panel first stated what their basic attitude was. Three groups of attitudes could be discerned. Some were sceptical about science and technology, and concerned that society was becoming increasingly directed by technology. This concern was based to a large extent on the fear that new technologies would only be accessible to the rich and that a one-dimensional emphasis would be placed on 'cure' at the expense of 'care'. Another group was optimistic about science and was afraid that if risky research was called to a halt then important developments would be missed. A third, 'in-between', group thought that scientific research often led to developments that were for the common good, but that unfortunately sometimes undesirable applications were possible, and we should make sure that the outcomes of research would be used only for the benefit of society.

This difference in basic attitudes led to many discussions and was one reason why 'never a position was taken that all agreed with *for the same reason*'.[115] This statement reveals the different interpretations that can be given to the concept of consensus. Consensus could either mean simply to agree on an outcome, and this could be agreement on the basis of compromise, or it could mean agreeing not only on the outcome, but also on the underlying reasons, as in Rawls' notion of the overlapping consensus. If we take consensus to mean agreement not only on outcomes but also on underlying arguments, obviously consensus was not reached by the Dutch lay panel, nor was this the purpose of the debates. On the other hand, the panellists did strive to make some recommendations that all could support, albeit for their own

---

[114] Interview with organiser of public debate on 2/4/2003.

[115] Emine Bozkurt (1999), page 5.

reasons. The fact that no real consensus was reached has been pointed out as one of the main reasons why no political commitments were attached to the conclusions of the lay panel.[116] What is striking in a comparison between the formulations of the Dutch and the Australian reports, is that the internal struggles the panel had experienced (even between one and the same panel member) became clear from the Dutch report, while the Australian report shows only the end results of the deliberations. The Dutch panellists state that many of them changed their opinion regarding animal cloning several times during the process. Also, the underlying arguments for their position are stated more explicitly than in the Australian case.

The contributions of several ethicists and a philosopher seems to have influenced the content of the discussions. The panellists were handed several theoretical frameworks (of consequentialist, deontological, and virtue ethics, or ethics of the good life) that seem to have been helpful. It turned out that consequentialist considerations were the least contentious and this might be the reason why many decisions were primarily taken on this basis. Deontological and life ethical considerations did persist, however, but these tended to be drawn into the private sphere.[117] It can be concluded that in the Dutch consensus conference there was less antagonism and more respect for differences than in Australia. The distinction between experts and lay persons was less strict, also because the conference was not as pre-structured as the Australian one, and informal communication between the two groups was not discouraged. The set-up of the conference created more opportunity to discuss the normative assumptions behind scientific expertise; this was mainly due to the presence of philosophers and ethicists who handed the lay panel a framework to deal with moral and ethical questions. The strategy of depoliticisation that Hisschemöller and Hoppe describe, is therefore, less prevalent in the Dutch situation than in the Australian. Still, to some extent viewpoints were excluded, because the bias towards utilitarian or extrinsic arguments that I also pointed out in Chapter 2 was also present here; a type of self-censorship occurred. Fallacies or self-interested motives were less of an issue in this conference than in the Australian one, which is likely the result in part of the different topics of the conferences and in part of the difference in political climate. Depoliticisation in the sense that Pettit describes, therefore, seems less important in the Dutch situation, while in the Australian situation it appears unsuccessful. I will come back to the issue of depoliticisation in the discussion of this chapter.

6.6 Participation

In Chapter 4, I argued that we should aim to be as inclusive as possible when unstructured problems like biotechnology are involved. This led to the formulation of six criteria, specifying that women and marginalised groups should not be

---

[116] Van der Bruggen (1999).

[117] Swierstra (2000).

6.6 Participation 217

disadvantaged in the debate, mechanisms should be in place to counter power differences, distorting group dynamics should be avoided, people with a wide variety of backgrounds and opinions should be included, participants should learn from the debate, and the outcomes of the debate should reach people other than the direct participants. I will now examine to what extent these criteria were met in the two consensus conferences under examination.

6.6.1 Australia: Massive Learning Effect

What role did the consensus conference play in stimulating wider public debate on GM foods? It appears that prior to the conference the organisers put little effort into involving the general public by encouraging them to be in the audience.[118] As a result the public gallery was not filled up and the majority of the audience consisted of stakeholders. On the other hand, several sources suggest that there was great interest from the public in the topic of GM foods and the process of the consensus conference during and immediately following the conference.[119] The conference was attended by 83 journalists and following the conference 173 media items were devoted to the conference nationally. Also, 114 news items appeared in regional media.[120] The Australian Broadcasting Corporation (ABC), who had exclusive access to the lay panel and broadcasted a series of radio programmes about the consensus conference, put up an internet site devoted to the conference, which included information and an interactive forum for citizens to discuss GM food and the conference. Particularly the interactive website provided an opportunity for involving a wider group of people in the public debate.[121] Analysis of the news coverage on biotechnology, and GM food in particular, in the year of the conference shows that the number of negative reports increased significantly.[122] According to McDonald, the consensus conference 'triggered, rather than facilitated, broad public debate'.[123] Nevertheless, the official evaluators of the consensus conference conclude that considering the limited resources available, the consensus conference significantly raised public awareness about GM food. This appears

---

[118] Two advertisements were placed in the Canberra Times, a newspaper read primarily by (part of) the citizens of Canberra and not by the rest of Australia. Mohr (2002). This judgment is supported by the official evaluation of Phase I. See McKay and Dawson (1999).

[119] Interviews with organisers of public debates on 10/4/2005 and 1/3/2005.

[120] Crombie and Ducker (2000).

[121] Members of the lay panel, the expert panel and the facilitator joined in the discussions on this website. The ACA also set up an internet forum about the conference. Davison and Schibeci (2002).

[122] Mohr (2002), pages 266–267. See also, Crombie and Ducker (2000).

[123] McDonald (1999), argues that despite the conference, gene technology has had little media coverage in Australia, but the numbers quoted by Mohr and Crombie seem to suggest otherwise.

to have led mainly to public suspicion towards GM food, but not so much to an increase in scientific understanding.[124]

Most of those involved consider the learning effect on the lay panel 'massive'.[125] From evaluations it appears that on a personal level they found it a rewarding experience. They particularly thought that the exchange with experts was an empowering experience and a privilege and they were pleasantly surprised about the respect that was accorded them.[126] One of the lay panel members in a radio broadcast stated 'it's been the most important thing I've ever been part of'.[127] They also gained a lot of knowledge that they could go on to share with other lay people and realised that as ordinary citizens they could in fact influence policy-decisions. On the other hand, some were quite cynical about the extent of this influence. This cynicism was exacerbated by the fact that there were no follow-up meetings or information update on the report's fate after the conference.[128] This is problematic, because as Archon Fung explains, the learning and socializing potentials of public deliberations are greatest when they are recurring or continuous.[129] Some participants, however, indicated that this experience made them more active in other fields of politics and more interested in other social issues as well.[130] It could be concluded that their awareness of, and interest in, the topic were raised and that they became more active citizens.

Tarja Cronberg points out two ways in which experts' views can be positively influenced by participating in a consensus conference.[131] Firstly, experts often work within a narrow framework of specialist knowledge, ignoring questions outside of this framework. Confrontation with lay panellists' views can alert them to the questions outside of this framework, including moral, social or political issues. This does not always have the effect of broadening the scope of the expert, however, as a common reaction is that such questions belong to the political sphere. Nevertheless, through a consensus conference, experts might start to realise that their own framework is not as neutral as they always thought. Secondly, experts who previously disputed each other's viewpoints might be drawn closer together as they are all challenged by the views of lay persons, which might be even more

---

[124] Crombie and Ducker (2000).

[125] Interview with organiser of public debate on 10/4/2005.

[126] Mohr (2002); McDonald (1999); Crombie and Ducker (2000).

[127] Australian Broadcasting Corporation radio recordings of 'Life Matters: Consensus Conference' about the consensus conference on Gene Technology in the Food Chain, 1999.

[128] Crombie and Ducker (2000).

[129] Fung (2003).

[130] In terms of spin-off of the consensus conference, two members of the lay panel were asked to sit on CSIRO Institutional Biosafety Committees, two members held presentations at a conference and a university, one member joined a focus group of Agrifood Awareness Australia, two panellists became members of the Gene Ethics Network, and one member entered discussions with her local council. See Mohr (2002).

[131] Cronberg (1995).

distant to their own views. These two influences can be discerned only to a limited extent in the Australian conference. The conference was a learning experience for some of the expert speakers, who gained a greater understanding of the importance of lay perspectives and the ability of lay people to become acquainted with a technical topic.[132] Also, some noted that the consensus conference had increased the dialogue between opponents, even after the conference. This did not lead to a reconciliation of opposing viewpoints, however. Even though the lay panel's views became less opposed and more moderate, this cannot be said of the participating stakeholders. Crombie and Ducker conclude in their post-conference evaluation that even though most stakeholders have gained a deeper understanding of the arguments and viewpoints of their opponents, 'the consensus conference process has not significantly softened or ameliorated the polarisation of beliefs and positions' about GM food.[133] Rather, they 'got to know the enemy' better. The adversarial character of the conference itself illustrated for many of those involved that the expert speakers and stakeholders have entrenched positions that they are not willing to depart from.[134] It does appear that the conference led to a consolidation of cooperation between the groups who were in one camp (either pro or contra GM food).[135] This finding supports Sunstein's point of the law of group polarisation, alluded to before.

The expert panellists did gain knowledge, particularly regarding the viewpoints of lay people, of reasons why many lay people were opposed to genetic engineering, and connected to this an understanding of the importance of communicating with citizens. Moreover, most were in awe of the ability of a group of citizens without prior knowledge of or formal education about the subject matter to come up with important questions and a balanced report. They and others involved lauded the lay panel for their common sense approach to the complex issue.[136] The lay panel also proved to be able to see through the rhetoric that was used by some of the expert presenters. For example, the panel were critical of arguments made by industry representatives about substantial equivalence and the impossibility of labelling, and of the claims of loss of international competitiveness.[137]

---

[132] For example, the representative of Monsanto, who first treated his presentation as a pr-exercise and misjudged the level of knowledge of the lay panel, during the later stages of the conference, after he was approached in a rather suspicious way by the panel, was absent. This increased the mistrust of the panel. After the conference he wrote a frank letter of apology for his absence, stating that 'the consensus conference made him realise that he really did have to listen to public opinion'. Mohr (2002), page 165. On the other hand, Hendriks suggests that this admission is 'another PR stunt'. Hendriks (2004).

[133] Crombie and Ducker (2000).

[134] This was especially illustrated by the two who were on both extremes of the GM debate, the representatives of Monsanto and of Gene Ethics.

[135] For a thoughtful interpretation of stakeholder roles in the consensus conference, see Hendriks (2004).

[136] Mohr (2002).

[137] McDonald (1999). Evidence of this criticism of rhetoric, for example can be found in the remark of one panellist who introduced himself as not wearing a tie, but still being 'substantially equivalent' to those who were.

Overall, in the Australian experience a lot of focus was on the question as to how bias could be avoided and whether undue influences on the lay panel had taken place, for example by the chairperson, conference coordinator, steering committee, conference facilitator, experts, professional writer (who helped to formulate the report), and audience. Especially regarding the steering committee the question of bias and undue influence appears to be valid. Many sponsors of the consensus conference demanded to be represented on the steering committee and thereby gained a measure of control over the process.[138] This casts doubt on the neutrality of the process. Also, the consensus conference was officially evaluated and this evaluation was overseen by a steering committee subcommittee that included two representatives of biotechnology industry interests, who had a stake in public acceptance of the technology.[139] A representative of the Gene Ethics Network (GEN), the main organization critical of genetic engineering in Australia, was rejected as a steering committee member, because he was judged to be too biased. Moreover, nobody from government or biosafety departments, and nobody representing ethics or religious views was invited to be a member of the steering committee. The final selection of expert speakers was also heavily influenced by the steering committee, partly due to the panel running out of time, but also partly because the committee did not agree with the initial selection. From this and from the fact that the panel's questions corresponded to the topics that were emphasised in the briefing paper, it can be concluded that the discussion was framed to a certain extent by the steering committee. Unlike the Danish tradition of staging a press conference after the event that would be attended by the whole lay panel, a selection of three spokespersons of the lay panel was made. Also, only half of the steering committee members were chosen to be present. Mohr suggests that this is another way in which the organisers tried to contain the process.

Did the consensus conference provide a power-free arena to let the power of the better argument prevail, unhindered by distorting group dynamics? Most panellists and observers reported that there were no power imbalances within the panel and that everyone had an equal opportunity to take part in discussions, although some pointed out that one panel member was trying to dominate discussions, creating tensions within the group.[140] Some members of the lay panel tended to take part in the discussions more actively than others. This was not determined by whether they were male or female, but by their level of education, personal passion, and a desire

---

[138] McKay and Dawson (1999). The Phase I evaluators recognize that some of those on the steering committee regarded themselves as representatives of their employers. Still, without further argument, they conclude that all members of the committee successfully attempted not to influence the conference outcomes.

[139] Mohr (2002), page 282. Indeed, phase 2 of the evaluation focused primarily on the degree of public acceptance of gene technology brought about by the conference.

[140] This emerged from several different sources, such as the audiotapes of the ABC programme on the consensus conference, my interview with an organiser of a consensus conference on 10/4/2005, and Mohr. Ibid.

6.6 Participation

to come to grips with the issue.[141] Particularly one panellist was very vocal and persuasive; he distributed the extra information materials and managed to moderate the views of those members who were initially optimistic about gene technology to become more cautious.[142] It was pointed out by many that the skills of the conference facilitator made all the difference in maintaining equality of power within the group. It can be concluded that 'equality of voice' can be stimulated by a skilled and neutral facilitator.[143] It should also be mentioned that the personal style and charisma of the individual expert speakers appeared to influence their believability and as a result some were questioned more rigorously than others.

It can be concluded that while marginalised groups were not disadvantaged and distorting group dynamics were countered, the influence of power differences was not completely avoided, because of uneven stakeholder presence in the steering group which framed the debate. The conference had a great educative effect on the lay panel, and the outcomes reached a larger group of people due to relatively extensive media coverage. Experts gained an understanding of the importance of lay person perspectives, but their own views remained stable and the polarisation between them was not moderated.

6.6.2 Netherlands: An Elite Affair

To what extent did the public debate in the Netherlands stimulate wider public debate? There was some media-attention, particularly during the public hearing and the closing session of the public debate, but this focussed primarily on the presence of Ministers or on incidents.[144] Also, most newspapers published the outcomes of the public survey, but not much attention was given to the lay panel report.[145] The audience during the public parts of the debate primarily consisted of stakeholders; there was not much participation from members of the wider public. This suggests that the Minister's goal of making as many citizens aware of the topic so that they could form their own opinion was reached only to a very limited extent. The politicians that were involved in the public debate did think it was so successful that they wanted to apply the same method to other controversial technological issues, such as xenotransplantation.[146]

What were the effects of the debate on the participants? The lay panel gained a lot of knowledge. Despite the criticisms that the lay panel's conclusions were based too much on emotions, according to Koos Van der Bruggen, 'the members of

---

[141] Interview with organiser of a consensus conference on 10/4/2005.

[142] McKay and Dawson (1999).

[143] This point is supported by Smith (2008).

[144] Interview with organiser of public debate on 2/4/2003.

[145] Biesboer (1999).

[146] Ibid.

the panel probably know more about the subject than members of parliament'.[147] Moreover, this criticism misses the point of the panel that emotions and intuitions have a legitimate place in ethics as a starting point of moral reasoning. One of the outcomes of the debate was that those who had previously held no opinion about cloning generally became more positive of its uses after they gained knowledge about it. Those who were negative about it to start with did not change this attitude. As the lay people gradually became less lay and more expert, their opinions became more moderate and subtle.

In their report, the lay panel members acknowledged that even though the lay panel was composed of persons with a diverse background they were not representative of wider society in at least one respect: they all possessed higher than average debating skills.[148] This can be attributed to the fact that they were self-selected and that most were highly educated. If we consider the list of the members and their backgrounds, it becomes clear that the term 'lay' is not completely appropriate either. It includes, for example, a dairy farmer, a theologian, an ethics teacher and a coordinator of quality assurance systems in animal production.[149] Other studies also show that often members of lay panels are not representative of views in wider society. A Danish study concludes, for example, that lay panels tend to be composed of persons with a higher than average concern about technological developments and social and moral aspects of these, which might explain why the lay reports tend to be quite critical of the proposed novel technology.[150] A study of the second Dutch consensus conference, about human genetics research, shows that compared to the general population, among those who responded to the advertisement there was an overrepresentation from higher educated people with a left wing political affiliation, and no religious background.[151] The Rathenau Institute did not intend the lay panel report to show politicians what ordinary citizens thought about cloning, because a group of 17 citizens, who were not elected, is not representative of the general public anyway. They viewed the lay panel conference as a type of laboratory experiment that would show what could happen when a group of lay people are well informed and deliberate about a complex issue.[152] This, however, raises questions about the status and implications of the lay panel report.

What about the distorting influence of power dynamics? The participants all had an open attitude to the debate and were willing to be convinced by arguments. Some people had a more dominant character and this was reflected in the debate, but the chair person tried to get everyone's input, for example, by sometimes choosing a structure in which one or two people had to prepare a small presentation on a specific question. Except for one notable exception, those who were the leaders in the

---

[147] Van der Bruggen (1999), page 17.

[148] Bozkurt (ed.) (1999).

[149] Ibid.

[150] Cronberg (1995).

[151] Igor Mayer, Jolanda De Vries, and Jac Geurts (1995).

[152] Interview with organiser of public debate on 2/4/2003.

discussion had enjoyed a high level of education. The men did not press their view-points stronger than the women; quite the opposite, the men tended to be more reserved.[153] This contrast with the research into the gendered character of juries discussed in Chapter 4 might be explained by the self-selected character of the lay panel. Distorting mechanisms such as group think or majority influence were not discerned by the participants, although, of course these mechanisms might be too subtle and implicit to realise if one is not particularly tuned in to them. It can be concluded that the representation of groups in society was not achieved very well and that there was less influence of the debate on people other than direct partici-pants; the Dutch lay panel debate could, therefore, be characterised as a somewhat elitist affair.

6.7 The Influence of Contexts: Discussion

What have we learnt from this comparative analysis of two consensus conferences? As I concluded in the case of the committee system, consensus conferences can be used to contain public debate regarding controversial issues. For example, during the selection of the lay panel members, those with prior knowledge or more extreme views were explicitly excluded. This means that the potential benefit from hearing experiential knowledge, or from stakeholder groups that have informed views on the topic, was lost. In Australia, more than in the Netherlands, the strict set-up of the conference was also used to contain the debate. Moreover, framing a debate can influence its outcomes. In both deliberative mini-publics, the problem definition and framing of the debate was influenced by the organisers. In the Dutch case, the agenda-setting public hearing was dominated by experts and in the Australian case it was dominated by the steering committee and its briefing paper. Such framing closes off other possible venues of inquiry and alternative views. On the other hand, the lay panels did manage to open up the debate about biotechnology to include a broader range of issues than had been the case up until then. Particularly, both panels felt that policy decisions should take more account of social, moral, and cultural issues; for example, they resisted the positivistic notion of risk and the exclusion of emotions. In this latter respect consensus conferences appear to be more inclusive of views than committee deliberations. Scientific uncertainties were addressed to a certain extent; scientific methodologies of risk assessment were challenged and it was understood that science is not value neutral. Nevertheless, there was still a strict separation between experts and lay people, exemplified by the fact that in both cases the lay panel was seated on one side of the room and the experts on the other side. Particularly in the Australian case the possibility of experiential lay knowledge was not recognized, because panel members were specifically selected on the basis that they had no knowledge of or experience with the topic.

---

[153] Interview with organiser of public debate on 2/4/2003.

Despite the influence of power imbalances in the conference set up and framing, during the actual panel discussions not many power imbalances or distorting group dynamics were experienced. This is most likely due to the skills of the conference facilitators. Regarding the quality of deliberation it also became clear from analysis of the Australian conference that the panellists changed their viewpoints most in the period prior to deliberation, but after reading the briefing material. This suggests that increased knowledge and the motivation to reflect on a topic already reap at least one of the results that deliberative democrats aim for: decisions based on information and reflection on the better argument, rather than a simple aggregation of preferences. This is supported by other research, such as that of Robert Goodin and Simon Niemeyer, who found out that 'much (maybe most) of the work of deliberation occurs well before the formal proceedings – before the "talking together" ever begins'.[154] If they are right, this means that the internal-reflective process of participants prior to the actual public deliberation is at least as important as the formal deliberating phase. However, the public part of deliberative mini-publics is very important, even if only to create the right conditions for such internal-reflective processes.

The members of the consensus conferences showed tremendous dedication and interest and most of them participated in ongoing initiatives and indicated that they would be willing to participate in more such policy processes.[155] This suggests that, given the right conditions, lay citizens are willing and able to take part in decision-making procedures. On the other hand, at both consensus conferences attendance from the general public was limited; the limited numbers in the audience were primarily composed of stakeholders and the media.[156] The fact that both panels were self-selecting in the initial phase casts doubt on their representativeness. In the Dutch selection procedure this was exacerbated by the fact that the topic of the conference was mentioned in the advertisement. The Dutch conference was more of an elitist affair than the Australian one, as the panel consisted mainly of well-educated people with a particular interest and skill in deliberating. This suggests that special care has to be taken to include participants from lower social-economic backgrounds, and minority groups. The lack of representativeness of a panel of around 14 members can be used by those opposed to the lay panel recommendations to discredit the outcomes and raises real problems for its status. As not each citizen has had an equal chance of participating in the conference, and the panel members were not chosen by the electorate, one can wonder how binding their recommendations can ever be. Furthermore, perhaps just as important as the distribution of views among the lay panel, is that among the expert panel. Rather than the dualistic way in which the Australian debate was framed, which led to polarization between the experts, it would be better to present the whole breadth of views on biotechnology that exists amongst experts, not only

---

[154] Goodin and Niemeyer (2003), page 629.

[155] Edna F. Einsiedel and Deborah L. Eastlick (2000).

[156] This also appears to be the experience in other consensus conferences. Ibid.

6.7 The Influence of Contexts: Discussion 225

technical but also ethical experts. One of the main advantages of consensus conferences seems to be that it alerts biotechnology representatives and legislators to the fact that the concerns of ordinary citizens about biotechnology are not necessarily based on ignorance or irrational fears and that citizens feel that many values besides absence of risks, such as cultural and moral values, should be included in decision-making.

6.7.1 Comparative Analysis

What similarities and differences were discerned between the Australian and Dutch consensus conferences and how can these be explained? In both consensus conferences the quality of argumentation was quite high and the lay panels proved to be very capable of exposing rhetoric and self-interested arguments. Also, both consensus conferences had a large educating effect on all participants, not only lay people, but also experts and organisers. A problem in both exercises was the lack of clarity about the aims of the lay panel conference, and confusion about the status of the lay panel reports. Both had three broad goals: stimulating wider public debate and creating more public awareness, corresponding to the liberal model of democracy; bringing to light the plurality of views, bridging the gap between lay and experts, and getting a better understanding of arguments behind considerations, conforming to the proceduralist-deliberative model; and finding shared values and influencing regulation, conforming to the republican model. Each of these goals has a different implication for the set-up of the deliberation and the status of the lay panel recommendations. Whether deliberative exercises are able to achieve their goals depends on the political cultural background of the countries in question.

In both countries, the lay panel recommendations had little influence on regulatory decisions, although this was for different reasons. In Australia, more than in the Netherlands, the legitimacy of a report drafted by only 14 unelected citizens was called into question. Politicians were reluctant to transfer their decision-making authority. This could be explained by arguing that Australia conforms to the liberal ideal of democracy. My findings suggest, however, that this is only part of the explanation. Specific countries correspond only partly to the Habermasian models, which are ideal types. The Dutch situation, with its focus on consensus building, appears to correspond more to the proceduralist-deliberative ideal, while the Australian situation appears closer to the liberal ideal. However, it is more accurate to say that both lie on a continuum between the liberal and proceduralist-deliberative ideals. The fact that the stated aims of both consensus conferences contained elements that correspond to all three democratic ideals already shows that real life democracies are hybrid models. The finding that some goals are approximated more closely in some countries than in others is very likely due to two other circumstances, which are probably more important than the background model of democracy: whether the state has an active or passive nature and whether the political climate is antagonistic or more cooperative.

In an antagonistic political climate, characterised by interest pluralism, it could be expected that different groups would try to influence the outcomes of the consensus conference and this explains the criticisms of bias and the need for the steering committee to control the process in the Australian experience. In the Netherlands, less problems with bias and antagonism were experienced and organisers did not appear to try to influence its outcomes. The process of the lay debate appears to have been more generally accepted, which could be explained against the background of the Dutch corporatist tradition. Moreover, the active nature of the Dutch state makes it more self-evident that input from citizens and social movements is sought than the passive nature of the Australian state, in which such involvement is not actively stimulated. Also, the fact that the Dutch have a more active neutrality conception makes it more likely that their government would stimulate political involvement from social movements. In the Netherlands, two main reasons for the lack of influence on policy decisions can be discerned; the recommendations were not conclusive enough to be directly applicable to regulation – which shows the downside of letting go of the aim of consensus – and the lay debate was only part of a series of debates. The lay panel, therefore, appears to have been viewed as one more stakeholder around the table. Even though the corporatist tradition creates more favourable conditions for consensus formation, it largely remains an elitist undertaking, as it excludes those groups that are not 'around the table'. This might also explain why the Dutch panel was less representative of the general population than the Australian one. It could be concluded that the stakes were not as high in the Dutch debate as in the Australian one and this would explain why the Dutch one was less contained and pre-structured and why there was less time pressure or pressure to come up with consensus. The stakes were higher in the Australian consensus conference, not so much because the lay report was going to exert direct influence, but because it could be used as fuel for opposing sides in a struggle fought via lobbying and the media. Similarly, the fact that experiments with consensus conferences in the United States and France are found to be less successful than in countries like Denmark, does not appear to be primarily due to the liberal nature of these countries – in fact, in practice the distinction between the different models of democracy is hard to make. More relevant is the point that both states have a passive attitude to social groups and have an antagonistic political climate, particularly the United States. As in Australia, both in France and the United States politics is dominated by one ruling party and one strong opposition party, rather than by a coalition of smaller parties.

6.7.2 Depoliticisation

What can we conclude from the foregoing discussion about the question to what extent deliberative mini-publics can be regarded as depoliticised fora and whether they were successful in this respect? As I explained before, the term depoliticisation has been used in two different senses: Pettit regards it as a positive force, as involving the public in order to counter self-interest and lobbying of formal political institutions. Pettit's proposal appears to be based on a view of (current, but not

6.7 The Influence of Contexts: Discussion 227

ideal) politics as simply a struggle between different interest groups. Hisschemöller regards depoliticisation in the negative connotation of the word as containing conflict by making a strict expert/lay division and relegating decisions to experts and by narrowing the problem definition. In his view, we need to counter technocracy and containment of debate by *re*politicisation. Even though he does not describe it as such, I think he views politics as contestation. We already saw in Chapter 5 that depoliticisation in the Hisschemöller sense of the word took place in both countries, but more so in Australia than in the Netherlands; in both countries there was a division between lay and experts, and in Australia this was even established by separating the two into different committees. Furthermore, both committees led to containment; while there was more room for public input in the Netherlands, the debate was still contained by the legalistic context and narrow framing of the debate. Also, particularly in Australia, bias, lobbying and self-interested motives were not kept out of the committee structure, and depoliticisation in the 'Pettit-sense', therefore, was not successful.

Regarding consensus conferences it is interesting to note that the Australian lay panel recommended to defer regulatory decisions to a body composed of different stakeholders, industry, and citizen groups, in order to counter pressures from lobbying and power imbalances and arrive at better informed decisions, which is reminiscent of Pettit's proposal of depoliticisation.[157] But what can we learn from my analysis of consensus conferences for the chances of this proposal? Even though consensus conferences are supposed to create a power free arena devoid of sectional interests, it became clear from my analysis that power imbalances and influences from stakeholders can not be completely avoided. In practice sectional interests do tend to come in and to what extent this happens is dependent on the political context in which the conference takes place. The lay panels were capable of recognizing rhetoric and self-interested motives dressed up as arguments for the common good, and consensus conferences, therefore, could in principle serve the function of countering sectional interests. Unfortunately, the status of the lay panel report was downplayed and the results were interpreted differently by different political actors. In an antagonistic political culture with an interest group model of decision-making, a depoliticised body could only hope to exert any influence if it were given enough power. This, however, is unlikely to happen precisely because of the vested interests of powerful players and the government's pro-biotechnology stance. In the Dutch debate sectional interests were not as influential. Groups with vested interests had other venues to influence the decision-making process; the lay debate was only one of a series of initiatives. Moreover, the Netherlands do not have the pluralist interest group model that informs Pettit's view, and therefore depoliticisation in the way he envisages it is not as relevant there.

As we have seen before, Hisschemöller and Hoppe argued that two strategies of depoliticisation were employed in order to treat problems as more structured and regain control over them; the policy problem was narrowly defined and a division between experts and lay was made while the decision-making remained the

---

[157] Mohr (2002).

prerogative of experts. In this chapter it has become clear that, like the committee-system, consensus conferences can also lead to a containment of public debate, rather than a stimulation of it. In the Australian case the consensus report gave the semblance of unity, glossing over any disagreements that exist between the panel members. Faced with a polarization between experts, the Australian panel tended to choose a position in between. These mechanisms, and the selection procedure for the panels, serve to exclude more vocal critics of biotechnology. Also, while consensus conferences are in principle meant to alter the balance of power between experts and lay persons, this occurred only to a limited extent. Experts were still seen as the ones providing 'objective' knowledge and lay persons as debating 'subjective' values. Like in the committee system, the interested public can come to feel excluded and not taken seriously and this can lead to a radicalisation of positions (of those who have been excluded) and, thus, politicisation rather than depoliticisation.

Whether or not *depoliticisation* works appears to be dependent on the political context in which a depoliticised body functions. Especially important is the amount of control over decisions that policy makers are willing to relinquish. In different countries the role of a lay panel advice will be interpreted differently. When policy decisions are regarded as legitimate because they are made by elected representatives, such as is the case in France – and which also seems to be the case in Australia – the influence of a lay panel report on policy decisions can only be minimal. Considering a consensus conference as a form of *depoliticisation* in such a context is illusory, because the consensus conference will either not have any influence (and be toothless) or it will become a strategic tool of politicians who interpret its findings to suit their own agenda. In countries where lay input in policy decisions is regarded as legitimate because lay people do not have self-interested motives and are therefore more likely to deliberate about the common good – as is the case in Denmark and to a certain extent the Netherlands – the consensus conference has more chance of influencing policy decisions. However, this already assumes a certain political culture, and one can wonder whether we can really make a sharp distinction between politicisation and *depoliticisation* in such cultures; the need for *depoliticisation* is less pressing here, because politics already involves more debate about the common good. What is regarded as depoliticisation, then, is in itself dependent on the context of the political culture that is being studied.

In my view, through public deliberation we should aim for both countering self-interested motives and for countering political control by the strategies of narrow problem definition and strict expert/lay division. Terming these functions of public deliberation 'depoliticisation' is confusing. Rather, they should be termed 'repoliticisation' of the public vis-à-vis self-interested motives of politicians and lobbying by powerful interest groups and vis-à-vis the special status of expert knowledge. This repoliticisation needs to take place in two directions. Both the scientific and the normative discussions need to be repoliticised; in scientific discussions there should be more eye for value, moral- and metaphysical disagreements, while in normative discussions there should be more eye for the normative assumptions of scientific methodology and the special status of expert knowledge should be challenged.

6.8 Conclusion

The central question of this chapter was to what extent consensus conferences in the Netherlands and Australia can deal with intractable disagreement. My hypothesis was that consensus conferences deal better with intractable disagreement the more they acknowledge scientific uncertainty and that this would be more the case in the Netherlands than Australia. Is my hypothesis correct? My conclusion from Chapter 3, that the discussion about the merits of biotechnology cannot be considered apart from broader moral, social, economic, and political views, and views about our relationship with nature, is supported by the deliberations during the consensus conferences. Still, the recommendations of both panels had a strong emphasis on regulatory issues, which could point to the fact that agreement on more fundamental issues could not easily be reached. As pointed out in the Dutch debate, consequentialist considerations appeared to be easiest to reach agreement on. Moral considerations of a deontological or life- or virtue ethical nature, tended to be drawn into the private sphere. Nevertheless, compared to the committee system, in deliberative mini-publics there appears to be more room to discuss issues of a fundamental value or metaphysical nature. This is in line with the idea that the committee system treats biotechnology as a moderately structured problem (goal) and consensus conferences as a moderately structured (means) problem. In my opinion, the Dutch debate allowed more discussion on scientific uncertainty as well as normative uncertainty. It included several philosophers and ethicists who raised questions about the normativity of expert knowledge and allowed more freedom in the selection of and discussion with experts. However, the lay panel in the Australian case also acknowledged that expert knowledge was value-laden, and in both countries a strict expert-lay distinction was maintained. Nevertheless, the Australian organisers tried to contain the debate more and created more polarisation between experts. Also, the Dutch debate was open-ended, while in the Australian one there was a strong sense of closure. In my view, rather than presenting a consensus report, a majority/minority report and an explanation of the underlying arguments – as was done by the Dutch panel – provides a more valuable basis for politicians to make a decision. In my proposed three-track approach decisions will have to be made in the political arena and this should feed off the other two (agonal and deliberative) tracks. This can only be done if all the arguments that are relevant to the decision are brought to light so that they can be weighed in the political arena. For these reasons, I think that the Dutch debate dealt better with intractable disagreement than the Australian one, although both could be improved.

While it became clear that depending on the context of the political culture in which a deliberative forum takes place, influences from powerful actors cannot always be avoided, consensus conferences could be viewed as a tool for opening up and exposing political climates that are elitist and untransparent. The fact that most of those involved regarded the lay panel reports as well-considered and were in awe of the lay panel's depth of understanding supports the argument that consensus conferences can function to counter claims of politicians, experts, or the industry that public concerns are merely based on irrational fears and ignorance.

230   6 Consensus Conferences: The Influence of Contexts

The stakes were higher in the Australian consensus conference, which led to more external control. These differences can be explained, in part, by the political culture of each country, and only to a lesser extent by their specific model of democracy. The Netherlands with its corporatist tradition and more active state is more used to having input from different groups in society, but deliberations do tend to be elitist. Australia, with its antagonistic political climate and its passive state, is less used to involving social groups in an equal debate, and more used to a struggle for influence by different lobby groups. In light of these differences, it is remarkable that the deliberative exercises did show some important similarities as well, such as the lack of power differences within lay panel deliberations, educative effect of the deliberations, interest and dedication from panel members, and minimal input from the wider public.

To conclude, the conditions of inclusiveness of arguments and lay input and of inclusive participation were met fairly well, but two problems were discerned that appear to be the result of the consensus conference model. Firstly, stakeholders, people with experiential knowledge, and people with extreme views tend to be excluded and this entails that not all relevant viewpoints are taken into account. This is not necessarily problematic, as long as other avenues of participation by these groups exist alongside consensus conferences. Secondly, the strict expert/lay distinction tended to be maintained, which means that challenges to the expert paradigm itself prove difficult. While lack of consensus regarding scientific knowledge was acknowledged by the organisers and lay panels, the conference did not lead to opinion transformation among experts. The condition of appropriate choice of goals was met better by the Dutch exercise. An important conclusion is, therefore, that in the organization of deliberative mini-publics more attention has to be given to their aims and their status within their specific political context. If the aim is to stimulate wider public debate, there should be less focus on reaching consensus and more on presenting the plurality of viewpoints that exist. If the aim is to have citizens gain direct influence in decision-making, one should ask how realistic this is within passive or antagonistic political climates. Finally, within corporate systems special care needs to be taken to create inclusive participation, in order to counter elitism. In the next chapter, I will discuss the implications of the results of my comparative case-studies for deliberative democratic theory.

References

Andersen, Ida-Elisabeth and Jaeger, Birgit (1999), 'Scenario Workshops and Consensus Conferences: Towards More Democratic Decision-Making', *Science and Public Policy*, 26 (5), 331–340.

Annison, Geoffrey (1999), 'Gene Technology Overview', paper given at First Australian Consensus Conference on Gene Technology in the Food Chain, Canberra, March.

Australian Museum (1999), 'First Australian Consensus Conference on Gene Technology in the Food Chain' (Canberra: The Australian Museum).

Biesboer, Frank (1998), 'Hoorzitting over Klonen en Kloneren (Public Hearing About Clones and Cloning)' (The Hague, Netherlands: Rathenau Institute).

References

Biesboer, Frank (1999), 'Clones and Cloning: The Dutch Debate', in Gert Van Dijk (ed.) (The Hague, Netherlands: Rathenau Institute).

Blamey, R.K., McCarthy, P., and Smith, R. (2000), *Citizens' Juries and Small Group Decision-Making* (Canberra: Australian National University).

Blowes, W.M. (1999), 'Speaker Presentation', paper given at Australian Consensus Conference on Gene Technology in the Food Chain, Canberra, March 10–12.

Bout, Henriëtte (ed.) (1998), *Allemaal Klonen. Feiten, meningen en vragen over kloneren (All Clones. Facts, Opinions, and Questions About Cloning)* (Amsterdam/Den Haag: Boom/Rathenau Institute).

Bozkurt, Emine (1999), *Het Burgerpanel Kloneren Zoekt naar Grenzen. Slotverklaring (The Lay Panel on Cloning Searches for Boundaries. Final Conclusions)* (The Hague, Netherlands: Rathenau Institute).

Crombie, Alastair and Ducker, Colin (2000), 'The First Australian Consensus Conference. Gene Technology in the Food Chain. Evaluation Phase 2'.

Cronberg, Tarja (1995), 'Do Marginal Voices Shape Technology?' in Lars Klüver, Simon Joss, and John Durant (eds.), *Public Participation in Science* (London: Science Museum), 125–134.

Crosby, Ned (1995), 'Citizens' Juries: One Solution for Difficult Environmental Questions', in Ortwin Renn, Thomas Webler, and Peter Wiedemann (eds.), *Fairness and Competence in Citizen Participation. Evaluating Models for Environmental Discourse* (Dordrecht: Kluwer), 157–174.

Davison, Aidan and Schibeci, Renato (2000), 'The Consensus Conference as a Mechanism for Community Responsive Technology Policy', *Melbourne Studies in Education*, 41 (2), 47–59.

Dryzek, John S. and Tucker, Aviezer (2008), 'Deliberative Innovation to Different Effect: Consensus Conferences in Denmark, France, and the United States', *Public Administration Review*, 68 (5–6), 864.

Einsiedel, Edna F. and Eastlick, Deborah L. (2000), 'Consensus Conferences as Deliberative Democracy. A Communications Perspective', *Science Communication*, 21 (4), 323–343.

Einsiedel, Edna F., Jelsøe, Erling, and Breck, Thomas (2001), 'Publics at the Technology Table: The Consensus Conference in Denmark, Canada and Australia', *Public Understanding of Science*, 10, 83–98.

Elgesem, Dag (2005), 'Deliberative Technology?' in May Thorseth and Charles Ess (eds.), *Technology in a Multicultural and Global Society* (Trondheim: NTNU University Press), 61–76.

Fishkin, James S. and Luskin, R. C. (2000), 'The Quest for Deliberative Democracy', in Michael Saward (ed.), *Democratic Innovation* (New York: Routledge), 17–28.

Fishkin, James S. and Luskin, Robert C. (2005), 'Experimenting with a Democratic Ideal: Deliberative Polling and Public Opinion', *Acta Politica*, 40, 284–298.

Fung, Archon (2003), 'Survey Article: Recipes for Public Spheres: Eight Institutional Design Choices and Their Consequences', *The Journal of Political Philosophy*, 11 (3), 338–367.

Goodin, Robert E. and Niemeyer, Simon J. (2003), 'When Does Deliberation Begin? Internal Reflection Versus Public Discussion in Deliberative Democracy', *Political Studies*, 51, 627–649.

Hendriks, Carolyn M. (2004), *Public Deliberation and Interest Organisations: A Study of Responses to Lay Citizen Engagement in Public Policy* (Australian National University).

Hindmarsh, Richard (1992), 'CSIRO's Genetic Engineering Exhibition: Public Acceptance or Public Awareness?' *Search*, 23 (7), 212–213.

Keller, Reinier and Poferl, Angelika (2000), 'Habermas Fightin' Waste. Problems of Alternative Dispute Resolution in the Risk Society', *Journal of Environmental Policy and Planning*, 2, 55–67.

Löfgren, Hans and Benner, Mats (2003), 'Biotechnology and Governance in Australia and Sweden: Path Dependency or Institutional Convergence?', *Australian Journal of Political Science,* 38 (1), 25–43.

Mayer, Igor, De Vries, Jolanda, and Geurts, Jac (1995), 'An Evaluation of the Effects of Participation in a Consensus Conference', in Lars Klüver, Simon Joss, and John Durant (eds.), *Public Participation in Science* (London: Science Museum), 109–124.

McDonald, Jan (1999), 'Mechanisms for Public Participation in Environmental Policy Development – Lessons from Australia's First Consensus Conference', *Environmental and Planning Law Journal*, 16 (3), 258–266.

McKay, Elaine and Dawson, Peter (1999), 'First Australian Consensus Conference, March 10–12, 1999: Gene Technology in the Foods Chain. Evaluation Report: phase 1' (Manuka, ACT (Australia): P.J. Dawson & Associates).

Mohr, Alison (2002), 'A New Policy Making Instrument? The First Australian Consensus Conference', Doctoral Thesis (Griffith University).

Parkinson, John (2004), 'Why Deliberate? The Encounter Between Deliberation and New Public Managers', *Public Administration*, 82 (2), 377–395.

Peacock, Jim (1999), 'Ethical Issues', paper given at First Australian Consensus Conference on Gene Technology in the Food Chain, Canberra, March.

Porsborg Nielsen, Annika, Lassen, Jesper, and Sandøe, Peter (2007), 'Democracy at its Best? The Consensus Conference in a Cross-National Perspective', *Journal of Agricultural and Environmental Ethics*, 20 (1), 13–35.

Purdue, Derrick (1995), 'Whose Knowledge Counts? "Experts", "Counter-Experts", and the "Lay" Public', *The Ecologist*, 25 (5), 170–173.

Rowe, Gene and Frewer, Lynn J. (2000), 'Public Participation Methods: A Framework for Evaluation', *Science, Technology, & Human Values*, 25 (1), 3–29.

Smith, Graham (2008), 'Deliberative Democracy and Mini-publics', paper given at Political Studies Association (PSA) Annual Conference, Swansea University.

Smith, Graham and Wales, Corinne (2000), 'Citizens' Juries and Deliberative Democracy', *Political Studies*, 48, 51–65.

Swierstra, Tsjalling (2000), 'Kloneren in de Polder. Het maatschappelijk debat over kloneren in Nederland, februari 1997–oktober 1999 (Cloning in the Polder. The Societal Debate about Cloning in the Netherlands, February 1997–October 1999)' (The Hague, Netherlands: Rathenau Institute).

Van der Bruggen, Koos (1999), 'Dolly and Polly in the Polder: Debating the Dutch Debate on Cloning', in Rinie van Est and Gert van Dijk (eds.), *The Public Debate on Cloning: International experiences* (The Hague, Netherlands: Rathenau Institute), 16–17.

Wills, Peter (1999), 'Gene Technology in the Food Chain: Science and Risk', paper given at First Australian Consensus Conference on Gene Technology in the Food Chain, Canberra, March.

Part IV
Conclusions: Deliberative Democracy Revisited

Chapter 7
Implications of Empirical Results for Deliberative Theory

The art of the novel came into the world as the echo of God's laughter. But why did God laugh at the sight of man thinking?....Because the more men think, the more one man's thought diverges from another's.
Milan Kundera, The Art of the Novel

Whenever people agree with me I always feel I must be wrong.
Oscar Wilde

7.1 Introduction

Disagreement is part of the human condition. We could call it a paradoxical force; for social and psychological reasons we try to avoid disagreement, while on a cognitive level we need it; the confrontation with other points of view can lead us to either reconsider or strengthen our own views and disagreement is a catalyst for change and progress. Even if we simply want to accept or even celebrate difference and disagreement, on a political level we still have to deal with it. This 'dealing with' can take different forms and I have argued that it does not need to mean avoiding conflict, nor reaching consensus. A government could avoid conflict by simply declaring one view to be correct or by remaining neutral between its citizen's views. In the first case it would not accord its citizens the equal respect they deserve and would fail to be democratic; in the second case it would fail to acknowledge the nonneutral consequences of doing nothing. While governments should participate as a non-party to conflicts for as long as possible, they ultimately need to make regulatory decisions, and it is my view that – paradoxically perhaps – they stand the best chance of being neutral when they stimulate public debate about disagreements. In cases of intractable disagreement the aim of such debate should not be to reach consensus, because this would be premature. In my opinion, then, intractable disagreement is best dealt with when the underlying sources of the disagreement come out in the open, when self-interested or fallacious arguments are exposed, and when the confrontation with opposing viewpoints leads people to acknowledge that others have legitimate reasons for their views as well. The aim of dealing with

B. Bovenkerk, *The Biotechnology Debate*, Library of Ethics and Applied Philosophy 29, DOI 10.1007/978-94-007-2691-8_7, © Springer Science+Business Media B.V. 2012

disagreement is, then, to create tolerance for citizens' views and to allow for the positive aspects of conflict whilst avoiding its escalation. I argued that on a theoretical level deliberative democracy could deal better with intractable disagreement than traditional variants of liberal democracy, such as political liberalism, because the emphasis on deliberation would enable us to address the real sources of disagreement and would enable opinion transformation. Governments, therefore, should enable rather than constrain public dialogue. But despite its theoretical attractions, a political theory ultimately gains its force from its practical application. For this reason my aim was to test deliberative democracy – and in particular the conditions that I argued any deliberative theory should fulfil in order to be able to deal with disagreement – through empirical case-studies. This test moved in two directions; on the one hand I aimed to develop insights about the strengths and weaknesses of specific deliberative fora by examining to what extent they conformed to deliberative theory, while on the other hand I aimed to refine my normative and sociological insights and make them more practically relevant. The question of this final chapter is, then: What are the implications of what I have learned in my comparative analysis for deliberative theory and practice?

7.2 Theory and Practice

In the Introduction I raised the question of whether we can ever test deliberative democracy. What does it mean to say that empirical research and political theory need to speak to one another? More particularly, does it mean that deliberative democrats need to formulate less stringent ideals? One of the main lessons arising from the obstacles we have encountered in Chapters 5 and 6 is that while deliberative democrats should hold on to their ideals, they also need to see the limits of deliberation; they need to become aware that deliberation is the right form of political interaction in some contexts, while in other contexts other forms – such as activism – are necessary. Moreover, they need to appreciate more keenly that deliberation can be applied in a variety of different ways and that the type of deliberation should be connected to particular types of problem. If we recall the distinction of problem types proposed by Hisschemöller, it can be noted that for structured problems deliberation is not necessary. For moderately structured problems some deliberation is necessary; in the case of scientific uncertainties deliberation is primarily necessary among experts about scientific matters, while in the case of dissensus on values deliberation will be necessary among stakeholders and citizens in general regarding values. While, as I argued, facts and values can often not be so neatly separated, it is often still possible to locate problems as primarily belonging to the scientific uncertainty camp, to be debated by experts, or as primarily belonging in the value uncertainty camp, to be debated in politics or by citizens. When both types of uncertainty are present, as is the case with biotechnology, we are dealing with unstructured problems that often lead to intractability. This intractability is augmented when we erroneously treat the problem as moderately structured, which is what was done by letting either expert committees or citizen consensus conferences

7.2 Theory and Practice

deal with them. This is where public deliberation is particularly needed, but where it at the same time has its limitations. It is for this reason that I proposed a three-track approach: The agonal track can function as an agenda-setting space in which anyone can raise concerns and make statements in whatever form they deem necessary. In the deliberative track, arguments are exchanged in a more organised and structured fashion, while still aiming to be as inclusive as possible. This is the track where committees and consensus conferences could play a role. The institutional track is informed by the former two tracks and here different options are weighed and decisions are made. While this track is less inclusive it should still function as transparently as possible.

If we accept that types of deliberation should be 'matched' to the right type of problem and that the aims of the particular deliberation has consequences for its specific set-up, we will be in a better position to test aspects of deliberative democracy empirically. Deliberative democrats, then, need to engage more in the type of empirical research that was carried out by Fung, who differentiated between different aims and functions of deliberative mini-publics and argued that these had consequences for their specific design choice.[1] For example, aims of a specific mini-public could be educative, the creation of more collaboration between citizens, direct citizen governance, empowerment of disadvantaged groups, etc. Its function could be informing officials, monitoring officials, mobilising the public, etc. The choice of aim and function determines who should take part and how large the mini-public should be and all of these factors have consequences for mini-public design. Deliberative polling, for example, is a good design choice when the aim is to educate a great number of people and when a function is also to inform officials. Since the number of participants is large, problems of representation are avoided. However, because they are single events, they do not play a monitoring function and do not lead to empowerment. Another example that Fung gives is that of participatory budgeting in the city of Porto Alegre in Brazil. Here one of the aims was the empowerment of the poorest citizens and, therefore, especially members of those groups were selected. It functioned to inform officials and monitor them in a bid to counter corruption. This meant that the participatory budgeting exercises had to have a continuous nature. While they led to a gain in democratic skills, their focus was on local problems and this did not lead to citizen concern for the greater good of the city or the country. In order for theorists of deliberative democracy to make the step to practical implementation, therefore, they need to differentiate more between different aims and functions of public deliberation and investigate the different conditions that are necessary for application in specific contexts. In order to examine whether the claims of deliberative democrats are feasible in practice we will necessarily have to focus on partial aspects of the theory and look for contexts where these aspects could be met in practice.

In sum, the different theories of deliberative democracy cannot all be reduced to one common denominator. This means that we can only test partial aspects of

---

[1] Fung (2003).

specific deliberative theories in practice in order to test whether certain claims that deliberative democrats make are realistic. I have investigated one such partial aspect, namely the ability to deal with intractable disagreement, which I narrowed down to three core conditions. I will now relate the results of my case-study to these conditions in order to see what the implications are of my empirical research for deliberative theory.

7.3 Implications of Comparative Analysis

Let us take another look at the conditions and more specific guiding criteria that I formulated at the end of Chapter 4 and see how they fared in my case-studies. A call for public debate is easily made, but is empty if it is not based on a clear vision about the goal of a specific debate. Only when the goal is explicitly stated it can be determined whether this goal is realistic and legitimate and what set-up will best reach such a goal. In the case of committees, I argued that the attempt to depoliticise the conflict about biotechnology only worked to a very limited extent. With Hisschemöller, I argue that one of the reasons for this failure is that biotechnology was treated as the wrong policy problem. Depoliticisation – in Pettit's sense – is not the right way to go when we are dealing with unstructured problems, because it assumes that a group of experts can reach agreement about contested facts and it downplays the extent of disagreement about values. There are also other reasons why depoliticisation was unsuccessful in my case-studies; first of all, depoliticisation assumes a certain picture of politics, which better fits the interest-group model, and the political culture of the Netherlands does not fit this model. Secondly, while the Australian political culture does fit this picture, the antagonism that we encountered here was so pervasive that neither expert committees nor consensus conferences could escape political influences.

In the case of consensus conferences it became clear that ambitious aims were stated, but that not much thought had gone into the question of what type of deliberation was appropriate for what aim. While the underlying idea of organising consensus conferences seemed to be an acknowledgment that both scientific and value-related uncertainties were present, a problem in both countries was that the distinction between experts and lay people – which I argued is problematic in the context of unstructured problems – was maintained, and even though there was open expert controversy the status of expert knowledge was not questioned. A notable exception occurred when the Dutch panel decided to ascribe a more prominent role to emotions, contrary to the experts' view that emotions were merely gut feelings. In general, the role of experts during the conferences was regarded as transferring knowledge to the lay members, whereas in a learning strategy such as proposed by Hisschemöller and Hoppe, experts have the same status as lay people and there is a dialogue between these different groups, in which the methods used by experts are also a topic of discussion. I argued that in the case of unstructured policy problems the aim should not be consensus; this argument was supported by my empirical results. In the Australian situation the focus on consensus led to pressure to conform and minority standpoints were not reported, lending credibility to the view that dissenting views are suppressed by aiming for consensus.

If we take a closer look at the guiding criteria, it became clear that opinion transformation did take place to a certain extent, but that this was mainly the case for people whose views were not ingrained yet, namely lay persons who were explicitly selected because they did not have strong views on the topic of debate. Experts were less likely to change their viewpoints. If, as I argued, one of the main motivations of deliberative democracy is preference or opinion transformation of all participants, more thought will have to be given to how the ideal of two-way transformative learning could be achieved. Apparently, the set-up of committees and consensus conferences does not invite experts to question their own views and methods. Moreover, if we want opinion transformation to occur in broader society, people with pre-existing views on the topic of debate should be allowed to take part in the deliberations as well. Not only will this bring more arguments to the table, but it may also lead to opinion transformation in these groups, including stakeholder groups. At the same time, deliberative democrats may have to tone down their optimistic expectations of the possibility of opinion transformation in groups with strong pre-existing views.

My second criterion was that deliberation should influence the decision-making process. A problem of letting go of the aim of consensus is that direct influence over the decision-making process will be limited, because no clear recommendations will be made. This was the case in the Dutch public debate, for example. However, in relation to the Australian conference the question was raised whether such direct influence would be legitimate in the first place, because the fourteen people on the panel were not elected nor were they completely representative of the general population. How much influence a particular public deliberation should have, then, will be dependent on the number of people that participate, the way these were selected, and the opportunity for those who did not participate directly to still have a say, for example through internet commentary. Moreover, it became clear from my case-studies that the extent of influence a public debate has in practice is dependent on the political culture of the country in question. A problem in both countries was that the status of the conference results was not clearly agreed on beforehand. This may partly be due to the fact that both formulated wide-ranging goals of the deliberation, encompassing quality, inclusiveness, and consensus, while I showed that a tension is present between these goals. As was argued convincingly by Fung in his comparison of five different deliberative mini-publics, different goals demand a different set-up and design of a public deliberation.[2] Despite the wide-ranging aims of the organisers, it was my impression that they did not have high expectations; the Dutch organisers, for example, regarded the lay panel debate as merely a 'laboratory experiment' unrepresentative of real-life contexts. What we can learn from these findings for the application of deliberative democracy is that the status of specific deliberative fora should be clear from the outset. This

---

[2] Ibid.

status will be dependent on the goals that the organisers want to reach and the amount of political power they can muster. Also, my condition of influence on the decision-making process should be understood in the context of the three-track approach as indirect influence.

Regarding the criterion of revisability I judged the Dutch public debate about cloning as better able to deal with intractable disagreement than the Australian one. The Dutch debate was part of a whole series of discussions between different parties, took place over a longer time span and was more open-ended. It conformed better to my three-track approach. Revisability, then, seems to be better able to achieve when the stakes are not so high.

My second condition stated that no relevant viewpoints should be excluded and this entailed that there should be lay person input and that fallacies and self-interested motives should be uncovered. I argued that the Dutch CAB dealt better with intractable disagreement, because it integrated scientific and value-related questions and it allowed for more public input. However, the terms of reference and assessment procedure did function to narrow down the debate and exclude certain viewpoints. Also, the legalistic context in which the CAB operated hindered an open discussion and public input remained limited to involvement of a group of stakeholders. These circumstances meant that the public input procedure acquired the character of a ritual dance and in effect the instalment of the committee meant that the debate was contained. The positive aspect of the Australian GTEC was that more fundamental issues could be discussed. However, the committee did play a self-censuring role and public input was entirely contingent on what the committee members decided to bring up for discussion. The main problem with the GTEC was its lack of power; I argued that this was due to the political culture in which lobbying plays an important role and in which groups with vested interests manage to contain the regulatory system. Public input could only be given regarding risk assessment and submissions were forced into a narrow range of topics. In the Australian case, therefore, the public debate about biotechnology was effectively contained by political actors and stakeholders.

In the case of consensus conferences my results are more varied and the debates in the two countries had more in common, while some important differences were also discerned. On the positive side, it became clear that the lay members that participated became very interested and involved in the topic and were enthusiastic about the exercise they were part of. Some even became more active and involved citizens after the debate. Contrary to popular belief among scientists and bureaucrats, the panel members turned out to be quite capable of digesting information about complex topics and writing a well-balanced report. They grasped many conceptual ideas, such as the view that facts and values are interrelated, and were able to uncover fallacies and expose self-interested arguments. Another positive but unexpected finding was that no power imbalances were discerned within the lay panels; it appeared that the presence of a neutral and capable facilitator does much to counter power differences. These findings suggest that there is definitely potential for involving citizens in decision-making procedures based on public deliberation and that my criteria are more realistic in the case of consensus conferences than of the committee-system. However, even in consensus conferences viewpoints were

7.3 Implications of Comparative Analysis

excluded. In the Netherlands, less effort was made to contain the debate than in Australia, where several attempts were made to influence the decisions of the lay panel and the results of the consensus conference and the report were downplayed by those who disagreed with the panel's findings. On the other hand, in the Australian case the organisers had succeeded better at getting representation from different groups in society, while the Dutch debate was dominated by highly educated persons. This latter problem is characteristic of the corporatist decision-making model used in the Netherlands, because it tends to lead to elitism. Moreover, it became clear that more emphasis was put on consequentialist arguments than on deontological ones and on extrinsic rather than intrinsic ones. Perhaps this could be explained by the fact that the Dutch model of democracy is based on liberal philosophy, with Mill as its spiritual father, which has always had a close attachment to utilitarian views. It can be concluded that efforts to contain public debate were present in both political cultures and that these can be detrimental for the condition of non-exclusion. An important question for deliberative democrats is, therefore, how containment of debate could be countered in a deliberative democracy.

My third condition stated that we should aim to be as inclusive as possible, which entails that marginalised groups should not be disadvantaged, power differences and distorting group dynamics should be avoided, a diversity of groups should be included or reached indirectly, there should be open expert contestation and participants should learn from the debate. Regarding the committee-system it can be concluded in general that this condition was not met very well. Especially in Australia, public input was limited, while the citizens that did become involved in the Netherlands did not have very diverse backgrounds, as almost all were affiliated with animal protection groups. While expert contestation took place to a certain extent within the committees, this was not open to the public. To the extent that participants learned from the debate, particularly in the Netherlands this was characterised as primarily a legal learning process. Finally, the debate stimulated by both the CAB and the GTEC did not gain much publicity and did not reach a broader public. It could be concluded that the committee-system is not the best avenue to stimulate broad, inclusive public deliberation and that its main role is to provide structured information and arguments. In other words, it could be regarded as one actor within a broader public debate, but should not be the be-all and end-all of public deliberation about novel technologies.

The consensus conferences that I analysed could be considered more inclusive. Women and marginalised groups were not disadvantaged and power differences and distorting group dynamics were not encountered. This is probably due to the organised nature of the conferences and the presence of a skilled facilitator. But perhaps more importantly, it could be explained through the selection procedure, which led to an exclusion of people with stronger views on the topic and, particularly in the Netherlands, mainly highly educated participants. As I argued, this is problematic with regards to representativeness, and it could give us a contrived picture of what deliberation would be like in less organised and selective mini-publics. The organised consensus conference is, therefore, perhaps not the best medium to achieve maximum inclusiveness and diversity. This condition may in fact be better met in the agonal track of my three-track approach.

While open expert contestation was aimed for in the set-up of the conferences, this aim somewhat missed the mark, because in Australia there was not enough room for middle-of-the-road positions and the debate became too polarised, and in both countries expert contestation did not actually lead to a reflection about expert knowledge per se. Still, in one important respect the consensus conference model did fulfil my criteria in that the participants learned a great deal from their experience. Lay people gained knowledge, debating skills, and an understanding of policy-making and their role in it; many had an experience of empowerment. Stakeholders, experts, and bureaucrats gained respect for the deliberating skills of lay people and an understanding about lay people perspectives. Finally, in Australia the outcomes of the lay panel deliberations reached a larger audience than in the Netherlands. This is reflective of the different political cultures; as Dryzek explained, in passive inclusive systems the role of the media needs to be greater and the media will be more politicised.[3] Moreover, this finding is reflective of the Dutch tendency towards elitism and backroom politics. Because of Dutch confidence in authorities the media in the Netherlands will be less likely to report on novel technologies as a politically charged topic.

7.4 Political Culture

What can we conclude in general from my comparative analysis for the potential of deliberative democracy? First of all, because many of my criteria were not, or only partially, met it has become apparent that when people come together in real-life deliberative settings they face more obstacles than would be expected on the basis of theoretical analysis. Reality is more complex than some deliberative democrats would have us believe. Nevertheless, some potential for public deliberation was encountered, mainly in the context of consensus conferences. While Mutz argued that in an everyday context people are generally reluctant to talk about politics with those who hold opposing views it appeared from my analysis that when you bring citizens together in a more formal context they are quite willing and able to engage in 'cross-cutting' debate and are prepared to hear the other side.[4]

Secondly, it has become clear that political culture has an important influence on deliberative practice. The deliberative fora that I studied functioned differently in the Netherlands and in Australia and these differences can be attributed to differences in political culture. This means that some of the conditions that I formulated will be better met in some countries than in others and if we want to deal with intractable disagreement by way of public deliberation, this should be approached differently in different countries. The specific design of deliberative fora and their connection to political institutions should take into consideration the limits and

---

[3] Dryzek and Tucker (2008).

[4] Mutz (2006).

opportunities of the country in question. To the extent that existing political culture hinders deliberation – which is the case with the Australian antagonism and the Dutch elitism – deliberative democrats will need to address the questions whether and how these political cultures can be changed. At the moment deliberative democracy as a theory does not pay sufficient attention to concrete political cultures. Because of this, deliberative democracy as an abstract theory does not generate sufficiently context-specific normative guides for action. In concrete circumstances the guides for action of deliberative theory appear to be without direction; the theory cannot at the moment inform our practice very well. Deliberative democrats should, in my view, not aim to present one all-encompassing sweeping theory of deliberative democracy for all times and places, but should make deliberative democracy more context-sensitive. Future research into deliberative democracy should make room for the question of how we could and should deal with the specificities of different cultures.

Finally, my research suggests that apparent theoretical disagreements between political theorists about deliberative democracy may result from implicit assumptions about political culture. For example, agonists start from empirical observations about political practices, which they interpret as being permeated by power. Within such a framework they necessarily disagree with Habermas' directive that decisions should be made on the basis of the forceless force of the better argument rather than political power resulting from status, lobbying, or log-rolling. They would deny that decisions could ever not be made on the basis of power. Also, deliberative democrats make assumptions about equality that are considered naïve by difference democrats, because the latter tend to define political relations between people in terms of difference, while the former tend to emphasise what people have in common, such as their common humanity or equal rights.

7.5 Conclusion

A lot can and has already been said about deliberative democracy by scholars before me. Their discussions have tended to focus primarily on the theoretical aspects of public deliberation or, more recently and scantly, on specific empirical practices with deliberation, but these two discussions have failed to speak to one another. In this book, I have made an attempt to link these two discussions in the context of one specific aspect of deliberative democracy, its potential to deal with intractable disagreement. After arguing that deliberative democracy could theoretically deal better with intractable disagreement than traditional liberal democratic theories, I formulated a set of conditions and guiding criteria in order to make my theoretical views more practical. I then analysed two specific deliberative practices in two different countries and, firstly, sought the implications of my findings for the feasibility of my theoretical position, and secondly, developed insights about the strengths and weaknesses of different types of mini-publics. My comparative empirical study has helped to refine my normative claims and has made deliberative democracy more practically relevant, while at the same time generating practical insights about the strengths and weaknesses of deliberative fora by comparing them against deliberative theory. My key insights should be useful to both deliberative democrats, who need take into account real-life deliberative contexts, and to organisers of deliberative experiments, who need to tailor their debate design according to their aims and pay attention to their cultural context. Simply taking one type of deliberative mini-public and applying it across a range of different cultures and contexts is like lumping together all human conduct under one universal code.

Because of the failure to relate theory to practice, deliberative democrats have not paid enough attention to differences in political culture and how these influence the feasibility of deliberative democracy. This has led to a lack of concrete guidelines for the practical application of deliberative democracy. Dryzek argues that discursive democracy calls for a drastic restructuring of society. He argues against liberalism, as 'the most effective vacuum cleaner in the history of political thought', meaning that liberalism incorporates or co-opts all theories that appear to criticise it, such as feminism or environmentalism, and thereby weakens their capacity to question existing political institutions.[5] However, in my view it is not the liberal framework that needs to be abandoned, but rather that firstly, more emphasis needs to be put on its deliberative elements, and secondly, that successful deliberation calls for careful attention to the political culture of countries. Particular challenges exist for countries that can be characterised as antagonistic, but where consensus democracies have elitist tendencies, these need to be dealt with as well. In my view, then, rather than aiming for 'a deliberative democratic state' we should look for possibilities within existing states to put more emphasis on public deliberation alongside other methods of political activity. This requires deliberative democrats to cultivate a greater context-sensitivity for the implications of political cultures. Tailoring deliberative democracy to specific contexts may be the next big challenge for deliberative democrats. Finally, to the extent that public deliberation has the potential to lead to opinion transformation, it appears that one of the main merits of deliberative democracy – particularly in the context of intractable disagreement – is its educative function and that the success of this theory should, therefore, be considered over time; its impact on society might only be felt in the distant future.

References

Dryzek, John S. (2000), *Deliberative Democracy and Beyond: Liberals, Critics, Contestations* (New York: Oxford University Press).
Dryzek, John S. and Tucker, Aviezer (2008), 'Deliberative Innovation to Different Effect: Consensus Conferences in Denmark, France, and the United States', *Public Administration Review*, 68, 5–6, 864.
Fung, Archon (2003), 'Survey Article: Recipes for Public Spheres: Eight Institutional Design Choices and Their Consequences', *The Journal of Political Philosophy*, 11 (3), 338–367.
Mutz, Diana C. (2006), *Hearing the Other Side. Deliberative Versus Participatory Democracy* (Cambridge: Cambridge University Press).

---

[5] Dryzek (2000), page 27.

Appendix A
Animal Biotechnology Debate

Animal biotechnology, or the application of genetic modification to animals, is carried out for two main purposes. Firstly, it serves a medical purpose. Genetically modified animals (GM animals or transgenic animals) are made for use in animal experiments, either in order to provide fundamental scientific knowledge about biological processes, or because they can serve as models for certain diseases; through genetic modification the propensity to develop certain human or animal diseases can be built into animals. A famous example is the oncomouse, designed to develop cancer. Animal biotechnology finds another medical application in xenotransplantation; the advantage of transplanting organs from genetically modified donor animals to humans is that the presence of a human gene in the animals can prevent organ rejection by the host.[1] Finally, animal biotechnology is used in the medical context for the production of vaccines and medicines, in particular medicines containing human proteins.[2] Secondly, animal biotechnology serves an agricultural purpose, namely in animal husbandry. The three main applications in this context are (1) increased production, (2) enhanced nutritional value, and (3) increased disease resistance of the animals. Examples of increased production are the injection of genetically modified Bovine Somatotropin (BST) – also labelled Bovine Growth Hormone (bGH) – into cows in order to enlarge milk production,[3] the introduction of a growth hormone-gene into pigs, sheep, and salmon in order to create faster growing animals, such as

---

[1] Verhoog (1998).

[2] Insuline is made with GM bacteria, but 'higher' animals are needed for other types of medicines. For example, GenePharming in the Netherlands has used rabbits that express human enzymes in their milk and that can be used as a medicine for patients with the metabolic 'Pompe' disease. See Jochemsen (2000), page 163.

[3] For a detailed account of the moral aspects of bGH, see Comstock (2000, chapter 1). The shift from the use of the term bGH to BST provides another example of the power of discourse framing. As Burkhardt points out, when GM bGH was first employed, it met with fierce public resistance due to fears for negative health effects after consumption of these hormones. Because of the negative connotations of the word 'hormone' scientists and the biotech industry decided to change the term into its purely technical name, thereby 'diffusing a significant amount of consumer activists' policy-affecting power' (Burkhardt 2003, page 338).

B. Bovenkerk, *The Biotechnology Debate*, Library of Ethics and Applied Philosophy
29, DOI 10.1007/978-94-007-2691-8, © Springer Science+Business Media B.V. 2012

the infamous Beltsville hogs, and the creation of sheep with faster growing wool.[4] An example of the application of GM-technology for increased nutritional value is the creation of animals that provide leaner meat, and examples of disease resistance are adding an 'anti-freeze' – gene to fish,[5] so that they can survive in colder water, or increasing the concentration of the anti-inflammatory lactoferrin in cow's udders in order to prevent mastitis.[6] On a global scale, but especially in the United States, the injection of genetically modified BST into dairy cows was the first application to raise concerns, because of negative health effects on the animals, because of the further 'instrumentalisation' of the animals that this technique was perceived to entail by some, and because of safety concerns for human consumption.[7] In the Netherlands, the genetic modification of animals was first put on the 'public map' in 1990 when a bull, named Herman, was genetically engineered in order for his female offspring to produce the anti-inflammatory human lactoferrin in their milk.[8]

Benefits of Animal Transgenesis

Most of the arguments in favour of animal biotechnology are quite straightforward and are directed at its favourable results for human beings, but also indirectly for animals. The discussion about the merits of animal biotechnology has largely focused on its applications in the field of medicine, where most transgenic animals are used. This is especially true for the Netherlands, where, at the time of writing, no licenses have yet been requested for animal biotechnology in the agricultural setting.[9] In Australia, research is being done into, for example, increasing the growth and meat quality of pigs and improving the wool production of sheep through genetic modification and cloning.[10] Also, researchers of the Australian Commonwealth Scientific and Industrial Research Organization (CSIRO) are doing tests in which a gene from

---

[4] Jochemsen (2000). The Beltsville hogs refer to a group of nineteen pigs genetically engineered with human growth hormones in 1985 in the town of Beltsville, Maryland, USA. The aim of the experiment was to create pigs that could convert grain into leaner meat more efficiently, but instead resulted in many negative effects in the pigs, such as physical deformations, arthritis, decreased immune function, and ulcers. The poor state of the pigs resulted in a public outcry (Comstock 2000). At present no licenses have been given yet for commercially exploiting GM fish, but research with GM fish is carried out. See Millar and Tomkins (2006).

[5] Rifkin (1998).

[6] Jochemsen (2000).

[7] Comstock (2000).

[8] Brom (1997a).

[9] Dutch researchers and animal breeders at this point in time reject cloning and genetic modification of animals for agricultural purposes, firstly for technical reasons, such as limited efficiency and reliability, and the expected erosion of genetic diversity, and secondly, because the social climate regarding these techniques in the Netherlands and Europe in general is negative (Regouin and Tillie 2003).

[10] Regouin and Tillie (2003), page 7.

Appendix A: Animal Biotechnology Debate   247

the tobacco plant is introduced into the genome of sheep in order for them to produce an insecticide against the parasitic sheep blowfly in their sweat glands.[11] In the agricultural field, the claimed benefits of animal biotechnology are primarily commercial and nutritional in nature, but defences of this type of animal biotechnology are less explicit. As a benefit to animals in agriculture, it is argued that animals can be created that are more robust and can therefore deal better with the farm environment. Moreover, animals with a narrower consciousness could be created, which would lead to less suffering. Most of the criticism of animal biotechnology (particularly in animal ethics literature), on the other hand, seems to focus on its agricultural purposes, suggesting that this type of animal use is more problematic, because its benefits are more disputable.

Proponents of animal biotechnology in the medical field point out that genetically modified animals are better and more precise 'models' for human and animal diseases, such as cancer. In their eyes, this means that more precise research can be carried out and that, potentially, less animals have to be used. As we have already seen, however, sceptics disagree, because they argue that genetic engineering is not as precise a science as many researchers purport it to be. Proponents only need to point to the many successful transgenic animal 'models' for diseases they have been able to create in order to counter this view. On the other hand, the extent to which insights gained from transgenic animal experiments can be extrapolated to the human situation remains disputed. The extrapolation issue plays a role in debates regarding the merits of conventional animal experiments as well as in transgenesis, but it could be argued that transgenic animals are even further removed from 'real life' circumstances than their conventional counterparts and this could influence the applicability of insights gained through these experiments.

One advantage of using GM technology over conventionally bred and crossed animals is that less generations are necessary in order to acquire the preferred traits in the animals. This enables more efficient research into causes and possible cures of human and animal diseases and leads to a reduction of animals used. Proponents appeal to the high value that is generally attached to health and most people appear to accept this position; opinion polls invariably reveal that the public has a more positive attitude about genetic modification when it involves medical purposes than agricultural or other purposes, especially when the medical benefits are clear and immediate.[12] Nevertheless, some commentators are critical of the emphasis on biotechnology (and other high technologies) in medicine and argue that its use diverts attention and resources away from what they deem to be the 'real causes' of ill-health.[13] These critics point to alternatives, such as better prevention strategies.

Finally, it has been argued that through genetic modification animal species that are extinct or are threatened with extinction could be brought back to life.[14] This

---

[11] http://www.gene.ch/genet/2000/Dec/msg00034.html, accessed on 25 October 2006.

[12] European Commission (2003).

[13] Such as poor diet (Ho 1999).

[14] For example by Pluhar (1986).

application of genetic modification is opposed by some, because firstly, it would attest to the human *hubris* referred to earlier, and secondly, because even if animals could be made that have the characteristics unique to a specific species, and even if these animals could function independently in an ecosystem, then still the species would lack an 'historical character'.[15]

Moral Status of Animals

Animal biotechnology has been widely debated by animal ethicists. One of the focal points of the field of animal ethics has been (and to a certain extent continues to be) the debate about the moral standing of animals. Some (e.g. Peter Singer) have argued that animals have moral standing in as far as they have interests, which in turn are based on the capacity to form preferences.[16] Others have argued that animals have moral rights,[17] and often this notion is based on the more general claim that animals have inherent value (e.g. Tom Regan)[18] or intrinsic value (e.g. Henk Verhoog).[19] Some (e.g. Mary Midgley) argue that we should attribute moral standing to animals because of the natural sympathy we extend to them.[20] While this is a lively and intriguing debate with many distinctive positions, the practical implications of the different positions have so far been remarkably similar. Both Singer and Regan, for instance, reject animal testing, meat consumption, wearing animal fur, and using animals in various forms of entertainment, such as circuses and cock fights. Some authors (such as Donald VandeVeer)[21] take a less absolute position and suggest that in moral dilemmas between the interests of humans and animals, or between different species of animals, a balance has to be struck based on the question whether the interests at issue are basic or trivial, and taking into account the psycho-social capacities of the animals involved. Others (such as Peter Wenz) argue that a difference should be made between wild and domesticated animals.[22] All of these divergent approaches and a few dissenting voices[23] notwithstanding, within the field of animal ethics there seems to be a general agreement that animals, particularly mammals, possess moral standing. This agreement is shared by a large section of the public – it has in fact become part of 'public morality'[24] – and has even been

---

[15] Van Staveren (1991).

[16] See Singer (1975, 1979).

[17] For a detailed and up to date exploration of the 'animal rights' discussion, see Sunstein and Nussbaum (2004).

[18] Regan (1984).

[19] Verhoog (1992).

[20] Midgley (1983).

[21] Vandeveer (1986).

[22] Wenz (1988).

[23] For example Frey (1980).

[24] See Brom (1997a).

Appendix A: Animal Biotechnology Debate

adopted by our legal systems. In the Netherlands, for example, the notion that animals possess intrinsic value, forms the basis of animal protection legislation.[25] In the debate about animal biotechnology, likewise, the idea that animals matter from a moral point of view appears to be a common assumption, and the question whether or not animals have moral status is bypassed. There seems to be common consensus that animals ought not to be harmed without good reasons. However, disagreement exists about the questions as to what constitute such good reasons, what constitutes harm to animals and what level of harm should be tolerated. These issues, however, are not unique to animal biotechnology. Animal biotechnology, however, has given rise to a particularly persistent disagreement, namely about the question whether animal biotechnology is only problematic when it results in animal suffering, or whether it is intrinsically problematic to genetically alter animals, even if it would not lead to suffering. First, I will turn to the question of animal welfare, and in the subsequent section I will discuss issues beyond animal welfare concerns.

Animal Welfare

The notion of animal welfare plays an important role in the debate about animal biotechnology, both in the public debate and in the debate between experts. What exactly is meant by animal welfare is disputed, however. Frans Brom analyses the meaning of the term 'animal welfare' and suggests that it refers to, firstly, the absence of negative experiences, secondly, the presence of positive experiences, and thirdly, the extent to which an animal is capable of functioning 'normally'.[26] The first two have been called the 'experiential notion' and the third the 'functional notion' of welfare.[27] In debates about animal welfare, suffering is most often referred to and this suggests that there is more consensus about the experiential notion than about the functional notion of animal welfare. Brom, therefore, proposes to take the experiential notion as the core concept of animal welfare. When we want to establish what the impact of, for example, animal housing in factory farming on an animal's welfare is, we can encounter a tension within the experiential notion; an indication of welfare can differ depending on whether we focus on the absence of suffering or on the presence of positive experiences. Narrowly focusing on the absence of suffering, as is the most common method for establishing the level of welfare an animal experiences, disregards the fact that sometimes a certain amount of stress can add to an animal's welfare in the second sense of the term, because it prevents animals from getting bored. Animal welfare should be distinguished from animal health. Even though unhealthy animals are often likely to suffer negative experiences, and animal health is usually a precondition for animal welfare, healthy animals are not guaranteed to experience welfare. This becomes clear in

---

[25] Nota Rijksoverheid en Dierenbescherming, Tweede Kamer II (1981).

[26] Brom (1997a).

[27] This distinction is derived from Frans Stafleu, and is cited in Brom (1997a), page 120.

factory farming, where animals are often kept in good health, for example through the administration of antibiotics, but they lack certain positive experiences, such as being nursed by their mothers, and they cannot function normally, for example because they cannot root in mud.

Proponents of animal biotechnology argue that some instances of animal transgenesis are more defensible from a moral point of view than conventional research and agricultural practices involving animals, because biotechnology would actually lead to less suffering.[28] This is firstly, because transgenic animals could be made that experience less pain, for example because they are genetically engineered to be more resistant to diseases, and secondly, because they think less animals will be used in the long run. Animal rights groups and many members of the public, on the contrary, are concerned about the animal suffering they think genetic modification brings about. Moreover, animal rights groups believe that animal biotechnology actually leads to more animal use because this new technology opens up many new possibilities for investigation. However, Paul Thomspson points out that not many reports about the impact of genetic modification on the welfare of (particularly farm-) animals can be found and the evidence supporting either standpoint, therefore, is inconclusive.[29] The few studies that have been conducted seem to indicate that there are generally no adverse impacts on animals from the biotechnological procedures themselves.[30] As most of the interventions take place on the embryonic level, one would not expect direct welfare impacts. However, most impacts associated with genetic modification of animals are not the direct result of the biotechnological procedures themselves, but are the result of the expression of the genes that have been altered or are the effect of the genes that have been knocked out. As the aim of many biotechnological procedures with animals, in particular those for medical purposes, is to create animals that are in some way diseased, one would expect the indirect results of the procedures to be at least as bad for the animals' welfare as conventional breeding methods. Nevertheless, molecular scientists point out that 'the effects of genetic change on animal welfare are usually trivial'.[31] This could be explained by the fact that many animal experimentation committees demand 'humane endpoints' to animal experiments, meaning that when an experiment causes too much suffering – although what constitutes too much suffering is debatable, of course – the animals should be 'euthanised'. This practice ensures that genetic modifications of animals that potentially cause severe harm are terminated before this harm can actually occur. This leads Paul Thompson and Bernard Rollin to conclude that 'it is commercial production that poses the most serious threats to animal welfare'.[32] The reasoning behind this conclusion is that the creation of dysfunctional animals

---

[28] Burkhardt (2003).

[29] Thompson (2005).

[30] Thompson (2005).

[31] Thompson (1997b). Thompson cites Ian Wilmut, the famous creator of Dolly the sheep (Ian Wilmut 1995).

[32] Thompson (1997b), page 6.

Appendix A: Animal Biotechnology Debate

might actually be commercially profitable and producers of such animals will have no economic motivation to stop animal suffering.

In the agricultural area, at least two cases of animal suffering due to genetic modification have been extensively documented. The first is the case of the Beltsville hogs, already alluded to earlier, in which the genetic modification of pigs resulted in many health and welfare problems, such as arthritis and lung problems, and which has led some to speak of 'cruelty' to animals through genetic engineering.[33] The second case of suffering centres on the detrimental health and welfare effects of injecting cows with genetically modified BST. Many are concerned that increasing the already artificially high milk production of cows by injecting them with GM growth hormones will exacerbate health and welfare problems already encountered in high milk yielding cows and that are associated with intensive farming practices.[34] They cite metabolic disorders and increased incidences of mastitis and other infections. In reply, it has been argued that the same could be said about other methods to increase cows' milk production, and in other words, that it is not the genetic modification per se that causes the problems. Moreover, it has been argued that the harmful effects of injecting cows with BST can be solved by veterinary treatments and good management practices.[35] However, these arguments do not hold up against all types of criticism; after all, pointing out that animal biotechnology for agricultural purposes does not cause more harm than already existing intensive farming practices elicits the response that perhaps existing farming practices have already gone too far in their disregard for animals' welfare. Indeed, a lot of the criticism of animal biotechnology should be viewed in the light of broader concerns surrounding animal use and it is, therefore, misleading to use conventional farming (or research) methods as the standard by which to judge new biotechnological techniques. Bernard Rollin is one of the most influential critics of industrialized agriculture and animal testing, because these practices lead to an enormous amount of animal suffering. For the area of animal biotechnology he has proposed the 'Principle of Conservation of Welfare', which states that genetic engineering should not be permitted if it makes an animal worse off than its non-genetically engineered counterpart who lives in similar circumstances.[36] In light of Rollin's general criticism of intensive animal husbandry practices, one could wonder what exactly he means by 'similar circumstances', and thus whether his principle holds up to the critique mentioned above. Nevertheless, his principle precludes the employment of 'a strict logic of comparing costs and benefits to humans and animals . . . to rationalize actions that make food animals worse off than they currently are'.[37] Presumably, this should also hold for animal testing, where the principle could provide a 'bottom line' that should not be transgressed.

---

[33] Thompson (1997b), page 1.

[34] Comstock (2000), Goldhorn (1990).

[35] Thompson (1997b), pages 4–5.

[36] Rollin (1995), page 179.

[37] Thompson (1997b), page 21.

As is exemplified in the above discussion, animal welfare is often understood narrowly as only referring to animal suffering. However, as Brom points out, in the debate about biotechnology, the other two senses of animal welfare – the presence of positive experiences and the capability to function normally – become more pressing, as it is (at least theoretically) possible to create transgenic animals that are well suited to intensive farming practices, in that they experience very little pain, but in the process have been stripped of their capacity to experience positive experiences and their species-specific behavioural traits.[38] However, the possibility of creating such an animal raises the question to what extent changing an animal's behaviour, so that it no longer functions 'normally', according to its species, should really count as an infringement of the animal's welfare. After all, it is the animal itself that has been changed, and the question is what it means for this changed animal to function 'normally'.[39] Rollin was one of the first animal ethicists to raise this question. He reintroduces the Aristotelian notion of *telos*, 'as a moral norm to guide animal use in the face of technological changes which allow for animal use that does not automatically meet the animals' requirements flowing from their natures'.[40] In Aristotle's' teleological notion of nature, the notion *telos* entails that natural phenomena can be explained according to 'final causes'; natural entities are the way they are in order to fulfil a goal or purpose that is essential for them. Rollin gives an alternative interpretation of *telos*, which he defines as 'the set of needs and interests, physical and psychological, genetically encoded and environmentally expressed, which make up the animal's nature'.[41] According to Rollin, we have a moral duty not to harm these interests – without good reason – but in his eyes it is not morally objectionable to change them. For if we change an animal's *telos*, don't we at the same time change its interests? As long as we respect these new interests – by not infringing on an animal's experiential welfare, both in the first and the second sense – we are acting in a morally unproblematic way. Rollin holds that changing an animal's *telos* is only wrong if this intervention leads to harmful consequences for the animal's welfare. For him, 'it is only wrong to change a *telos* if the individual animals of that sort are likely to be more unhappy or suffer more after the change than before'.[42]

Animal Integrity and Dignity

Many ethicists who employ a more deontological approach also use the notion of *telos*, but interpret this differently, which leads them to reach the opposite conclusion, namely that changing an animal's *telos is* morally wrong. One of the main issues in the debate about the moral justification of animal biotechnology centres on

---

[38] Brom (1997a).

[39] This question is raised by Brom (1997a), page 129.

[40] Rollin (1998), page 161.

[41] Rollin (1986), page 295.

[42] Rollin (1998), page 88–89.

Appendix A: Animal Biotechnology Debate

this different interpretation and on the question whether or not it is wrong to adjust an animal's make-up to its (farm or laboratory) environment instead of the other way around. This issue comes up in the public debate as well and seems to express a moral intuition that animals should not be treated as if they were mere things that can be changed as if they were parts of a production process. Even though, as I mentioned before, in many practical issues animal ethicists of different 'schools' reach the same conclusion, this point is where their theoretical differences do lead them in separate ways. Welfarists, like Rollin, take animals to have interests because, and only insofar as, they are sentient. In other words, Rollin holds that because animals have interests by virtue of their sentience, only welfare matters from a moral point of view. Rollin does take a broad view of animal welfare, however, as he includes the fulfilment of needs, wants, desires and goals into the category of interests. The point that Rollin emphatically makes, however, is that even though we have a prima facie duty not to interfere with an animal's interests, there is nothing wrong with changing these interests themselves. On the contrary, changing an animal's interests could even benefit the animal. In his influential book *The Frankenstein Syndrome*, he argues that

> when we [genetically] engineer the new kind of chicken that prefers laying in a cage and we eliminate the nesting urge, we have removed a source of suffering ... the new chicken is now suffering less than its predecessor and is thus closer to being happy, that is, satisfying the dictates of its nature.[43]

Other welfarists, such as Nils Holtug, do make an objection to changing animals into 'senseless machines', by arguing that attention to welfare should not be limited to the prevention of suffering, but should also be directed to the promotion of positive experiences. By making animals senseless we would deny animals the possibility to enjoy positive experiences.[44] However, welfarists that argue this need to invoke an extra premise, not reducible to mere sentience, to explain why positive experiences matter to the animal. They need to show that it is not justified to take away positive experiences that are due to the animal, as these experiences should be part of the animal's good life.[45] Such an extra premise can be given by preference utilitarians like Peter Singer, because animals have preferences that go beyond the mere absence of suffering. Thompson argues that Singer should support Rollin's view, however, because the overall preference calculation would be more positive if animals were made to suffer less.[46] On the other hand, the capacity to enjoy positive experiences might also be removed from these animals and this would bring down the total amount of preference satisfaction. Nevertheless, if we compare a world where the

---

[43] Rollin (1995), page 172.

[44] Holtung (1996).

[45] Brom (1997b).

[46] Thompson (1997b). However, Thompson also explains some fundamental differences between the utilitarian approach and that of Rollin, most importantly that Rollin does embrace the notion of animal rights, albeit in a less strict form than Regan. Moreover, Rollin thinks that Singer defines suffering too narrowly.

circumstances of a modified and a non-modified animal would be the same, it would be preferable from a utilitarian point of view to choose the animal that is adapted to its circumstances. This also seems to be the background of Rollin's view; considering the circumstances in which animals are kept, he sees changing an animal's *telos* as the lesser of two evils: 'while it is certainly a poor alternative to alter animals to fit questionable environments, rather than alter the environments to suit the animals, few would deny that an animal that does mesh with a poor environment is better off than one that does not'.[47] However, Rollin does not give a clear basis for his idea that this is a 'poor alternative'. This assertion seems to acknowledge the moral intuition that changing an animal's nature is objectionable, even though the circumstances may deem it necessary. Clearly, however, suffering is not the main issue here. In other words, Rollin's concept of interest is too narrow to analyse this moral intuition.

Surprisingly, even Tom Regan's more inclusive rights view does not offer a sound basis on which to reject genetically engineering animals to fit their circumstances rather than the other way around. Tom Regan argues that, for example, raising animals for food is wrong not primarily because it causes animal suffering, but that it is wrong in principle. This is the case because animals, like humans, are valuable in themselves and not only by virtue of their value to others. In other words, they possess inherent value and therefore have moral standing. According to Regan, the basis for this inherent value is that animals are subjects-of-a-life.[48] Regan takes only mammals that possess a certain amount of awareness as being subjects-of-a-life. However, animals that have been genetically engineered in order to suffer no pain and even be completely unconscious, rather like animal machines, do not fit Regan's description of subject-of-a-life. It seems to follow that it is not wrong in principle to change, for example, chickens into living egg-laying machines. At most, this causes moral problems in the transitionary phase, when the chickens still possess awareness. It seems that the proponents of animal rights cannot adequately deal with the commonly held intuition that changing animals into living machines is morally problematic either. Moreover, if such animals could be made, we would be dealing with animals that are as yet non-existent. This raises the question how one can attribute rights to future creatures. David Cooper suggests that the conventional approaches to animal ethics, namely Singer-style utilitarianism and Regan-style deontology, are deficient because they fail to appreciate that biotechnology 'commits new wrongs' (rather than old wrongs in a new context).[49]

In the Netherlands, animal ethicists and veterinarians – particularly those from the 'Utrecht school' – have proposed to use the notion of *animal integrity*, rather

---

[47] Rollin (1995), page 172.

[48] Regan (1984), page 243. To be a subject-of-a-life is to 'have beliefs and desires; perception, memory, and a sense of the future, including their own future; an emotional life together with feelings of pleasure and pain; preference and welfare-interests; the ability to initiate action in pursuit of their desires and goals; a psychophysical identity over time; and an individual welfare in the sense that their experiential life fares well or ill for them, independently of their utility to others'.

[49] Cooper (1998), page 147.

Appendix A: Animal Biotechnology Debate

than interests or rights, to give voice to and analyse the moral intuition that we should adjust the farm or laboratory environment to the animal and not vice versa.[50] Bart Rutgers describes integrity as 'the wholeness and intactness of the animal and its species specific balance, as well as the capacity to sustain itself in an environment suitable to the species'.[51] This notion has also been adopted by the Dutch Committee for Animal Biotechnology, that reviews applications for biotechnological procedures with animals. As I will explain in a more detailed way in chapter 5, this committee gives advice on the basis of health and welfare considerations for the animals involved, but also determines whether the extent to which the animals' integrity has been violated is acceptable. There are a few important aspects of the notion of integrity. The first is that integrity refers to an animal's intactness and this could give the impression that any animal that is in some way no longer intact – for example because it has been ill – has lost its integrity. However, integrity is never invoked on its own, but always in relation to a *violation* of integrity and this presupposes a link to human action.[52] This in turn entails that the purpose of changing the animal's intactness influences whether its integrity has been violated. Rutgers holds that, for example, the docking of a dog's tail for aesthetic reasons constitutes a violation of the dog's integrity. When the dog's tail needs to be docked for medical reasons, however, he claims that its integrity has not been violated.[53] The docking of a dog's tail for these two reasons could be regarded as two different actions, depending on the intention with which the action is carried out. However, as the physiological result is the same, it seems that integrity is not a biological aspect of the animal itself, but should be understood as a moral notion.[54] The second aspect is that a violation of an animal's integrity should be understood as a notion that permits of gradations, in order for this moral category to be helpful in a balancing context, such as that of the Committee for Animal Biotechnology. In other words, an animal's integrity can be violated to a higher or a lower degree. Finally, integrity applies to different units of concern. Brom distinguishes between integrity of the individual animal, of the animal's genome, and of the species.[55] Regarding the imaginary animal as a senseless living machine, the question rises whether a potential violation of integrity has taken place on the level of the individual animal or only on the level of its genome or even of its species. Rutger's definition does not seem

---

[50] For instance Heeger (1997), Grommers et al. (1995), Vorstenbosch (1993), Brom (1997a). The concept of integrity is generally understood to be a deontological concept, although De Vries argues that it is essentially consequentialist (De Vries 2006, pages 479–481).

[51] Grommers et al. (1995), Rutgers and Heeger (1999).

[52] Brom (1997a).

[53] Grommers et al. (1995).

[54] For a more in-depth discussion of this issue, see Bovenkerk et al. (2002). See also Brom (1997a), pages 131–140.

[55] Brom (1997a), page 132. Rob De Vries argues, however, that the concept of integrity should not apply to the levels of the species and the genome, but only to that of the individual animal. See De Vries (2006).

suitable for genomic integrity. Moreover, any genetic modification would automatically constitute a violation of genomic integrity. For this reason, the Committee for Animal Biotechnology only considers changes on the phenotypical level to be relevant. Rutger's definition could apply both to the individual animal and the species it belongs to. However, regarding the individual animal it remains problematic to speak of the violation of its integrity if the actual infringement takes place before the animal is born – as is usually the case with genetic modification –, and we are therefore dealing with a future animal. Integrity seems to refer to an intuition that we should not change the way an animal is, or its 'species being', for our benefit: we should not tamper with the characteristics that make a chicken a chicken or a pig a pig. It, therefore, seems to apply mainly on the species level. However, this opens a whole new debate, namely about the question whether a species has a good of its own and possesses moral standing independent from its individual members.[56]

Opponents of the notion of animal integrity often base their rejection on the claim that an appeal to integrity is based on mere emotions, or in other words on unwarranted subjectivism. It is argued, in reply, that emotional responses to issues like animal biotechnology are actually important intuitions that signal that something might be wrong, something that needs to be more precisely articulated. Cooper claims, for example, that 'a way to identify what is distinctively wrong with bioengineering of animals is to reflect on the revulsion felt by many ordinary people'.[57] More generally, Mary Midgley states that 'the sense of disgust and outrage is in itself no sign of irrationality. Feeling is an essential part of our moral life'.[58] Another criticism of integrity does not deny the role of intuition, but rather centres on the idea that integrity is not the best notion to give voice to our intuition. Cooper, for example, argues that the notion of integrity carries an unwarranted pretension of describing a scientific fact, whereas in his eyes, in reality the intuition is about the fact that humans do not display a proper sense of 'humility', in the Murdochian sense of a 'selfless respect for reality'.[59] Others argue that instead of integrity, the concept of 'animal dignity' gives a better basis for the moral intuition in question.[60] The concept of animal dignity is analogous to the concept of human dignity, except that the latter is based on respect for autonomy, whereas the former is based on respect for an animal's 'own good', defined as 'the uninhibited development of certain specific functions'.[61] The concept of dignity, then, seems to focus on one aspect of the notion of integrity, namely on the idea that an animal should retain the capacity to sustain itself in an environment suitable to its species. Ortiz argues that this makes

---

[56] Robin Attfield, for example, doubts whether we could meaningfully speak of species conceived as individual, as is proposed by Holmes Rolston. See Attfield (1998), Rolston (2002). Lawrence Johnson, on the other hand, does argue that species and even ecosystems have a good of their own and possess moral standing (Johnson 1992).

[57] Cooper (1998), page 151.

[58] Midgley (2000), page 9.

[59] Cooper (1998), pages 153–155.

[60] Balzer et al. (2000), Ortiz (2004).

[61] Ortiz (2004), page 112. Ortiz bases her account on that of Balzer (see previous footnote).

Appendix A: Animal Biotechnology Debate

more sense than the idea that an animal's 'species specific balance' should not be violated, as it is unclear how such a balance could be determined.[62] Balzer et al. defend the use of animal dignity, a concept which has in fact been adopted by referendum in the Swiss Constitution, with reference to the inherent value of animals, and their theory can, therefore, be regarded as an attempt to apply Regan's theory to the field of animal biotechnology.

As mentioned before, other authors invoke the concept of *telos* in order to argue that animal biotechnology is problematic. It is not clear in what respect the idea of *telos* differs from that of integrity, however. Michael Fox seems to conflate the two, when he argues that genetic modification of animals violates 'genetic integrity' or *telos* of organisms or species; he even refers to the 'wolfness' of a 'wolf'.[63] Thompson suggests that permitting a radical change of an animal's *telos*, – thereby actually creating a different creature altogether – while not permitting this in the case of humans – as in Brave New World, where a class of 'subhumans' is created – would be morally indefensible.[64] Aldous Huxley's Brave New World is often referred to in this debate, because of the analogy between creating animals whose 'nature' is artificially conditioned to suit their surroundings, and Huxley's human beings that through drugs and gene technology are happier in the world they live in, but at the same time are 'alienated' from their true human nature. Regardless of what terms authors have employed in this debate, they all connote the idea that animals deserve to be treated with respect and that this entails that we should take into account more than just considerations of animal welfare.[65] The appeal to integrity, dignity, and *telos* all reflect the intuition that an animal is not just a thing, or an instrument, and should not be treated as such.

This latter point is also encountered in the debate about animal biotechnology by reference to another group of related terms: the 'objectification', 'commodification', or 'mechanization' of animals.[66] Brom distinguishes two senses of the equivalent Dutch term *verdinglijking*. The first conveys the idea that animals are used *as if* they are things and the second conveys the idea that through genetic modification animals are changed to the extent that they actually *become* things. The first sense has to do with our attitude towards animals, and in a broader sense also with the tendency in society in general, lamented by some, to regard more and more goods as commodities. As an example of the second sense, we could think of the senseless living machines alluded to above, and which have also been termed 'animal microencephalic lumps'.[67] According to Wouter Achterberg, such animals completely merge into their status as instruments for human benefit.[68] In

---

[62] Ortiz (2004), page 108.

[63] Fox (1990a).

[64] Thompson (1997b).

[65] See Brom (1997a).

[66] For example Rifkin (1998).

[67] Ortiz (2004), page 95.

[68] Achterberg (1989).

the debate we also encounter the term 'instrumentalisation' of animals. It is argued that even though conventional agriculture and even conventional domestication processes already entail an instrumentalisation of animals, genetic modification takes this a drastic step further by creating the animals as if they were products that roll of a conveyor belt. The appeal to instrumentalisation also refers to Kant's notion of treating humans never merely as means but also as ends in themselves, and that has been adopted by Regan regarding animals.

The concern over instrumentalisation or mechanization also connects to a set of broader concerns with animal biotechnology in agriculture. There are fears that introducing transgenic animals in agriculture will intensify the tendency to make agricultural practices high-tech and this might lead to social injustice. For example, in order to administer BST, high tech management practices and inputs are needed, which will disadvantage smaller sized 'family farms' and will not easily integrate into developing country farming practices.[69] There seems to be a 'cultural' concern behind this argument as well. Wolfgang Goldhorn, for instance, regrets that 'family farms, where cows still have names and are almost regarded as members of the family, will be replaced by 'factory farms' with 'animal machines' for whom veterinarians will become their 'maintenance technicians'.[70]

Conclusion

Animals have been subjected to genetic modification for two main purposes: a medical and an agricultural purpose. The potential benefits of using GM animals in medical research and for the production of medicines for the enhancement of human health are great, especially in light of the high value that is commonly attached to health – although some regret the emphasis on high technology in healthcare. The potential benefits of using GM in animal husbandry for humans are mainly of a commercial and a nutritional nature, are more controversial, and therefore seem to be left out of the debate. The benefits for animals themselves of genetic modification in both medicine and agriculture are more disputed. It is held that the precision of genetic modification leads to more reliable experiments that cause less suffering and that animals that experience less suffering themselves can be created, so that they can deal better with the laboratory or farm environment. In response it is argued that biotechnological procedures have led to a sharp increase in the amount of animals used for experimentation. These claims have led to two broad discussions; the

---

[69] Comstock (2000). Comstock argues that from an economic point of view the risks associated with BST, such as possible harm to cows and to consumers, are not justified. It is not warranted to increase milk yields in this way if in fact 'in developed countries, there is too much milk, not too little'. Moreover, in developing countries introduction of BST will lead to a competitive advantage for big plantation-style farms to the detriment of smaller family-farms. See pages 16–17. See also Fox (1990b).

[70] Goldhorn (1990), page 85.

first centres on the effects of genetic modification on animals' welfare and the second centres on concerns that go beyond welfare issues. The evidence for animal welfare problems is inconclusive and needs to be assessed on a case-by-case basis. However, the determination of animal welfare has centered too much on the absence of negative experiences and not on other indicators, such as the presence of positive experiences or on the extent to which an animal can function normally. Regarding the latter, the question rises whether this should really be taken as a welfare-factor, or whether it has more to do with respect for animal integrity. It has been argued that traditional animal ethics approaches cannot deal with concerns that go beyond purely animal welfare concerns. The moral intuition that changing an animal's *telos* or nature, or that changing an animal's make-up to fit its environment instead of the other way around, is wrong, cannot be expressed by reference to either Singer's utilitarianism or Regan's rights-based view. Other ways to express this intuition have been discussed, in particular the notion that it violates an animal's integrity, but also the appeal to animal dignity, *telos* (in a deontological sense), commodification, and instrumentalisation. This overview of the animal biotechnology debate shows that there are several different levels of disagreement: there is disagreement over whether animals have moral standing, and if it is granted that they do, how much this matters when their interests have to be balanced against human interests. There is disagreement about the question whether genetic modification entails an infringement or an improvement of animal welfare and about how this welfare should be measured in the first place. Finally, there is disagreement about the question whether only animal welfare concerns are legitimate or whether there are considerations beyond animal welfare. If it is agreed that there are broader concerns, disagreements still exist about the basis for these concerns. In the next chapter, I will map the different viewpoints and arguments in the debate regarding the genetic engineering of plants, and the underlying values these adhere to.

Appendix B
Plant Biotechnology Debate

While recombinant-DNA techniques had already been successfully employed in 1973, it would be another decade until the technique could successfully be applied to plants.[71] It would be more than another decade until, in 1996, the first transgenic crops were commercially planted on a large scale.[72] Some of the goals for which plants are genetically modified, or are expected to be so in the future, are enhancement of flavour, colour or nutrition of fruits, improved shelf-life of, for example, tomatoes, creating herbicide-resistant crops, creating crops that are resistant to pests, creating crops that can deal better with environmental stresses such as drought, heat, cold, or high salinity in soils, production of drugs and vaccines in plants ('biopharming'), creating plants that can clean up environmental pollution, or that contribute to sustainability, such as genetically modified tries for sustainable wood-production, and even manufacturing plastics from genetically modified plants and micro-organisms instead of petroleum.[73] Transgenic plants have so far primarily been employed for agricultural purposes. The most common application of genetic engineering in plants is for herbicide tolerance of crops and the second most common use is for insect resistance. The most common transgenic crops are soybeans (60%), maize (24%), cotton (11%), and oilseed rape (or canola, which is one particular cultivar of oilseed rape: 5%).[74] In 2005, 10 years after their commercialisation, 90 million hectares of GM crops were grown worldwide; this entails an increase of 11% compared to 2004. These crops are grown by 8.5 million farmers in 21 countries, mostly in the United States (55% of the global area), Canada, Argentina, Brazil, and China. In Europe (in Spain, France, Portugal, Czech Republic, and Germany), the most commonly grown transgenic crop is *Bt*-corn. In

---

[71] Nuffield Council on Bioethics (1999).

[72] Vasil (2003).

[73] Hughes and Bryant (2002). Huges and Bryant give quite an extensive list of (possible) applications of plant genetic modification. See also: Nuffield Council on Bioethics (1999), Straughan (1992), Rifkin (1998), Hughes and Bryant (2002), Vasil (2003), Paula and Birrer (2006).

[74] International Service for the Acquisition of AgroBiotech Applications. See http://www.isaaa. org, accessed on 4 July 2006; Nuffield Council on Bioethics (1999), page 31; Brookes and Coghlan (1998).

the United States, in 2003, 80% of soybeans and 70% of cotton grown was transgenic.[75] The global market value of GM crops was estimated to be 5.25 billion US dollars in 2005.[76]

One of the reasons why many breeders seem eager to adopt genetic engineering is that particular traits can be isolated more quickly and be manipulated more precisely than through traditional breeding programmes, in which unintended genes could 'hitchhike' along with targeted genes.[77] Steve Hughes and John Bryant explain that on the one hand genetic engineering is a precise technology, because it concentrates on the transfer of one or a group of specific genes, but on the other hand, it is imprecise, because the position the inserted gene will take on the chromosome cannot be controlled. The latter problem entails that in the first generation after gene transfer the plant has to be monitored closely and the plants with desirable characteristics have to be selected and bred on with. However, this process is still quicker than traditional breeding methods. In their eyes, 'the main advantage of this technique is that it increases hugely the genetic variety available to the plant breeder whilst avoiding the problem of bringing in unwanted genes'.[78] A possible reason why farmers are willing to accept GM crops is that they fear to lose a competitive advantage if they do not. This fear seems to be fostered by farmers' media, such as the *Australian Farm Journal*, which published an article about genetic engineering, entitled 'Farmers Risk Irrelevancy if they Fail to Involve Themselves'.[79]

At the moment, most applications of agricultural biotechnology do not seem to confer a direct benefit to consumers, but only to biotechnology companies and certain groups of farmers.[80] This could provide one explanation for consumer rejection of transgenic foods, which is particularly strong in Europe. Proponents of biotechnology believe that in the future more benefits to the consumer, such as improved nutritive value and pharmaceutical applications, will be achieved and they think this will lead to less public resistance.[81] Whether this is the case remains to be seen; from the start of the biotechnology era there have always been groups that question the need to have genetic engineering in the first place and it is unlikely that they

---

[75] Vasil (2003).

[76] http://www.isaaa.org: 'In 2005, the 21 countries growing biotech crops included 11 developing countries, and 10 industrial countries; they were, in order of hectarage, USA, Argentina, Brazil, Canada, China, Paraguay, India, South Africa, Uruguay, Australia, Mexico, Romania, the Philippines, Spain, Colombia, Iran, Honduras, Portugal, Germany, France and the Czech Republic'.

[77] According to Snow and Palma, 'historically, genes coding for economically important traits have been obtained from related taxa by hybridisation and several generations of backcrossing, with little knowledge of the identity of nontarget genes that "hitchhike" along due to genetic linkage. Now, however, the use of recombinant DNA techniques allows for precise transfer of only the gene(s) of interest without repeated backcrossing' (Snow and Palma 1997, page 88).

[78] Hughes and Bryant (2002), page 122.

[79] Smith (2003).

[80] Lindhout and Danial (2006).

[81] Conway (2000).

would change their mind. If they do not experience problems in the areas targeted for genetic modification – for example, if they deem conventional foods as nutritious enough – they are unlikely to look at biotechnology as a solution to any problem.[82] Besides questioning the potential benefits of the application of genetic engineering, critics also perceive many potential problems in the areas of interhuman relationships, and human and environmental health. While these topics are often conflated in the biotechnology debate, I will here treat them one by one, in order to get a more precise picture of the nature of the existing disagreements: first, I will sketch the context of modern agriculture in which crop biotechnology is introduced ; I will also discuss arguments about the (in)justice of the socio-economic relations flowing from crop biotechnology, and, closely related to concerns about justice, the discussion about the merits of the patenting system; the next section deals with concerns surrounding human safety, with particular attention to the discussions about risks and labelling; after that, I will turn to environmental concerns, including the issues of chemical use, possible hybridisation, and loss of biodiversity; I will end with a short evaluation and an analysis and conclusion.

From Green Revolution to Gene Revolution

The application of genetic engineering to crops is widely regarded as a continuation of the Green Revolution, which refers to developments in agriculture from the 1950's onward in which scientific inputs, that enabled the creation of 'high yielding varieties' of crops, mechanization, widespread use of chemical herbicides and pesticides, and more intensive irrigation were employed in order to increase yields for a growing world population.[83] The Green Revolution has led to high-input, intensive forms of large-scale agriculture and to the farming of monocultures. According to Gordon Conway, president of the Rockefeller Foundation,[84] 'the Green Revolution was one of the great technological success stories of the second half of the 20th century. Food production in the developing countries kept pace with population

---

[82] Qualitative research into public concerns about genetically modified food has shown that many have doubts about the necessity of GM food, because current food is 'fine as it is' (Mayer 2002, page 142).

[83] Since 1950, the global yield of cereals has increased to three times its original volume and the world population has reached six billion (Trewavas 2002, page 668).

[84] This Foundation is a charitable organisation that played a crucial role in the Green Revolution by funding the research that led to the high-yielding (dwarfed) crop varieties. It should be noted that the methods employed by the Rockefeller Foundation have been criticized for creating Third World dependency on biotechnology and for advancing corporate interests from the North. See Hindmarsh and Hindmarsh (2002) and Hindmarsh (1994, especially chapter 5). According to Hindmarsh 'the location of Mexico as a starting point for the Green Revolution may or may not have had something to do with the "fact that Rockefeller holdings in Mexico had recently been nationalised and that the climate for US private investment was in definite need of improvement" ' (Hindmarsh quotes Susan George here).

growth, with both more than doubling over the past forty years'.[85] Still, even Conway acknowledges that 'it was in some respects flawed'.[86] It is now widely acknowledged that the Green Revolution has led to environmental degradation. Moreover, the rate of increase in crop yields is in decline; it seems that the success of the Green Revolution is coming to an end. This leads some to call for a second Green Revolution based on biotechnology, or as Conway puts it a 'Doubly Green Revolution', that avoids the environmental problems of the first, and is therefore 'green' in a real sense, and that leads to even higher productivity.[87]

Critics of the Green Revolution argue, on the other hand, that we should not use a 'technical fix' for problems that were caused by modern technology in the first place. They tend to look at its flaws in a broader perspective, not simply as side-effects of an otherwise benign development, but rather as symptomatic of a 'business-like' approach that is not sustainable and does not suit the special nature of agriculture and food production. With the production of higher yields the baby was thrown out with the bathwater, because some of the benefits of traditional farming practices were lost; less straw was left over to feed animals, the practice of intercropping was given up, and the choice to grow primarily grains meant that cheap and nutritious sources of protein, such as pulses, were lost to poor farmers and consumers in developing countries.[88] Ironically, traditional plant varieties that were well suited to local circumstances and that contained many of the now highly sought after traits by biotechnologists – such as drought resistance – were forgotten in the drive to plant monocultures.[89] The environmental damage resulting from the Green Revolution is not limited to the leaching of chemical herbicides and pesticides into soils and waterways, but also includes soil erosion, loss of soil fertility, increased disease-proneness of crops and livestock, the increased release of greenhouse gasses, and the loss of genetic diversity.[90] Furthermore, even though one of the main targets of the Green Revolution was to increase food security by creating a larger food supply and bringing down the costs of food, its socio-economic consequences have been criticized. Comstock, for example, concludes that 'the technology of the green revolution seems to have led to a concentration of land in the hands of a few large farms producing crops for export, displacing peasants from farms and apparently

---

[85] Conway (2000), page 5. Similarly, a special working group on genetic modification of the Nuffield Council on Bioethics argues that the Green Revolution has led to great increases in income from labour and access to food for people in developing countries; especially in the 1970s and 1980s hunger was greatly reduced (Nuffield Council on Bioethics 1999, page 59).

[86] Conway (2000), page 5.

[87] Conway and Toenniessen (1999).

[88] Levidow (2001). This has also been argued by Vandana Shiva, who looks at the issue from an even broader ecofeminist perspective. In her eyes, the increased yields are not sustainable and 'the gain in "yields" of industrially produced crops is based on a theft of food from other species and the rural poor in the Third World' (Shiva 2000, pages 12–13).

[89] For example Knud Vilby in Meyer (2001).

[90] Tilman (1998). Moreover, the reliance on agro-chemical inputs also caused damage to the health of farmers applying the chemicals (Bharathan et al. 2002).

Appendix B: Plant Biotechnology Debate

decreasing the availability of low cost food'.[91] Peter Rosset argues, similarly, that it is short-sighted to think that yield increases and corresponding low food prices increase food security and reduce poverty. Overproduction leads to low crop prices, which actually tends to cause poverty in rural areas.[92] The Nuffield Council on Bioethics, on the other hand, points out that during the Green Revolution 'food production increased faster than food prices fell', making farmers better off and giving them an incentive to employ more people.[93]

Clearly, the success of the Green Revolution has been interpreted differently. Many of those who consider the Green Revolution to be a success-story, also expect great benefits from the 'Gene Revolution'. They think that genetic modification will increase yields, because it can avoid losses caused by weeds, pests, and pathogens. This means that existing farmland can be used more efficiently and less new land will have to be exploited in order to be able to feed an increasing world population. They also believe that through biotechnology not only will valuable land, which could be set aside for wildlife refuges, be conserved, but also water, energy and other resources.[94] Some even point out that we cannot afford to reject biotechnology. Many paint dark scenario's for the future, including disconcerting figures of population growth.[95] Plant biologist and staunch defender of plant biotechnology, Anthony Trewavas, even warns that

the Future is threatened by global warming and unpredictable climate change. The old enemies of locusts, floods, disease, drought and pests still exist. In the face of these adversaries, diversity in technology becomes a strength and a necessity, not a luxury. We have developed genetic manipulation of food and plants only just in time. Companies and scientists may fumble in its use, but now is the time to experiment, not when a holocaust is upon us.[96]

---

[91] Comstock (2000), page 157. Comstock later changed his mind about supporting the global case against agricultural biotechnology. One of the reasons he cites is that reality has caught up with his view that small to medium sized family farms were preferable, as most farms, at least in the US, are now large corporate style farms. Regarding developing countries, Comstock is optimistic about the potential of tailor-made biotechnologies. His change of mind in favor of agricultural biotechnology is a qualified one, however. Whether or not a specific biotechnological application is desirable depends on its goal (he is against making Beltsville pigs or injecting cows with genetically modified BST, but for Golden Rice, for example). Jack Kloppenburg explains in more detail what the causes of these developments are and cites issues such as 'exacerbation of regional inequalities, …specialization of production, displacement of labor, accelerating mechanization,… rising land prices,… and agrichemical dependence' (Kloppenburg 1988, page 6).

[92] Rosset (2006).

[93] Nuffield Council on Bioethics (1999), page 62.

[94] Vasil (2003), Conway and Toenniessen (1999).

[95] Vasil, for example states that, 'in the hope that the world population can be stabilized at 11 billion by about 2050, the challenge for the agricultural sector is to double food production by 2025, and triple it by 2050, on less per capita land, with less water, and under increasingly challenging environmental conditions' (Vasil 1998, page 399).

[96] Trewavas (1999), page 231.

Trewavas cites evidence of an already increasing yield of food and fibre and of increased farm incomes and reduced pesticide use as a result of genetic engineering.[97]

Critics of this view agree that there are many environmental and economic problems that need to be solved, but they do not share the pro-biotech camp's optimism about the potential of genetic engineering to solve them. They argue that the benefits of biotechnology are being exaggerated: it has not lived up to its economic promise so far and the herbicide and pest-tolerant crops that have been grown until now do not contribute in any real way to the famine problem. According to Buiatti, despite 'prophecies' of solving mass starvation, 'in over twenty years of research carried out with considerable investment by thousands of groups in many countries, only two new characters (resistance to herbicides and to insects) have been inserted into only four species, yielding a very limited number of productive cultivars'.[98] A 2002 report by the UK's organic certification association, the Soil Association, even concludes that the introduction of transgenic crops into the United States and Canada has amounted to an 'economic disaster', due to 'high subsidies, lower crop prices, loss of major export orders and product recalls'.[99] Contrary to Trewavas' claims, it found that in general the aim of increased yields was not realized.[100] Trewavas' claims are, on the other hand, supported by a study into the global socio-economic and environmental effects of ten years of growing genetically engineered crops; Graham Brookes and Peter Barfoot conclude in this study that over this ten year period of commercial planting of GE-crops global farm income net gains have been around 27 billon US dollars. Moreover, the majority of this gain accrued to developing countries (although they do not mention whether mainly rich farmers or poor subsistence farmers gained most).[101]

It is generally accepted that the Green Revolution has been more successful in Asia and Latin America than in Africa.[102] African scientist Florence Wambugu hopes that 'the biotechnology revolution will not pass by [Africa] (as the Green Revolution did) due to a lack of resources and unrealistic controversial arguments from the North, based on imagined risks'.[103] Despite Wambugu's optimism about the potential of biotechnology for African countries, there is one important difference between the Green and the Gene Revolutions. The first was funded by the public sector and charitable organizations, and its success depended in part upon the fact that genetic diversity and information were freely exchanged.[104] The

---

[97] Trewavas (2002), page 669.

[98] Buiatti (2005), page 20.

[99] Meziani and Warwick (2002).

[100] Meziani and Warwick (2002). Only Bt corn yields were found to be slightly increased.

[101] Brookes and Barfoot (2006).

[102] One reason why it has been more successful in Asia than in Africa is that access to water was not a problem in Asia. See Meyer (2001).

[103] Wambugu (1998). Quoted in Bharathan et al. (2002), page 182.

[104] Conway and Toenniessen (1999).

Appendix B: Plant Biotechnology Debate 267

Gene Revolution, on the other hand, is mainly directed by the private (corporate) sector.[105] This takes us to the next topic of debate, which focuses on the socio-economic consequences of the Gene Revolution.

Justice

The question of who 'owns' biotechnology; who benefits, who gets disadvantaged, and who gets left behind with its introduction, has been at the heart of the biotechnology controversy. Proponents of biotechnology list many economic benefits of growing GM crops, resulting from higher yields, longer shelf-life, tolerance to early maturing of plants, and tolerance to environmental stress. Critics argue that these applications primarily benefit the biotech industry and rich farmers in the West and that they stand to increase the gap between the rich and the poor, and therefore lead to injustice.[106] This criticism is rejected by biotech advocates, who argue that one of the most important aims of biotechnology is to reduce poverty.[107] Still, many commentators, both of the pro-GM and the anti-GM camp lament the concentration of power in the biotechnology field: 'Through a spate of mergers amongst their own kind and through purchase of seed companies, the number of major players on the world biotechnology stage has been reduced to about six, including Monsanto (USA), DuPont (USA), Novartis (Switzerland), AstraZeneca (UK/Sweden) and Aventis (Germany/France)'.[108] Many originally pharmaceutical companies are now also involved in biotechnology, both in its medical and agricultural applications. Biotech companies appear to be gaining power over an ever more diverse range of human practices. In the words of a former Monsanto executive: 'What you're seeing is not just a consolidation of the seed companies, it's really a consolidation of the entire food chain'.[109] Critics think this is a problematic development, because firstly, a monopoly in the production and distribution of food can lead to a lack of diversity and choice. Secondly, these companies arguably do not have the interests of poor farmers at heart: 'poor farmers do not make an attractive market'.[110] Thirdly, there is a concern that the development of biotechnology will merely benefit these big multinationals and not consumers, farmers, or researchers.[111] Particularly contested is the question what consequences biotechnology will have for developing countries.

---

[105] Bharathan et al. (2002).

[106] For example, Vint (2002).

[107] Trewavas, for example, argues that, indeed, biotechnology is already benefiting the poorest farmers. He names the example of Bt cotton, which has increased poor farmers' incomes, cut their costs and cut down on expensive inputs of chemical pesticides (Trewavas 2002, page 669).

[108] Bharathan et al. (2002), page 186.

[109] Robert Fraley quoted in Hindmarsh and Hindmarsh (2002), page 1.

[110] Barry (2001), page 28.

[111] Nuffield Council on Bioethics (1999).

268 Appendix B: Plant Biotechnology Debate

Will Biotechnology Feed the Poor?

Publication materials from the association of European biotechnology companies, EuropaBio, state that 'biotechnology is a key factor in the fight against famine ... biotechnology will help increase the yield on limited land'.[112] Proponents argue that it would be irresponsible not to use these 'humanitarian' technological advances in order to feed the world and alleviate poverty.[113] The Nuffield Council on Bioethics (hereafter: the Nuffield Council) even believes that we have a moral duty to develop GM crops in order to enhance food security in developing countries.[114] Critics of the 'feed the world' argument claim that this aim could only be reached if biotechnology research and development were to focus on applications that are useful for third world farmers, which is currently not the case.[115] Not much research has been done into crop varieties that are relevant for developing countries, such as the staples of Africa's poor – millet, sorghum and yams – or into ways to incorporate biotechnology into developing country agricultural systems.[116] The main transgenic crops that are grown in developing countries are cash crops, such as *Bt*-cotton.[117] *Bt*-corn, which could become a staple of the poor in South America, is instead used primarily in developed countries as a source of animal feed.[118] Biotechnology companies argue that they are interested in the developing world, and indeed, are looking into market openings there. Nevertheless, critics think that even if developing countries were to benefit from GM crops, it would be the rich farmers in these countries that stand to gain from their introduction, and not the 'starving millions'.[119] This is because the technical inputs necessary for growing GM crops are better suited to large-scale mechanized farms than to small-scale subsistence farms, meaning that only wealthy farmers can gain a comparative advantage through biotechnology and will ultimately be able to buy out smaller farmers. This process of monopolization is thought to stand in the way of food security, for as the Nuffield Council argues, poverty and hunger should be countered by raising the productivity on small farms, because this will not only lead to an increased food supply, but also to productive employment.[120]

---

[112] Mack (1998).

[113] For example, Vasil (2003).

[114] Nuffield Council on Bioethics (1999). The World Food Summit has given the following definition of food security: 'Food security exists when all people, at all times, have physical and economic access to sufficient, safe and nutritious food to meet their dietary needs and food preferences for an active and healthy life' (Bharathan et al. 2002, page 171).

[115] Meyer (2001).

[116] Bharathan et al. (2002), page 194. For lack of a generally accepted politically correct term for the poor countries of the world, I will use the terms 'developing countries/world', 'less developed countries/world', 'Third World', 'poor countries', and '(countries of) the South' interchangeably.

[117] Conway and Toenniessen (1999).

[118] Nuffield Council on Bioethics (1999).

[119] Mack (1998).

[120] Nuffield Council on Bioethics (2003).

Appendix B: Plant Biotechnology Debate 269

Some biotech enthusiasts accuse European consumers, the majority of which have rejected genetically modified foods, of being elitist. In their eyes, Europeans enjoy the luxury of being able to choose or reject certain types of food, but they are destroying a viable market for poor Third World farmers who do not have this luxury. Critics of biotechnology, however, seek to expose the self-interested nature of this argument. As Mara Bün from the Australian Consumers Association point-edly formulates it: 'GMO's won't feed the world – they'll feed the wealthy'.[121] Moreover, even though some groups in the developing world are hopeful that biotechnology will lead to greater prosperity,[122] resentment is also expressed by African nations at being used as an 'excuse' by the biotechnology industry, as the following quote from a letter by African Food and Agriculture representatives from all African states (except South Africa) demonstrates:

> We strongly object that the image of the poor and hungry from our countries is being used by giant multinational corporations to push a technology that is neither safe, environmentally friendly, nor economically beneficial to us ... we think it will destroy diversity, the local knowledge and the sustainable agricultural systems that our farmers have developed for millennia, and that it will thus undermine our capacity to feed ourselves.[123]

Many of these accounts refer to an episode in 2002, when the US government offered African countries that were stricken with famine, food aid in the form of genetically modified grains, which the governments of Zambia and Zimbabwe declined. Many have attributed these countries' reactions to a fear that their products would be rejected by European trade partners.[124] Others argue that these coun-tries had many legitimate, independent, reasons to want to stay GE-free, such as the fact that crops with antibiotic resistance genes are particularly risky for African citizens.[125] Critics claim that the US government had ulterior motives to offer GM food in the form of assistance; they think the US is using Africa as a 'dumping ground', thereby creating further dependency on foreign aid. In their eyes, rather than food relief to Africa, the donation constituted a subsidy to US farmers, who were left with surpluses because their genetically modified goods were rejected in

---

[121] http://www.austmus.gov.au/consensus/ (accessed on 17 November 2005).

[122] Meyer (2001).

[123] Vint (2002), page 1.

[124] The Nuffield Council, for instance, argues that 'the freedom of choice of farmers in developing countries is being severely challenged by the agricultural policy of the European Union (EU). Developing countries might well be reluctant to approve GM crop varieties because of fears of jeopardising their current and future export markets' (Nuffield Council on Bioethics 2003, page xvii). The Council also thinks that developing countries might lack the infrastructure to be able to comply with EU labelling requirements, because it will be hard for them to segregate GM and non-GM crops.

[125] According to UK National Coordinator of Genetic Food Alert, Robert Vint, this is because firstly, due to the AIDS epidemic many people there suffer from lowered immune systems, which places them in a higher risk category, secondly, in Africa bacterial diseases are common and, thirdly, outdated antibiotics are commonly used there (Vint 2002).

the global market.[126] Critics also consider American aid part of a push to open up new markets for GM crops in Africa.[127] In fact, biotech companies themselves have admitted that they are looking to expand their market in the developing world.[128]

Some African spokespersons do believe that biotechnology could help to control the viral diseases that damage so many of Africa's crops, so that Africa could grow enough grain and would no longer rely on grain imports from the developed world.[129] Nevertheless, in their eyes, African countries need to strengthen their knowledge base and must avoid becoming the victim of multinational interests.[130] Many spokespersons from organizations in developing countries, aid agencies and environmental groups, on the other hand, are sceptical of the claim that biotechnology will increase food security. An often-heard counterargument is that hunger and poverty are not the cause of a supply but of a distribution problem. This argument is marked as naïve by biotech proponents, who say that solving distribution problems amounts to wishful thinking and is, therefore, 'purely academic'.[131] This 'slogan' has many different facets, however, and it is worthwhile to shed some light on the different arguments put forward under this banner. Some simply point out that globally, more than enough food is produced to feed the whole world population. Countries that are suffering from famines are known to export food to rich countries who use it as cattle feed. The problem is that many people in these countries are just too poor to purchase food.[132] Contrary to this point of view, it is put forward that redistribution is not a sustainable solution in the future, because without

---

[126] This is for instance argued by Amadou Kanoute, the African regional director for Consumers International. See Kanoute (2003). This view would be supported by the fact that the Bush Administration refused to supply Zambia and Zimbabwe with non-GM grains after their resistance; in other words, they were given the choice between GM grain or nothing ('GM or Death'). It should be noted that many of those in favour of biotechnology nevertheless hold that developing countries should be able to make up their own mind about whether food aid to them is GM or non-GM derived. See for example Nuffield Council on Bioethics (2003). The Council argues that aid agencies should comply with the wishes of developing countries regarding GM food aid, and that 'it would be unacceptable to introduce a GM crop into any country against its will by this means'. However, if these countries' wishes are based solely on environmental concerns, it could be justified to offer GM food, but then in milled form, so that the grains cannot be planted.

[127] Kanoute (2003). Aid agencies, environmental organizations and some farmers unions in developing countries seem to share this view and they are also wary of poor countries becoming a 'dumping ground' or test case for GM crops. South Africa's Biowatch, for instance, is suspicious of the US government's motives: 'Africa is treated as the dustbin of the world. To donate untested food and seed to Africa is not an act of kindness but an attempt to lure Africa into further dependence on foreign aid'. Quoted in Vint (2002), page 2. In response to this view, it has been pointed out, however, that most GM crops are still grown and tested in developed countries.

[128] Bharathan et al. (2002).

[129] Wambugu (1999). Wambugu suggests that the concerns of many European citizens that GM crops would be harmful for Africa and that this continent will be used as a dumping ground for GMO's is misplaced and paternalistic.

[130] Wambugu (1999).

[131] For example Per Pinstrup-Andersen in Meyer (2001).

[132] Vint (2002).

Appendix B: Plant Biotechnology Debate

biotechnology, by the year 2030 there would not be enough food available, even if we were able to distribute it equally.[133] Others point out that the real causes of food insecurity are not technical but political. The fact that US and European farmers get subsidies creates an unfair advantage and means that small-holders in the developing world cannot compete on an equal footing with them on the international market, because prices are kept artificially low.[134] These critics are doubtful that the introduction of GM crops will do anything to ameliorate this situation. Others on the other hand, most notably the Nuffield Council, recognize the political problems facing agriculture in the developing world, but argue that GM crops will provide positive changes more quickly than socio-political measures.[135] Those who reject biotechnology believe that increasing yields through novel technologies rather than using socio-political measures is misguided and will not offer a sustainable solution to world hunger. In fact, overproduction of food actually causes poverty, because it keeps down the price of the crops that poor farmers grow. Rosset argues that

> Third World food producers demonstrate lagging productivity not because they lack 'miracle' seeds that contain their own insecticide or tolerate massive doses of herbicide, but because they have been displaced onto marginal, rainfed lands, and face structures and macroeconomic policies that are increasingly inimical to food production by small farmers.[136]

In his eyes, the problems of poverty and environmental destruction are an inheritance from colonialism, which is continued by postcolonial liberal capitalism. He argues that structural adjustment policies of organizations such as the World Bank and the International Monetary Fund (IMF), that operate within the paradigm of privatisation and continuous growth, exacerbate poor farmers' problems. As a result of these institutions' policies, instead of growing food for their own populations, developing countries are compelled to use a great amount of their agricultural land to grow cash crops to export to the West, such as cotton, coffee, tobacco, and flowers.[137] As the World Bank recommends all countries to shift to the production of the same goods for the international market, the competition between these countries keeps the prices they receive for their goods low. Nevertheless, the Nuffield Council believes that Third World farmers cannot afford to forego the yield enhancing benefits of genetic engineering, because they have to compete with farmers from the developed world, where GM crops will be paramount before long.[138]

---

[133] Mack (1998).

[134] For example Kirsten Brandt in Meyer (2001). Also, many developing countries lack infrastructure such as roads and irrigation systems, which makes it difficult for third world farmers to compete as well.

[135] Nuffield Council on Bioethics (2003, section 4.1). It argues that in order to feed the world with the food that is available today would require for everyone to become vegan and for an equal distribution of all the world's farmland; two measure that one could expect to meet with fierce resistance.

[136] Rosset (2006), page 84.

[137] Vint (2002).

[138] Nuffield Council on Bioethics (2003, section 4.11).

Others argue, in contrast, that it might actually be a good marketing strategy for developing countries to remain GE-free zones. The Food Ethics Council argues in this context, that food security is complex and should not only be measured by its cost-effectiveness as expressed by a narrow range of variables, as seems to be the method of the World Bank and IMF.[139] Les Levidow similarly argues that the focus on agricultural efficiency discounts informal, non-marketed, produce, 'food which never reaches the market and thus tends to be omitted from official figures of production'.[140] Agricultural biotechnology limits these informal food networks, because small farmers are less efficient and go out of business. Levidow criticizes the Nuffield Council report for adopting a one-dimensional view of the market as essentially efficient and benign. In her book *Stolen Harvest*, Shiva also argues that genetic engineering is not compatible with the type of small-scale, low-input farming practices that are prevalent in Third World countries.[141]

Godfred Frempong, who has studied agricultural practices in Ghana, on the other hand, places the cause of poverty amongst Third World farmers, and the genuine obstacles to increased food production, in the fact that they still use outdated farming methods and bad crop varieties.[142] He thinks that science and technology could improve this situation, but as genetic engineering has met with a lot of resistance, he proposes to use the less controversial genomics. While genetic engineering actually alters the genetic make-up of plants or animals, genomics merely enables scientists to gain more insight into which genes are at work in a particular genome, which means that favourable traits can be selected and passed on in a much more precise way than through conventional breeding methods.[143] Frempong – a clear representative of Ruivenkamp's 'redesign' category – thinks that the main problem with applying genetic modification in developing countries is that the primary actors in the biotechnology field are multinational companies upon which farmers

---

[139] Food Ethics Council (2003). Similarly, Vandana Shiva argues that the free trade ideology championed by these organizations assumes that food security should be measured in terms of the power to buy food from international markets, rather than food self-sufficiency, or the ability to grow food locally for local consumption. In the international standards of food security, the fact that small subsistence farms manage to feed an extended family is not taken into consideration; this does not show up in economic calculations and therefore does not 'exist' in the eyes of international economists (Shiva 2000).

[140] Levidow (2001), page 51.

[141] Third world farmers traditionally grew a diversity of crops on small plots of land. The advantages were that the symbiotic relationship between different plants stimulates higher yields and that many of the crops generate a higher income than the monocultural crops that farmers were required to grow during the Green Revolution. Moreover, using plants genetically engineered to withstand herbicides or pesticides are only more cost effective on large monocultural farms. For small and diverse farms the chemical inputs make them too expensive (Shiva 2000, page 112).

[142] According to him, they use: 'poor planting materials, crops with poor genetic characteristics (low yielding, poor in food nutrients, long gestation periods)... Other militating factors include small holder farming (subsistence farming), poor agronomic practices and continuous dependence on hoe and cutlass (as the main farm implements)' (Frempong 2006, page 51).

[143] Frempong 2006, page 52.

Appendix B: Plant Biotechnology Debate

become dependent. This means that they can no longer participate in the development of seeds and that they can't choose their own specific method of farming. Also, industrial biotechnology has no regard for the diversity and heterogeneity of local farming practices.[144] Some applications of crop biotechnology are simply not relevant to smaller farms; herbicide-resistance, for example, is not useful for farmers who do not spray herbicide in the first place, because they use hand-weeding and use the 'weeds' thus obtained as nutritional supplements to their meals.[145] Frempong objects to biotechnology being developed by scientists in laboratories and then given to farmers as a 'quick fix', without farmer input. The new process of 'tailoring biotechnologies', on the other hand, builds on the experience of local and indigenous farmers; they play an active role in the development of new seed varieties. This means that the technology is made to respond to local farmers' specific requirements and problems, and aids the adoption of the new technology.[146] Local and indigenous farmers have unique knowledge not possessed by scientists – they could be considered experts in their own right – and, therefore, the two groups should work together, already at the stage of formulating research.[147] The problem with this approach, however, is a lack of incentive for private biotech companies, as the resulting crops by definition will have a small, specialized market that might not raise enough return on investment. The Nuffield Council, therefore, recommends that the UK spend a considerable part of its foreign aid money on public initiatives to tailor biotechnology to poor farmer's needs.[148] However, this recommendation can expect to meet with fierce opposition from biotech critics and aid agencies who think that other aid projects have a higher priority and that a mix of different farming methods would serve the interests of Third World farmers better than just the planting of GM crops.[149] Kirsten Brandt, for example, thinks that organic farming, with

---

[144] This view is supported by Bharathan et al., who explain that there are large variations in farm size in less developed countries, and that the uniformity that biotechnology seems to require will therefore not work there (Bharathan et al. 2002).

[145] Bharathan et al. (2002), page 184.

[146] An example of such tailor-made biotechnology is offered by Lindhout and Danial with regard to quinoa production in Bolivia. Genetic modification is combined with other methods, such as 'improved crop management'. The Andean region is characterized by highly variable environmental and climatic conditions and the preferences of farmers is also very variable. These complex conditions and preferences are insignificant on a global scale, 'and thus difficult to address by breeding programs with a global mandate'. '…Farmer participation is a prerequisite to ascertain that varieties are developed that are demand driven' (Lindhout and Danial 2006, page 37).

[147] Frempong (2006). Many proponents of biotechnology who are genuinely concerned about this technology's potential to feed the poor also recommend a closer consultation relationship between researchers and local farmers in order to create crop varieties that will be relevant to local circumstances. For example, Nuffield Council on Bioethics (1999) and Conway and Toenniessen (1999).

[148] Nuffield Council on Bioethics (2003).

[149] The Food Ethics Council, for example, thinks that research funding should not be spent on the development of GM crops, but on alternative projects directed at small-scale farmers who should have a say in all stages of development, from planning to implementation. It proposes to

274   Appendix B: Plant Biotechnology Debate

its low level of inputs, and therefore investment, and its labour intensive farming style is better suited to small farm holders in the Third World.[150]

Despite its overall optimistic stance towards the potential of biotechnology to feed the poor, the Nuffield Council acknowledges that the current research and development priorities of the biotech industry and research institutes is unlikely to fulfil this potential. It fears that research will continue to focus on quality enhancement of crops rather than on drought tolerance or increased yields, as is needed in developing countries, and that biotechnology innovation will be directed to saving labour-costs – herbicide-use, for example, replaces the need for hand weeding – which will run counter to the objective of creating productive employment. Moreover, yield enhancement in the developed world might reduce the need to import food from developing countries. Another danger of introducing GM crops in the developing world is that safety regulations might not be complied with as strictly as in the developed world.[151] In order to counter these considerable problems the Council recommends joint ventures between the private and the public sector, mainly through the Consultative Group on International Agricultural Research (CGIAR), which prioritise biotechnology research and development relevant to the developing world. However, according to critics there is no reason to think that private companies will be interested in these ventures. Some commentators are not optimistic either about the role of public institutions in addressing inequalities arising from the introduction of GM crops; in their eyes, certain international agencies are working towards the same goals as multinational biotech companies.[152] This is supported by the 'increasing convergence of private & public sector' research and development, which is amongst other things, a result of the fact that public institutions have to rely to an increasing extent on private funding.[153] One last potential problem of introducing GM crops in developing countries is that it is unclear who would bear the costs if the introduction of GM crops has unintended adverse consequences. It is questionable whether either private companies or the governments of developing countries could be relied upon to counter any damage to the environment or to compensate the losses of farmers.[154]

Golden Rice

The benefits of genetic modification for the poor are not only sought in increased yields, but also in the potential added nutritive value of crops, particularly Third world staple crops. As mentioned before, no high quality foods have reached the

---

use 'multi-dimensional strategies based on already-available knowledge and tools' (Food Ethics Council 2003, page 2).

[150] Meyer (2001).

[151] Nuffield Council on Bioethics (1999), pages 66–67.

[152] Bharathan et al. (2002). The authors point at Vandana Shiva as someone who holds this view.

[153] Hindmarsh (1994), page 2.

[154] Hindmarsh and Hindmarsh (2002).

Appendix B: Plant Biotechnology Debate

dinner table as of yet, so its impact on improved health (or increased consumer acceptance) is not yet known. Examples of high quality GM foods that could counter malnutrition are corn and rice with increased bioavailability of iron and Andean potatoes that block their own bitter tasting glycoalkaloids.[155] But the most famous GM crop that has been developed with the interests of the poor in developing countries in mind is Golden Rice, with its enhanced ß-carotene levels, which the human body can convert into vitamin A (ß-carotene is therefore also called provitamin A).[156] Vitamin A deficiency is estimated to be suffered by 180 million people globally and is particularly problematic in children. Sufferers have a higher chance of going blind and contracting (more severe) infections.[157] Golden Rice is the flagship of the biotech industry; it provides the industry with a moral justification for developing GM crops.[158] Its name is derived from the golden colour of the rice grains, but also, perhaps not coincidentally, carries an association with 'miracle' or 'valuable' grains.

Critics of Vitamin A-enriched rice argue that it does not address the real problems of Vitamin A deficiency: 'Vitamin A deficiency is not best characterized as a problem, but rather as a *symptom* . . . people do not present Vitamin A deficiency because rice contains too little Vitamin A, or beta-carotene, but rather because their diet has been reduced to rice and almost nothing else'.[159] They think it would be preferable to reintroduce other sources of Vitamin A (such as green leafy vegetables) into poor people's diet. Moreover, opponents argue that even if Golden Rice would fulfil its nutritional promise, consumers in countries with high levels of vitamin A deficiency will have no access to it, because their farmers do not have the means to purchase the seeds and their consumers cannot afford to buy the rice, which they think will be more expensive.[160] According to McAfee, the focus on the 'technical fix' of

---

[155] Genetic Resources Action International (GRAIN) (2001).

[156] This type of rice was developed in 1999 by Ingo Potrykus of the Swiss Federal Institute of Technology and Peter Beyer of the University of Freiburg. See Food Ethics Council (2003). Usually, ß-carotene is only expressed in the photosynthetic (green) tissues of plants, but in Golden Rice a small number (I have read the claim that it involves, 2, 3, or 4) genes (derived from a daffodil and from a bacteria) for ß-carotene enzymes have been added, so that it is also expressed in the non-photosynthetic tissues, such as the rice grains. See Conway and Toenniessen (1999), Genetic Resources Action International (GRAIN) (2001). Vitamin A deficiency is most common in Asia and hence rice, Asia's staple crop, was earmarked as the crop to produce higher provitamin A levels. Even though some, such as the Food Ethics Council, claim that it would have been more cost-effective in other crops, such as potatoes (Food Ethics Council 2003).

[157] Conway (2000).

[158] After Golden Rice had been developed, as Sarah and Richard Hindmarsh phrase it, 'it was immediately hailed as proof that biotech was there to help the poor' (Hindmarsh and Hindmarsh 2002). According to a group of Third World organisations critical of biotechnology, Golden Rice 'provided a much-needed public relations boost for the biotech industry at a time when genetic engineering is under siege in Europe, Japan, Brazil and other developing countries' (Genetic Resources Action International (GRAIN) 2001).

[159] Rosset (2006), page 87.

[160] Hindmarsh and Hindmarsh (2002).

276   Appendix B: Plant Biotechnology Debate

Golden Rice, and the positive image it provides for biotechnology, causes society
to overlook the question of why poor people have lost their original sources of vitamin A.[161] Comstock suggests that before the change in the developing world from
subsistence to export farming, and thus modern agriculture, the problem of 'lysine
deficiency' in children was not present, because children were complementing their
diet of rice with legumes that contained this amino acid. Rather than looking for a
technical fix to this problem, all we need to do is look at alternatives that are already
present in traditional farming methods.[162]

Buiatti writes that despite years of discussion about the supposed nutritional benefits of Golden Rice, it has finally been withdrawn, because the level of expression
of the transgene turned out to be too low.[163] This claim stands in stark contrast to
the observation Jorge Mayer made in the same year, that a new type of Golden Rice
has been created that expresses 23 times more provitamin A than the earlier version. Mayer argues that the suggested alternative solution to vitamin A deficiency of
encouraging poor farmers to grow vitamin A rich crops so that they can have a more
varied diet is doomed to fail, because these foods are not available all year round
and, unlike rice, are perishable. Moreover, the poor cannot afford to pay for such
a varied diet.[164] It is also argued that a lot of servings of these alternative sources,
such as green leafy vegetables, are necessary in order to uptake the recommended
daily amount (RDA) of vitamins.[165] Others argue, on the other hand, that either
two tablespoons of yellow sweet potatoes, half a cup of green leafy vegetables, or a
small mango already satisfy the RDA of a young child.[166]

Critics argue that the multiple factors underlying malnutrition, such as poverty,
lack of purchasing power, bad public health systems and lack of public education,
cannot be solved by Golden Rice.[167] However, the pro-biotech camp is not arguing that it can offer a single solution to all these problems; many agree that it
offers only a short-term solution. What they do argue is that the nutritional value
of Golden Rice is higher than that of conventional rice varieties. This, however,
is disputed as well. Disagreement exists about the amount of rice that would need

---

[161] McAfee (2003), page 213. Also, she wonders how Golden rice will be available to people to
whom at the moment other sources of vitamin A or cheap vitamin A supplements are not.

[162] Comstock (2000), page 161. The Food Ethics Council, similarly, points out that vegetable
garden schemes in developing countries have proven to result in more varied diets and increased
provitamin A levels already one year after their implementation (Food Ethics Council 2003). Not
only does this address vitamin A deficiency, but it provides a whole range of other nutrients as well
and it, therefore, takes a broader approach rather than the narrow focus on one specific problem
(Genetic Resources Action International (GRAIN) 2001).

[163] Buiatti (2005), page 23. In fact, according to Buiatti, most applications of biotechnology never
make it out of the laboratory because they just don't work. The few varieties that are successful are
repeated over and over in a variety of different contexts, as if there would be 'one optimal plant'
for all circumstances.

[164] Mayer (2005).

[165] Nuffield Council on Bioethics (2003, section 4.22).

[166] Genetic Resources Action International (GRAIN) (2001).

[167] Genetic Resources Action International (GRAIN) (2001).

to be consumed in order to counter the negative effects of vitamin A deficiency. Widely divergent estimates are given, from 3 kilos (Greenpeace) to 200 grams (the developers of Golden Rice) of uncooked rice a day. These differences rest on how much ß-carotene a certain amount of Golden Rice is assumed to yield, on what one takes to be the conversion rate of ß-carotene to vitamin A, and on what percentage of the recommended daily allowance is estimated to counter the effects of vitamin A deficiency.[168] Apparently, no scientific consensus exists on these questions. Furthermore, it is argued that in order for human bodies to be able to absorb ß-carotene, it needs to be consumed in combination with fat and other nutrients and minerals, which poor people can often not afford.[169]

The question whether the supposed benefits of Golden Rice will reach those who need it most is disputed. The research into Golden Rice is publicly funded, by the European Union and by the Rockefeller Foundation and even though as many as 70 patents apply to it, this appears not to have hindered its application in developing countries.[170] Syngenta has apparently given free Golden Rice seeds to subsistence farmers who earn less than $10,000 annually from rice production, provided that they do not export it. Moreover, a benefit of vitamin A enriched rice is that this trait is monogenic and can, therefore, be bred into local rice varieties.[171] Still, this does not automatically mean that the rice will reach poor farmers or will be affordable by poor consumers. Moreover, critics are suspicious of Syngenta's motives and think that it is a venue for entrenching its technology in developing countries.[172]

Terminator Technology

One particular topic in the debate about biotechnology's effects on poor farmers, which has received a lot of media attention and has met with fierce opposition, is the development of genetic-use restriction technologies (GURTs), labelled 'terminator technology' by opponents.[173] This technology consists of inserting genes into crops that make them infertile, so that germination of the next generation of seeds does not occur and farmers will no longer be able to save seeds for next year's sowing. This means that farmers need to purchase new seeds from biotech companies every year

---

[168] Nuffield Council on Bioethics (2003, section 4.24). BIOTHAI and other organisations argue that Golden Rice could only provide 10% of a child's RDA of vitamin A (Genetic Resources Action International (GRAIN) 2001).

[169] Genetic Resources Action International (GRAIN) (2001). Not only are the benefits of Golden Rice doubted, some, like Mae Wan-Ho also think there are risks, such as allergic reactions (due to the fact that a daffodil gene was inserted) and vitamin A poisoning from an overdose of ß-carotene.

[170] Bharathan et al. (2002).

[171] Mayer (2005), Van den Belt (2003).

[172] Hindmarsh and Hindmarsh (2002).

[173] A similar result can be produced with so called 'traitor crops', which are engineered to stop growing or ripening if they are not sprayed with a particular chemical sold by the same company that sold the seed. See Vint (2002).

and some argue that this leads to injustice and is part of a treadmill of farmer dependency on the biotech industry.[174] Terminator technology is advantageous especially in countries where patent laws are weak – which are usually developing countries – and whose governments can, therefore, not be relied upon by companies to protect their patents.[175] McAfee, for this reason, describes this practice as 'hard-wiring property rights into plant genomes'.[176] A USDA scientist describes it as a way to 'self-police your technology, other than trying to put on laws and legal barriers to farmers saving seed, and to try and stop foreign interests from stealing the technology'.[177] In other words, proponents regard terminator technology merely as a way to recover their investments. Vandana Shiva argues against this view that not all means are warranted to recover investment; after all, weapons manufacturers could just as well argue that they should be allowed to sell weapons for this reason.[178] Critics are fearful that the biotech industry will use this technology as an instrument to gain ever more control over the world's food supply.[179] Moreover, they fear that seeds from 'terminator crops' might escape and hybridise with wild relatives and local non-transgenic crops, making these infertile as well.[180] Proponents, on the other hand, argue that terminator technology will not harm poor farmers because they can still rely on seeds engineered by the public sector. However, this is not true if companies use patenting law to 'tie up enabling technologies' and to restrict further breeding and developing initiatives, so that the public sector's power to create plant varieties that will aid the poor is restricted.[181] For these reasons, many who are otherwise positive about biotechnology also object to terminator technology. Global protests against this technology have been successful in preventing their commercial release.[182]

---

[174] For example McAfee (2003), Bharathan et al. (2002), Shiva (2000).

[175] Bharathan et al. (2002), page 191.

[176] McAfee (2003), page 215.

[177] Quoted in Shiva (2000), page 82.

[178] Shiva (2000).

[179] Shiva (2000).

[180] Even though at first this possibility was denied by scientists and spokespersons from the biotech industry, when it turned out such hybridisations had in fact occurred, they argued that terminator technology could actually be used in order to stop the evolution of unwanted genetic constructs in the wild (McAfee 2003).

[181] Conway and Toenniessen (1999).

[182] Conway (2000). It is interesting to note in this context that non-transgenic hybrid corn, which does not germinate either, has been available for 75 years already, and that this corn was also rejected by many African countries (Bharathan et al. 2002, page 191).

Appendix B: Plant Biotechnology Debate

Patenting

Another topic in the biotechnology debate that is related to justice, is the question to what extent patenting of genetically modified organisms or constructs should be allowed. This is relevant in the context of the gap between rich and poor countries, because almost all patent holders are from the developed world and if developing countries infringe on patent legislation for certain products or technologies, they cannot sell their products in the global marketplace.[183] But the justice of the patent-system is also questioned in the national context of developed countries themselves, because of the concentration of patents in the private sphere. Again, the central question is who benefits and who pays the costs of this new technology. Moreover, some object to patenting for intrinsic reasons as well.

Biotechnological knowledge can be owned in the sense that it can be considered intellectual property. Three requirements have to be fulfilled before a patent is granted; firstly, the technique or the gene construct under patent application has to be original, or in other words involve an inventive step, secondly, it should not be obvious; and thirdly, it should be capable of industrial application.[184] Patents give patent-holders exclusive rights to commercial exploitation of their invention for 17–20 years.[185] Patent holders can charge a fee for licensing others to use their invention, but they are not obliged to exploit their invention. Patents give holders the right to practice a certain invention and to exclude others from doing so, but this does not mean they are actually granted ownership of the subject matter under patent.[186] There are exceptions to patents, which legally allow for the practice of someone else's invention without the requirement to obtain a license. This exception is often made for research, in order to stimulate further advances in biotechnology.[187] However, the researchers in question are not allowed to commercialise the outcomes of their experiments. The patent-system is influenced by two main international agreements. Firstly, through the 1995 Trade Related Aspects of Intellectual Property Rights (TRIPS) agreement, the WTO encourages all its trading countries to institute intellectual property rights over micro-organisms and some plants and to observe the patent rights of other countries.[188] This puts pressure especially on developing countries, many of whom do not recognize patent rights at the moment and can often not afford to pay license fees. Secondly, the 1993 Convention on Biological Diversity (CBD), which was not signed by the United States, is mainly concerned with the preservation of biodiversity and with the aspect of ownership of

---

[183] Nuffield Council on Bioethics (1999).

[184] Hughes (2002), Meyer (2001).

[185] Mayer (2002), page 146.

[186] Hughes (2002), page 155.

[187] Hughes (2002), page 157.

[188] Hughes (2002), page 164, Christie (2001). Member states are allowed to exclude animals from patent laws, but in practice parts of animals and humans are treated as micro-organisms.

the genetic resources that might be exploited by biotechnology institutions. As could be expected, there have been tensions between these two international approaches.

The aim of the patent-system is threefold; firstly, it rewards inventors, who have invested time and money in order to develop their invention, secondly, it is meant to encourage further innovation by making investments worthwhile financially, and thirdly, it aims at stimulating further inventions by requiring disclosure of knowledge.[189] Proponents of the patent-system argue that patents stimulate inventions which ultimately benefit the general public, because they protect the intellectual property rights without which companies and research centres would have no motivation to invest in biotechnology research and development. Critics, on the other hand, argue that patents only benefit large multinational companies, who hold around '90% of the new patents on products and technologies'.[190] They are of the opinion that patents limit innovation, stifle scientific research and decrease biodiversity, because patents exclude others from using the knowledge or source material to develop new varieties.[191] In this view, the purpose of patents is not to share knowledge, but to limit it, in order for the patent-holder to get maximum benefit out of his or her invention. Moreover, the system is so complex and so many patents can apply to a single set of genes or a specific technology that it is not only costly to obtain several different licenses, but it is also difficult to figure out which patents apply and who are the patent holders, and this could function as a disincentive to further innovation.[192]

A contentious issue in the debate over patenting in the field of biotechnology is whether the (parts of) organisms that patents are applied for should be considered as discoveries or as inventions. Many argue that living creatures should not be patentable, because they can only be discovered, but not invented. There have been several landmark Supreme Court rulings, on the other hand, that argue that the issue is not whether the organism is alive, but whether it is a product of nature or human-made.[193] Opponents argue that an organism can hardly be considered human-made when only one or a few genes have been added to its thousands of genes. Despite the heatedness of this debate, in reality this distinction is not as important as it seems, because under the TRIPS agreement many patents that are deemed to be discoveries are granted anyway. Moreover, as we saw before, patent holders are not actually the owners of organisms, but, rather, have the right to exploit certain uses of genes of the organisms.[194] Another contentious issue in the patent debate is the granting of

---

[189] The last two of these aims are held to serve the 'common good' (Brom 2003).

[190] Meyer (2001), page 25.

[191] Sterckx (2004), pages 4–5.

[192] Hughes (2002), page 162. Research among North American public sector plant breeders has revealed that many experienced difficulties carrying out their work because of corporate patents (Mayer 2002). Mayer bases this on findings by S.C. Price (1999).

[193] A famous decision was in the 1980 Diamond v. Chakrabarty case, where a genetically engineered micro-organism that could break down oil and would be used for treating oil spills was deemed by the Supreme Court patentable.

[194] Hughes (2002).

Appendix B: Plant Biotechnology Debate 281

broadbased patents. In order to maximise the value of their patent, applicants prefer to formulate the area to which their invention pertains as broadly as possible, which can lead to patents that can give companies monopoly control over a whole transgenic plant species, leading to a lack of competition resulting in high prices and restricted choice.[195]

Another type of patent which may act as a barrier to further biotechnology research and development is the patent granted for enabling technologies that function as tools of genetic modification.[196] On the one hand, these technologies help researchers to acquire patents, because they enable them to make the step from a gene to an invention, but on the other hand, if a new invention involving newly discovered genes involves the use of an enabling technology, a licence has to be obtained.[197] According to the Nuffield Council, patents on enabling technologies are controlled by only five big companies.[198] Even GM crop varieties that are the result of publicly funded biotechnology research are often not freely available because they make us of enabling technologies that are covered by private industry patents.[199] The role that publicly funded research has played in the development of biotechnology in general, and of enabling technologies in particular, is often discounted when a patent is given to companies that only become involved in the later stages of the research process.[200] This raises the issues of fair distribution of benefits, equality of opportunity, and dependence on the biotech industry. Within this patenting climate, public sector research institutions, such as universities, are also under pressure to apply for patents in order to ensure returns for their investments and no longer share all the new information and technology they develop.[201] Also, public sector institutions sometimes need to resort to 'defensive patenting' to make sure that their products are not patented by private companies, locking public institutions out of their use.[202] Furthermore, patents are regularly given for goods that used to be part of the public domain.[203]

Another problem caused by the patent-system is that farmers can be held liable for their unintended use of patented GM seeds after their crop has accidentally crossbred with a nearby GM crop. Not only will they not be compensated when they can no longer sell their crop as 'GE-free', but they can actually be held in breach of

---

[195] Hindmarsh (1994). For example, initially, a patent was granted that covered all the genetic transformations of the cotton plant, but this was contested by competitors (Hughes 2002).

[196] An example is the 'construction of a vector system for plant transformation, based on the *Agrobacterium tumefaciens* tumour-inducing plasmid', which causes localized growth and can function as a 'natural genetic engineer' (Hughes 2002, page 160).

[197] Hughes (2002).

[198] Nuffield Council on Bioethics (1999).

[199] Bharathan et al. (2002), page 192.

[200] Hughes (2002).

[201] Conway and Toenniessen (1999).

[202] Van den Belt (2003).

[203] Sterckx (2004).

licensing conditions of a biotechnology company.[204] In this context, critics speak of 'corporate feudalism' and they often use one particular highly publicized case as an example: Percy Schmeiser, a Canadian farmer, was sued by Monsanto after it found its Round-Up Ready herbicide resistant canola on his farm while he had not paid Monsanto royalties. Schmeiser claims that his crop had been contaminated by genes from a neighbouring farmer's crop.[205]

Owning Life

An intrinsic objection to patenting that is often heard is that it amounts to owning life, which is deemed to be morally objectionable. Patenting is viewed as licensing and legitimising scientists to change 'life as we know it'.[206] This objection is often put forward by religious groups, because biopatenting is considered to constitute a denial of the sacredness of life.[207] In response to this objection it is argued that it is not the organism itself that is patented, but merely something that is derived from the organism. And again, patents only bestow a negative right on patent holders, but do not give them ownership of life.[208] Frans Brom argues, however, that these replies do not take into account the real issue, which is the widely held belief that is actually constitutive of a society's identity, that 'society should not neglect the fundamental difference between living beings and human inventions'.[209] Brom goes on to criticize the assumptions underlying the arguments that biopatenting rewards the inventor and encourages research and development for the common good. The first relies on a Lockean view of ownership which can be contested, because it paints an oversimplified picture of complex societies with their division of labour. In complex societies there is no straightforward answer to the question of how ownership

---

[204] The Soil Association gives an example of a non-GM farmer who was sued by Monsanto for infringing on patent rights to the amount of $400,000 after his crop was contaminated by the GM crop of a neighbouring farmer (Meziani and Warwick 2002). Rogers argues that the Australian law works with double standards, because GE seeds are regarded as a company's private property, but traditionally bred plants are not, otherwise farmers would be compensated for their loss due to contamination (Rogers 2002).

[205] Monsanto won the Supreme Court case after the Court decided that it was irrelevant whether or not Schmeiser knew that GM canola was growing in his fields. However, Schmeiser did not have to pay a technology fee, because he had not harvested or made a profit from the canola. See Eckersley (2004). Eckersley refers to the following website: http://www.percyschmeiser.com. In the media it has been claimed that Monsanto flew planes over Schmeiser's fields, dropping Round-Up Ready herbicide in order to determine whether the crop would survive, in which case it must be Round-Up Ready canola.

[206] Hindmarsh (1994, chapter 1).

[207] A group representing 80 religious organizations, for example, stated in 1995: 'We believe that humans and animals are creations of God, not humans, and as such should not be patented as human inventions' (Brom 2003, page 120). Brom in turn quotes K. Lebacqz (2001).

[208] Brom (2003), Hughes (2002).

[209] Brom (2003), page 121.

Appendix B: Plant Biotechnology Debate 283

should be conferred to someone. Scientific knowledge relies on a process of co-operation that takes place in large, hierarchical, institutions; who gets ownership of an invention in this context will to a certain extent always be contingent. Moreover, the Lockean view does not take into account the fact that natural resources are a scarce good and that, therefore, the raw material that workers have to combine their labour with in order to be granted ownership, is usually already owned by somebody else. In the case of biotechnology, what some consider to be such raw material is often in fact the result of the cumulative efforts by generations of farmers and breeders. The argument that biopatenting serves the common good can also be disputed, for reasons outlined above. In the words of Brom, 'biopatents are tools for building and protecting market power in a free market, therefore biopatents reproduce and magnify existing inequalities'.[210] Many groups, such as indigenous communities, environmentalists, and religious groups, furthermore, argue that patenting life is problematic because it turns living organisms and parts of nature into saleable commodities. Like in the debate about animal biotechnology, here we encounter again the objection to commodification and instrumentalisation of living beings.[211]

Bioprospecting Versus Biopiracy

The discovery of useful genes is the subject of a heated controversy about the question who should have access to, and in effect control over, the world's biological resources and that centres on the divergence of interests of the North, where most of the knowledge, technology, and financial resources reside and the South, especially the tropics, where most of the world's genetic resources can be found.[212] Critics of biotechnology tend to call the collecting of genetic resources from developing countries by developed world biotech companies and public sector institutions 'biopiracy', whereas advocates term it ' bioprospecting'. According to the Nuffield Council on Bioethics these are two different practices, however. Bioprospecting is the legal transfer of resources which is 'commercially fair' because the countries of origin have given their informed consent and are compensated, whereas biopiracy is the unauthorized and uncompensated removal of resources.[213] Bharathan et al. take a wider definition of biopiracy, which may or may not be authorized, as 'the exploitation (including the use of intellectual property rights) by organizations usually based in developed countries of biological (including agricultural) resources from less developed countries'.[214] An often cited example of biopiracy is the

---

[210] Brom (2003), page 124.

[211] Protests on the basis of this objection have led to the revocation or retraction of some patents. For example, after worldwide opposition the US Centre for Disease Control revoked its application for patenting a cell line from indigenous women from Panama (Ho 1999, page 31).

[212] Christie (2001), page 173.

[213] Nuffield Council on Bioethics (1999), page 74.

[214] Bharathan et al. (2002), page 190.

284   Appendix B: Plant Biotechnology Debate

Rosy Periwinkle-case, which concerns a plant from the rainforests of Madagascar, that was found to have characteristics that could be used to create an anti-cancer medicine. The pharmaceutical company Eli Lilly developed the medicine and is making great profits, while the people of Madagascar have never been compensated for the extraction of the genetic material of the Rosy Periwinkle.[215] Another famous example is that of the Indian Neem Tree, which was used by locals for centuries for all sorts of purposes and the seeds of which provide a powerful pesticide. U.S. company W.R. Grace patented the seed emulsion extraction process, claiming to have invented a new process, and subsequently sued Indian companies for extracting Neem seed emulsion. The price of the pesticide rose sharply, which had a devastating impact on Indian poor or subsistence farmers.[216] W.R. Grace spokespersons argued that the company was actually helping the Indian economy by making Neem seeds available all over the world. Spokespersons of the biotech industry in general claim that genetic resources only get market value after they have been engineered into useful products or applications and that, therefore, the people or the countries who supply the genetic material need not be compensated.[217] It has been noted, however, that the rush to patent inventions in biotechnology can easily backfire on companies in the developed world, as most centres of genetic biodiversity located in developing countries, are rethinking their policies regarding access to their genetic resources.[218] It is also pointed out that today's wealthy countries lacked a strict patent system in their industrial development stages, which aided their economic development.[219]

Jeremy Rifkin traces the 'worldwide race to patent the gene pool of the planet' back to the movements of 'enclosures' of the common lands, in Britain in the 1500's and 1800's. In his eyes, patenting the genepool is like privatising the commons, which resulted in the displacement of millions of peasants from their farms.[220] In the case of biopiracy, the main concern is with the interests of indigenous and traditional farmers communities. Sigrid Sterckx discusses the implications of international patenting regulations for indigenous communities' access to and benefit sharing of the results of biotechnological research and development, and for the preservation of biodiversity.[221] It is difficult to regulate biodiversity, especially in the agricultural field, because the origins of the resources in this field are 'typically

---

[215] Rifkin (1998), page 38.

[216] See http://filer.case.edu/~ijd3/authorship/neem.html, accessed on 16 May 2007.

[217] Rifkin (1998), page 37.

[218] Conway and Toenniessen (1999).

[219] Meyer (2001), Van den Belt (2003), page 230. According to Van den Belt companies that are now pushing for the protection of intellectual property even admit that the 'lack of patent protection has been a decisive factor in [their own] development'.

[220] Rifkin (1998), pages 38–41. Besides pastoral lands and the global genepool, other 'commons' that have been enclosed are parts of the oceans, the atmosphere, and the electromagnetic spectrum.

[221] The CBD defines biological diversity as 'the variability among living organisms from all sources ....this includes diversity within species and of ecosystems', and genetic diversity'. Convention on Biological Diversity, article 2, note 1.

Appendix B: Plant Biotechnology Debate

obscure and certainly ancient'.[222] The TRIPS agreement has broadened the range of patentable goods, especially in the pharmaceutical and biotechnology domains.[223] The CBD does not counter the patenting of genetic resources, which subjects them to private property rights, but it does state that prior informed consent has to be acquired from the country where a genetic resource originates. Sterckx shows that there are several problems with this principle: Firstly, many resources of agricultural biotechnology have already 'escaped' and are now freely accessible, which means that no country can claim them. Secondly, one could wonder why national governments should be the decision-making body. Indigenous and local communities have often contributed a lot to the preservation of biodiversity, the breeding of new plant varieties and the discovery of agricultural and pharmaceutical uses of plants and other organisms. Christie remarks that 'quite literally, the gene pool that world agriculture must rely on is kept alive and developed by the daily work of small family farmers of the "developing" countries'.[224] However, whether their informed consent is sought, or whether they are compensated for the appropriation of their traditional knowledge is left up to national governments, which in the past have often not protected their rights and interests very well; there is little reason to assume that the financial rewards obtained by states for sharing knowledge or resources will be used to benefit the communities that have fostered them.[225] Thirdly, many people argue that genetic resources should be considered 'the common heritage of humankind' and from this point of view the fact that a genetic resource originated in one country arbitrary.[226]

According to Sterckx, the argument often used to defend patenting, that inventors should be rewarded for their contribution to society, is based on the principle of distributive justice and can also be used by local and indigenous communities to argue for benefit sharing of the results of biotechnological inventions.[227] She argues that it is in a sense arbitrary who counts as the inventor of something, because their invention often builds on ideas of many others before them. The invention can be seen as the 'finishing touch to the realizations of others'.[228] Like many others, she argues that the international patent system is biased in favour of Western scientists and corporations.[229] Rogers, for example, claims that 'the status of intellectual property rights as private rather than public rights renders them unsuited to the process of traditional plant breeding, which is a collective, generational and cumulative development'.[230] Granting patents to the inventions or discoveries of indigenous

---

[222] Sterckx (2004), page 1.

[223] Sterckx (2004), page 3.

[224] Christie (2001), page 176.

[225] Roht-Arriaza (1996), page 948.

[226] Sterckx (2004), page 4.

[227] Sterckx (2004), page 5.

[228] Sterckx (2004), page 6.

[229] See, for example, also Rogers (2002).

[230] Rogers (2002), page 13.

communities would be complicated, because applying for patents is very expensive, indigenous communities often do not recognize property rights and if they do, the knowledge is regarded as the property of the whole community, whereas patents require an application by an individual or group of individuals. Moreover, traditional knowledge[231] is often classified as belonging to the category of discoveries rather than inventions and is often not considered new or nonobvious and is not transmitted in written form.[232] Roht-Arriaza, furthermore, points out that resources that indigenous and local groups have preserved and improved over the course of centuries are often characterized as 'wild species' by Western scientists, that are, in their eyes, free for the taking.[233] Also, only formal knowledge systems are recognized as giving rise to intellectual property, whereas informal knowledge systems are considered the common heritage of humankind, not worthy of protection by intellectual property rights. Roht-Arriaza argues that traditional 'plant breeders' are often not acknowledged because they are peasant women who lack power and status.[234] Similarly, Shiva argues that this group is the most disadvantaged by genetic engineering.[235] Shiva also claims that not only are traditional farmers' efforts often not recognized and their knowledge and resources appropriated, but through patent laws they are often required to pay royalties to the very same companies who pirated their knowledge.[236]

Within the context of this thesis it is interesting to consider the role of Australia. Christie argues that while Australia is in the unique position that it is both a location of high genetic diversity and of financial and technological resources, its government sides with the 'plunderers' of the industrialised North. Nevertheless, in her estimation Australia is more likely to be plundered, because it has opened its doors to 'corporate commercial exploitation' by bioprospectors from Japan, Europe, and the US.[237] One famous example is the smokebush, which has been used by Aboriginal people for centuries for medicinal purposes and which was patented by the US Department of Health and Human Services, because it turned

---

[231] Traditional knowledge is information that people have acquired over time and often passed on orally. It includes information about which trees or plants 'grow well together, and indicator plants, such as plants that show the soil salinity..it includes practices and technologies, such as seed treatment and storage methods and tools used for planting and harvesting' (Sterckx 2004, page 7). The term 'traditional' is being referred to because this knowledge is based on traditions, not because the knowledge would be primitive or untechnological.

[232] Because such knowledge is often transmitted orally through stories and songs, they are dismissed by Western researchers as superstition or folklore (Roht-Arriaza 1996, page 933).

[233] Recall also Brom's remarks in the previous section criticising the Lockean assumption that ownership is conferred when raw material is mixed with labor, because there is not much raw material left.

[234] Roht-Arriaza (1996).

[235] Shiva (2000).

[236] For example, Indian farmers have been punished for infringing on Rice Tec patents on a type of basmati rice that was derived from Indian basmati (Shiva 2000).

[237] Christie (2001).

Appendix B: Plant Biotechnology Debate

out to inactivate the HIV virus.[238] In Australia, Aboriginal people provide tours for visitors explaining the medicinal purposes of plants. This knowledge, which they have built up through collective effort of studying, describing, and testing over centuries is used by so-called 'gene hunters' who attend these tours, without any acknowledgment or compensation to the Aboriginal people who led them to this knowledge.[239]

Food Safety and Health Risks

The question whether genetically modified food is safe or whether there are unintended health risks involved in their consumption is at the heart of the biotechnology debate.[240] The perceived health risk associated with GM food is one of the main reasons why many European consumers have rejected it. Proponents of biotechnology claim that it is safe to eat GM foods[241]; in fact, they point out that more health problems are associated with the consumption of conventionally produced food. Tony Conner of the New Zealand Institute for Crops and Food Research, for instance, claims that 'many nightmares predicted for genetically engineered crops have already happened ... [but] not many people noticed or cared', because they were the result of conventional breeding and not of transgenesis.[242] Similarly, Klaus Ammann, curator of the botanical garden at the University of Bern in Switzerland, argues that

> When we eat wheat, we consume varieties mutated by nuclear radiation. It is not known what happened with the genomes, but we have been eating this wheat for decades, without any type of problem. Today, with more extensive knowledge ... ... resulting from genetic engineering, we are faced with a new system where control is greater, more precise, and less risky than that of the old systems.[243]

Proponents of agricultural biotechnology often point out that consumers, particularly in the United States, Canada, Australia, and Mexico have eaten GM food for decades, without experiencing any health problems.[244] In Australia, for example,

---

[238] Christie (2001), page 181.

[239] Christie (2001), page 185. Similarly, the collections of botanical gardens all over the world are researched by biotech companies. This is possible because the CBD contains a 'loophole' clause which states that biological collections that were put together before the CBD came into force are not covered by it.

[240] It should be noted that GM food can refer to food obtained from GM crops or to food that has been processed using genetic modification techniques (such as 'vegetarian cheese' that uses a genetically engineered enzyme instead of animal rennet), or from animals that have been fed GM feed. In this chapter the focus is on GM crops, but in the debate about food safety, the two methods are often conflated.

[241] For example, Nuffield Council on Bioethics (1999).

[242] Quoted in Cohen (1998), page 42.

[243] Quoted in Fedoroff and Brown (2004), pages 9–10.

[244] For example, Wambugu (1999).

GM soy, corn, potato, canola, and sugarbeet have been approved for sale and these products are present in food such as breads, pastries, snacks, oils, confectionary, and soft drinks.[245] Proponents assume that the fact that consumers from these countries have eaten these foods shows that they had no qualms about biotechnology. However, critics cite opinion polls in order to demonstrate that these consumers did so out of apathy and ignorance rather than on the basis of informed choice or even indifference about food safety.[246] Also, a high degree of scientific consensus about safety of new technologies in the past has sometimes proven unsubstantiated and consumers are, therefore, justified in being suspicious of experts' claims.[247] Judy Carman argues that it is a fallacy to think that the fact that no cases of illness after eating GM foods have been documented must mean it is safe. She thinks that it would be near impossible to determine whether a particular GM food was causing illness in the community. If there is no surveillance system, as there is in the case of cancer for example, nobody will be alerted to an increase in a certain disease and the problem may go unnoticed. In the case of GM foods there is no surveillance system, because we do not even know what disease to look for.[248] Moreover, even if an increased incidence of a particular illness was discovered, then it would still be hard to trace the cause back to GM food, especially when many people are not aware that they are consuming GM products.[249] Furthermore, Kanoute suggests that even though GMO's have been consumed without apparent problems in the US, consumers in Africa stand a higher risk of adverse health effects from consuming GMO's as a much larger portion of their diet consists of grains, which are the principal transgenic crops.[250]

The perceived unsafety of GM food in the first place appears to be related to its potential to cause allergic reactions. This is especially the case when genes from known allergens, such as the brazil nut, are inserted into different food types.[251] Up till now, producers have refrained from using genes from such allergens, however. Nevertheless, some biotechnology advocates defend the position that even if minor allergic reactions would occur in some people after eating GM foods, the benefits

---

[245] Carman (2004), page 83.

[246] They ascribe it to the lack of labelling of GM foods in these countries and to the fact that many developments have been kept out of public view (Barry 2001). Barry also concedes that, as the Food Ethics Council suggests, the difference between the US and Europe can in part be explained by the traditionally more instrumental approach towards food in the US and the more cultural and moral values attached to food in Europe.

[247] Thompson (1997a).

[248] Carman (2004). She gives the example of the HIV/AIDS epidemic that nobody noticed for decades. Apparently the HIV/AIDS epidemic was discovered by coincidence, due to the fact that records were kept of a particular drug that showed up an unusually high number of rare pneumonia cases. There were already thousands of infected people by then.

[249] Carman also argues that even if it was found that a particular GM food caused increased illness it would take many years to remove it from the food supply; biotech companies could be expected to contest the findings, as tobacco companies did when a relation between smoking and lung cancer.

[250] Kanoute (2003).

[251] Cohen (1998).

Appendix B: Plant Biotechnology Debate

still outweigh the risks. After all, we also accept this in other cases, such as the use of penicillin that causes an allergic reaction in some people.[252] Worries exist about other unintended effects as well. A much cited case is that of research by Dr. Arpad Pusztai, who fed rats genetically engineered potatoes and found abnormal growth rates of the rats' internal organs, suggesting that transgenic potatoes would be unfit for human consumption as well. The 'Pusztai case' represents a particularly grim episode in the history of the biotechnology debate, with biotech advocates doubting the truth of Pusztai's findings and biotech critics claiming that a propaganda war was launched to discredit Pusztai and his work.[253] Another concern with the consumption of GM foods focuses on the use of DNA sequences coding for antibiotic resistance which are used for selecting GM plants; critics believe that this will worsen the already common problems experienced today with antibiotic resistance bacteria.[254] Finally, concerns arise from the fact that DNA from plants that are normally not consumed, such as petunias, are inserted into the genomes of GM plants, as well as genes from bacteria, animals, and viruses.[255] Critics argue that as components are added to food that have not previously been used for human consumption, it will be difficult for standard allergy testing procedures to identify possible new allergens.[256]

According to critics, much of the safety assessments have not been carried out by independent scientists, but by industry scientists who are likely to be biased. Carman points out that even government watchdog agencies such as the Food Standards Australia New Zealand (FSANZ) do not carry out their own tests, but rely on the results of company experiments.[257] She analysed the safety reviews of FSANZ and found that none of the crops that had been approved for human consumption had been tested on people in feeding trials and one had not even been tested on animals. Moreover, she argues that the duration of the animal tests was too short; one to two weeks of testing is not long enough especially for cancer studies or for studies to determine effects on offspring. Usually not the whole food was tested but only the part that had been genetically modified, meaning that possible interactions of the GM component with the rest of the foodstuff could not be revealed. If the whole food

---

[252] Wambugu (1999).

[253] For example, Wambugu points out that an independent review of Pusztai's experiment has been done, which has found that its outcomes were misleading and data were misinterpreted Wambugu (1999). Shiva, on the other hand, cites independent research supporting Pusztai's findings. Moreover, as Dr. Puzstai confirmed during a public lecture I attended, before his experiments could be published, Pusztai was suspended by his lab, was accused of scientific fraud, forbidden to talk to the media, and his computers were confiscated, causing Shiva to proclaim that Pusztai was 'sacrificed to protect corporate control and profits' (Shiva 2000, page 110).

[254] Carman (2004), Hindmarsh and Lawrence (2004), Conway and Toenniessen (1999). Conway, an otherwise enthusiastic supporter of GM crops, for this reason calls for an end of using antibiotic-resistance genes in crop biotechnology.

[255] Carman (2004).

[256] Mayer (2002), page 145.

[257] Carman (2004).

had been fed to animals, the sample sizes were very low, making it easy to not find significant differences with the control group. Another anomaly she encountered was that uncommon animal models had been used, such as chickens, cows and trout, which raises questions about the possibility of extrapolation to human health. Also, the measurements that were taken from the animal tests did not give an adequate reflection of human health; for example 'abdominal fat pad weight, total de-boned breast meat yield, and milk production'.[258] These measurements suggest that the real reason for carrying out the animal tests were not related to human health, but rather to increased yield and economic benefit after feeding animals GM crops. The results of the tests focused on body weight and sometimes organ weight of the dead animals, but biochemical, immunology, tissue pathology and microscopy results were not given and had probably not been tested for.[259] These observations cause Carman to proclaim that the assumptions underlying the experiments 'make a mockery of GM proponents' claims that the risk assessment of GM foods is based on sound science'.[260] Moreover, even the limited tests in Carman's eyes do not justify the claim that the GM foods were safe, because adverse effects, such as increased liver weight, were in fact found.[261] Another problem of the tests Carman analysed was that no detailed statistical data were given, so that the results could not be reviewed by others.[262]

Substantial Equivalence

Proponents of biotechnology tend to argue that more extensive safety testing is not necessary for GM foods, because there have not been significant changes to GM crops and there is no 'conceptual distinction' between organisms that have been traditionally bred and those that have been genetically modified.[263] The term that has been used to describe this view is 'substantial equivalence'. This term was first adopted to safety assessments of food by the Organization for Economic Cooperation and Development (OECD), which held that the safety of GM foods

---

[258] Carman (2004), page 87.

[259] Carman argues that these limited tests would be fine as a starting point of a long series of tests, but subsequent tests were not carried out. It should be noted in this context that the call for more animal experiments would raise objections from an animal ethics point of view; this could be another reason why critics think the benefits do not outweigh the costs of GM food.

[260] Carman (2004), pages 87–88.

[261] Carman (2004). Also, in a test with GM potatoes on rats abnormal findings were noted, but still FSANZ considered them safe for consumption, because in the control rats also showed problems. Carman argues that this could either indicate that rats are not the appropriate animal model for these tests or that something uncommon happened, such as a virus that affected all the rats. In any case, the conclusion that GM potatoes were safe for human consumption was not warranted.

[262] Carman (2004). Carman's view that much of the safety assessment that is carried out is supported by organizations such as the Royal Society of Canada, the American National Academy of Sciences and the Royal Society of London.

[263] Miller (1999).

Appendix B: Plant Biotechnology Debate

could be assessed using generally the same methods as conventional foods. In fact, the OECD suggests that less risks are associated with GM food than with conventional food, because of the greater predictability and precision of biotechnology.[264] Substantial equivalence is a comparative term based on the idea that the conventionally produced counterparts of GM foods can function as a reference for evaluating how safe GM foods are. Saying that a certain GM food is substantially equivalent to its conventional counterpart is not a safety assessment in itself, but merely establishes whether there are significant differences between a GM and a non-GM food product, and whether extra safety testing needs to be carried out. This concept, therefore, provides a starting point and not an end point of safety assessment.[265]

Critics argue that testing only the composition of GM-foods to determine whether they are substantially equivalent to conventional foods is insufficient.[266] For example, GM-soybeans are tested in isolation, whereas their chemical composition should be tested after exposition to herbicides.[267] Millstone et al. think that the term substantial equivalence is misleading; it gives the impression of being a scientific term, but it is not defined clearly enough: 'the degree of difference between a natural food and its GM alternative before its 'substance' ceases to be acceptably 'equivalent' is not defined anywhere'.[268] They recommend toxicological, biological and immunological tests on novel GM-foods, rather than just chemical tests, and the establishment of levels of 'acceptable daily intakes'. In their eyes, regulators and the biotech industry have come up with the term 'substantial equivalence' because such comprehensive testing would be too time-consuming and expensive and because many GM-crops are staple crops, which make up a large part of consumers' diets and would quickly exceed acceptable daily intake levels. They claim that, strangely enough, many GM foods are classified substantially equivalent enough to their conventional counterparts to limit safety tests, but new enough to justify granting patents to their inventors.[269]

Miller points out that contrary to what Millstone suggests, substantial equivalence was never meant to be a scientific term, but rather a tool for regulators to determine whether new foods have to be tested more rigorously. He also thinks it is disproportional to require costly biological, toxicological and immunological testing of GM foods but not of foods of plants that have been improved through traditional breeding methods, such as hybridisation, which in his eyes, are in fact much less

---

[264] OECD (1986) quoted in Miller (1999).

[265] Kuiper et al. (2002).

[266] For example Mayer (2002).

[267] Millstone et al. (1999). Ho, similarly, argues that because substantial equivalence is not clearly defined, there is a loophole in safety assessment (Ho 1999, page 35).

[268] Millstone et al. (1999), page 525.

[269] Millstone et al. (1999).

predictable.[270] Carman, on the other hand, points out that products that have a high level of substantial equivalence to their non-GM counterpart can still turn out to be dangerous.[271]

In general, the 'anti-GM camp' argues that the use of substantial equivalence is misguided and overly atomistic, in that it limits the comparison between two food-stuffs to only their components on a genetic level, on which there is little difference between humans, animals, plants or bacteria.[272] Also, they argue that there is no test actually establishing substantial equivalence, meaning that substantial equivalence can just be declared in advance, which makes further testing redundant.[273] The 'pro-GM' camp, on the other hand, argues that their opponents' focus on the process of biotechnology rather than on the actual products achieved by it, is seriously misguided, because it carries the assumption that 'somehow gene-splicing systematically introduces into organisms (and the foods derived from them) greater uncertainty or risk than other, older, less precise techniques'.[274] They think that one should not assess the risks associated with genetic engineering as such, 'horizontally', as if all applications would be the same, but one should analyse the characteristics of specific organisms and their interaction with the environment into which they are introduced.[275] They are, therefore, also opposed to blanket policy responses to genetic engineering such as moratoria on all GM crops. It appears that the question whether or not one regards biotechnology as a precise and predictable technology determines one's opinion of the merits of substantial equivalence in risk assessment. Now let us take a closer look at the controversies surrounding the methods of risk assessment.

Risks

In order to determine whether the consumption of GM foods and the introduction of GMO's into the environment are safe, scientists make a risk analysis. They measure risk as the probability of an adverse effect to happen combined with the seriousness of the consequences if it does. Risk analysis entails three phases: first a

---

[270] Miller (1999).

[271] Carman (2004), page 86. She points to the tryptophan scandal, in which a Japanese company made a GM dietary supplement which killed 37 people and disabled many more, involved a product that was 99.6% pure, and therefore, 'much more substantially equivalent in its chemical composition to pure tryptophan than products of GM crops are to their non-GM counterparts'. In her eyes, this shows that the argument that 'substantially equivalent means safe' is dangerous. Carman also argues that even though biotech industry spokespersons have argued that the poisoning was a result of cost cutting by the Japanese company Showa Denko KK, rather than of the fact that it involved a GMO, this cannot be proven conclusively, because these two factors were present at the same time.

[272] For example, Barry (2001).

[273] Ho (1999), page 149.

[274] Miller (1999).

[275] Miller and Gunary (1993), page 1500.

Appendix B: Plant Biotechnology Debate 293

risk assessment is made, then a plan is made to manage this risk, and finally information regarding the risks is communicated to affected parties.[276] The conventional approach to risk analysis and its underlying assumptions on all three of these levels have come under increasing scrutiny, particularly after episodes such as the BSE crisis in Britain. Turner and Wynne point out that there are three broad approaches to risk analysis. The first is the conventional cognitive approach, which is still favoured by many scientists and adopted by policy makers, and treats risk as an 'objective probability of harm'.[277] In this approach, the phase of risk assessment is understood to be value neutral, while risk management focuses on objective aspects such as mortality rate, and risk communication is regarded as simply the transfer of information to passive receivers aimed at gaining public acceptance. This approach has been criticized for neglecting the social and cultural dimensions of risk, and for being simplistic and ineffective.[278] More subjective elements of risk perception, such as the extent to which a risk is taken voluntarily, the controllability of the risk, and the immediacy of the effects, are not taken into account in this approach.[279] Moreover, the idea that any of the phases are value free, or involve value judgments only to a limited extent, has been attacked.

Before one can make a risk assessment one has to determine what constitutes a risk and since a risk can be described as the chance that unwanted consequences will occur, it has to be decided first what constitutes an unwanted consequence. This is based on a value judgment.[280] Moreover, the conflict about the risks involved in GM foods already starts at the level of the definition of risk. As Levidow puts it, 'the controversy is not simply about how to 'balance' risks and benefits, but also about how to define risk'.[281] Similarly, in risk management a decision about the acceptability of certain risks is implicitly made and this rests on value judgments. What risks one is willing to accept depends, amongst other things, on the perceived benefits involved.[282] According to the second and third approaches to risk analysis – based on the cultural and social theories of risk – a wider set of values needs to be included in risk management. Cultural theories, in short, view beliefs about risks as socially constructed.[283] A good example of cultural influences in risk management in the case of food safety is the fact that the French are willing to accept the risks involved in consuming unpasteurised cheese, because of the cultural value

---

[276] Brom (2004).

[277] Turner and Wynne (1992), page 111.

[278] Turner and Wynne (1992).

[279] See also Roeser (2006).

[280] Brom (2004).

[281] Levidow and Carr (1997), page 33.

[282] Brom (2004).

[283] Turner and Wynne (1992). This theory is attributed to the work of Mary Douglas, who observed that different social groups who apparently were discussing the same problem tended to focus on different types of risk. She traced this back to these groups' different social organisation and political culture. See for example Mary Douglas (1986).

they attach to this food. Social theories of risk look at risk perception from a sociological perspective and argue that people define risks 'primarily according to their perceived threat to familiar social relationships and practices, and not by numerical magnitudes of physical harm'.[284] In other words, the purely physical risks that scientists point out are subordinate to other meanings people attach to risks.[285] The insights from this theory are particularly relevant for the phase of risk communication, because risk communication has proven to be ineffective and even to have contrary effects if social dimensions are overlooked. People tend to define risks on the basis of the level of trust they have in the institutions that are managing and communicating risks. The 'one-way sender/receiver model' as employed in the cognitive approach often does not lead to public trust. Especially when the public perceives these institutions as functioning along lines of social dominance and control – as is the case with biotechnology – in order to be effective risk communication needs to take a more interactive shape in which information flows in both directions.[286]

In an Amicus Curiae Brief to the WTO regarding biotechnology, five of the most prominent scholars in the field of the sociology of science warn that 'risk assessment is neither a single methodology, nor a 'science'', and that 'judgments about the same hazard, based on the same scientific knowledge and evidence, do not always lead to the same estimates of possible harm', because risk analysis is influenced by the political and cultural context in which it takes place.[287] Similarly, Millstone argues that

> in science-based risk debates, one of the main reasons why different groups of experts reach different conclusions is not because they reach conflicting interpretations of shared bodies of evidence but because they adopt differing framing assumptions about the categories of risks that they should address and those that they should discount or ignore. In other words, they reach different conclusions because they are answering different questions.[288]

What makes risk analysis in gene technology especially problematic is that it is characterized by low levels of certainty – we are dealing not only with known uncertainties, which is what the cognitive approach to risk analysis is based on, but also with 'unknown unknowns'[289]– and low levels of 'consensus with respect to the parameters of the scientific issues to be addressed'.[290] Research shows that lay citizens tend to take into account the existence of unknown unknowns while scientists and policy makers tend to ignore them.[291] Sheila Jasanoff notes in this

---

[284] Turner and Wynne (1992), page 122.

[285] Turner and Wynne (1992).

[286] Turner and Wynne (1992), Brom (2004).

[287] Bush et al. (2004), pages 5 and 115.

[288] Millstone (2006), page 45.

[289] Wynne (2001).

[290] Bush et al. (2004), page 5.

[291] Wynne (2001). Wynne bases himself on the findings of several research groups from the United States FDA, the UK Biotechnology and Biological Sciences Research Council, and a European research project, which all corroborate each other.

Appendix B: Plant Biotechnology Debate

context that the relationship between science, society and government are increasingly being questioned. Science is no longer generally regarded as separated from values and citizens have become sceptic of the confidence of scientific predictions and policy decisions based on them. Moreover, the idea that there is one objective sound science that is agreed on by all scientists, is questioned. The fact that much of the scientific research has not been open for public scrutiny has caused special concerns about the so-called undemocratic character of science. Governments and the biotech industry have been criticized for employing a ''top-down' model of the science-society interaction' in which the results of 'sound science' are simply disclosed to the public, who if they do not accept the findings, are accused of being ignorant or overly emotional.[292] Differences between risk perceptions of experts and lay people have traditionally been seen as stemming from lay persons' irrational fears, whereas they can also be regarded as the result of a different interpretation of the social implications of technology and the different framing of policy issues.[293] What is dubbed a fear of technology may in fact be based on intrinsic moral concerns, such as described earlier in this chapter.[294] As already noted before, more knowledge of genetic engineering does not automatically lead to acceptance of this technology; quite to the contrary, it tends to lead to more rejection. Ulrich Beck, who coined the concept of the 'risk society', argues for more democracy in science through open expert controversy and a role of scientists in public debate as consultants rather than as authoritative decision-makers. Wynne goes even further and argues that we do not simply need what Beck calls 'deviant experts' to show us that certain risks are out there. Rather, we need to question the priority given to the scientific model that is used by experts. The values that are inherent in risk analysis have to be made more explicit. It is then open for public discussion what outcomes of risk analysis are realistic and what risks people are willing to accept.[295] Wynne, furthermore, argues that lay people can provide experiential knowledge that forms an indispensable contribution to the risk analysis of experts.

Levidow and Carr argue that the term 'risk' itself has become ambiguous. For some members of the public, it 'denoted an ethical challenge to the further industrialized control over nature.'[296] From the point of view of biotechnology advocates, however, society is at risk of not benefiting from biotechnology's potential if it is overly regulated. Discussions about risks, then, do not only have normative aspects simply because values are involved in risk analysis, but more fundamental moral issues are also raised when risk terminology is used. Grove-White and Szerszynski in this context argue that 'many social conflicts, overtly about the technical determination of environmental risks, can be more usefully seen as conflicts between

---

[292] Barry (2001).

[293] Jasanoff (2001).

[294] This point is also made by Bruce (2001).

[295] Wynne (1996).

[296] Levidow and Carr (1997), page 32.

commitments to certain models of how society is – or should be – ordered'.[297] According to Levidow and Carr, the biotech industry and many government agencies have ignored the normative aspects of risk and have actively sought to separate risk from ethics and to reduce both to matters that could be dealt with by experts. The next step has been to declare that risks are a technical matter that could be more objectively determined than ethics. They criticize the way in which many governments have dealt with moral concerns, for example by setting up separate technical and ethical advisory committees: 'the issues of biotechnological control were fragmented – into an expert risk assessment, a free consumer choice, and a residual ethics, for ensuring proper "balance".[298]

Precautionary Principle

Because there are so many scientific uncertainties involved in gene technology and the possibility of unknown consequences is perceived to be high, many propose to use the so-called 'precautionary principle' when making decisions in this field. In contrast to the US, the European approach to risk analysis is based on the precautionary principle.[299] It should be noted, however, that some critics also question the true precautionary nature of the European approach to risk assessment.[300] Generally, the Netherlands are perceived to employ a strong version of the precautionary principle, whereas Australia is claimed by critics to employ a weak precautionary approach.[301] The standard definition of the precautionary principle is from the 1992 Rio Declaration on Environment and Development: 'Where there are threats of serious or irreversible damage, lack of full scientific certainty shall not be used as a reason for postponing cost-effective measures to prevent environmental degradation'. In the discussions about biotechnology, this is implicitly supplemented by 'or harms to human health'.[302] The definition is vague though, and it is open to widely varying interpretations. For example, what do we mean by 'serious' or by 'irreversible' or by 'damage' or 'cost-effective'? How we interpret these terms is, again, based on value judgments. One problem is that the principle does not specify 'a minimal threshold of scientific certainty warranting preventive

---

[297] Grove-White and Szerszynski (1992), page 292.

[298] Levidow and Carr (1997), page 40.

[299] Brom (2004).

[300] Mayer and Stirling (2002).

[301] Hindmarsh and Lawrence (2004).

[302] Rio Declaration. This principle was first proposed in the context of global warming. Skeptic scientists argued that you could not conclusively prove that the atmosphere was warming up and even if you could, you could not prove that this was caused by humans. Some companies used this argument to claim that they did not have to reduce greenhouse gas emissions. The precautionary principle was used to counter this claim. So even though we could not prove it, if it was true that we caused global warming the consequences of not taking action to prevent it would be very bad and so we should not take any risks.

Appendix B: Plant Biotechnology Debate 297

action'.[303] Donald Bruce argues that the precautionary principle, like the concept of sustainable development, has been re-interpreted and re-defined in different ways to suit different groups' interests or beliefs. He identifies three broad interpretations, a strict one, a more nuanced one, and one that rejects it outright. The latter is typified by the US dismissal of the concept. Those who reject the principle altogether often argue that it is not based on sound science.[304] The nuanced one is the one proposed by the Rio Declaration. The strict interpretation is favoured by some environmental groups, which argue that if there is even a remote possibility of harm from a technology or intervention, we should not go ahead.[305] One facet of this strict interpretation is that the burden of proof is reversed; instead of forcing opponents of, in this case, biotechnology to prove that there is clear evidence of risk, it lays the burden of proving that no risks exist with proponents. However, this is an uneven distribution, because it is much harder – it is in fact logically impossible – to prove that there are no unwanted effects than to prove that there are.[306] Another problem that some note with the strict interpretation of the principle, is that it precludes any weighing of risks and benefits,[307] According to the strict interpretation there is no such thing as sound science and science cannot deal with the uncertainties associated with biotechnology; therefore, the principle should be based primarily on moral concerns rather than scientific ones.[308]

Hughes and Bryant, on the other hand, argue that even though risks can never be brought to zero, scientists possess enough information and knowledge to enable adequate risk assessments to be made, and to preclude the need for the use of the precautionary principle.[309] Many proponents of biotechnology appear to assume that critics of biotechnology demand the complete absence of all risk before a new technology can be introduced and point out that if this would have happened in the past, we would not now be able to enjoy the benefits of aeroplanes or mobile phones. They argue that invoking this principle amounts to doing nothing, which in itself is a risky option; the precautionary principle can work both ways – genetic modification could just as well lead to environmental catastrophes as it could help to prevent it.[310] According to Turner and Wynne, however, research shows that assumptions by industry and governments that the public is naïve and wants a zero-risk society are unfounded.[311] As mentioned before, a more nuanced interpretation of the precautionary principle is also possible. Some proponents, for example, point out that

---

[303] Van den Belt and Gremmen (2002), page 105.

[304] For example, Henry Miller. See Van den Belt and Gremmen (2002).

[305] Bruce (2001).

[306] Bruce (2001).

[307] Van den Belt and Gremmen (2002).

[308] Van den Belt and Gremmen (2002).

[309] Hughes and Bryant (2002), page 135.

[310] Nuffield Council on Bioethics (2003), Comstock (2000).

[311] Turner and Wynne (1992).

originally, it called for careful planning and risk avoidance and not just for determining how much risk is acceptable.[312] The principle can be understood as demanding that steps should be taken to minimise the risks when a novel technology is introduced, rather than calling for an a priori rejection of the technology. Some also think that the principle commends us to recognize the complexity, variability, and vulnerability of nature and to examine all the available alternatives to the risk generating activity.[313] Van den Belt and Gremmen argue that the principle should not simply be used to reverse the burden of proof, but that it should be carefully decided how the burden of proof should be divided amongst proponents and opponents of biotechnology. They explain that there are type-I errors, in which the assumption that a harmful effect will occur turns out to be false, and type-II errors, in which it is wrongfully assumed that no harmful effects will occur. There will always be a trade off between these two types of error. If we are very concerned to avoid type-II errors because we do not want to risk any environmental damage, we will probably make a lot of type-I errors, or in other words, we will 'produce a lot of false ecological alarms'. We have to decide in each case what the costs are of making each type of error. Even though we might be raising many false alarms and preclude certain benefits, it could be argued that we are morally more obliged to protect the public from grave harms (such as creating antibiotic resistance) than to improve welfare (such as creating cheaper food).[314]

Labelling

Many consumers feel that the regulatory process does not deal well with scientific uncertainty and that they lack control over the direction that biotechnology developments are taking.[315] One way in which they feel they can at least exert some control is through their consumption choices, but this requires that they can inform themselves about the products they are purchasing, and hence it requires labelling of GM foods. The question of whether and how GM foods should be labelled has been one of the most discussed and most contentious topics in the biotechnology debate.[316] In Europe and Australia, when a product contains more than 1% GMO's, it has to be labelled. However, according to Mayer, labelling requirements are side-stepped in many cases, because products containing derivatives of GM crops, such as vegetable oils and processed foods, often do not have to be labelled. The reason for this 'discrepancy' is that the labelling rule applies to the final product and not to the process used to make the product, even though consumers often want labelling because

---

[312] Hindmarsh and Lawrence (2004).

[313] Mayer and Stirling (2002).

[314] Van den Belt and Gremmen (2002). Van den Belt and Gremmen base themselves in part on Kristin Shrader-Frechette here.

[315] Mayer (2002).

[316] Mayer (2002).

Appendix B: Plant Biotechnology Debate

they are concerned about the process.[317] Advocates of the 'product approach' argue that analysing the product is the only way to measure whether there are GMO's in food and that health risks can only be based on the components of the food consumed and not on the process by which the food was made.[318] Biotech companies go on to argue that there is no need for labelling, because the chance of allergic reactions is very low.[319] It is also pointed out that a precondition of labelling is that GM and non-GM grains can be segregated effectively, which can be problematic, because it is unreliable and because GMO's can easily become intermingled with non-GMO's.[320] Nevertheless, McHughen thinks that segregation can be achieved as long as the price people are willing to pay for it is high enough, a price that will be passed on to consumers.[321] This raises the question of who should bear these costs, those who sell and buy GM foods – as these are the new addition to the market – or those that wish not to, and, therefore, demand labelling.

Thompson argues that lax labelling laws violate consumer sovereignty and religious liberty by preventing people from making informed food choices on the basis of their own moral or religious convictions.[322] According to Thompson, it is enough to show that some of the convictions that lie behind people's personal rejection of GM food are reasonable, to justify labelling. One need not agree with the substantive concerns people have to accept that they are reasonable. If people are not able to act according to their reasonable convictions, their liberty of conscience is violated. In other words, when it is decided that labelling is required on the basis of the product, the legitimate concerns that some have with the process by which the product was made – for instance environmental concerns –, are not taken into consideration, whereas according to the principle of consumer sovereignty they should. The Nuffield Council on Bioethics acknowledges the value of consumer choice, but argues that if the demand for choice to eat non-GM food is not based on safety concerns, labelling needs to be explained and justified, because otherwise it could be interpreted, as indeed US producers have, as simply a restraint on free trade.

---

[317] Mayer argues that opinion polls the world over have shown that consumers prefer labelling because they are concerned about the process and not only the product of genetic modification (Mayer (2002).

[318] Mayer (2002).

[319] Cohen (1998).

[320] Alan McHughen explains that segregation between certain products already occurs, for example when different grades of wheat are segregated from one another. Segregating GM from non-GM crops is more difficult, however, because it is not possible for the receiving party to determine whether a particular product is non-GM, and the system is, therefore, be open to abuse and might not be reliable. Also, McHughen argues that GM and non-GM varieties of crops can easily become intermingled, through pollen which blows from one field to the next and cross-pollinates plants, through the spilling of seeds from tractors, through the blending of different loads of grain with grains left behind in farm machinery, and through the accidental mislabelling of seed bags (McHughen 2000). Brookes, similarly, points out that transgenic crops can just not be grown in isolation; some of them will inevitably get out. Brookes (1998).

[321] McHughen (2000).

[322] Thompson (1997a).

The Council thus assumes that we know that GM food is safe and that demands for labelling, therefore, have to be based on other than safety concerns. As I will argue below, they indeed are often based on other concerns, but this does not alter the fact that many calls for labelling are actually based on safety concerns as well. The Council proceeds to argue that a 'balance has to be struck' between the 'costs to producers of offering the choice and the cost to consumers of foregoing it',[323] thereby assuming that financial costs to producers and autonomy of consumers are two similar categories that can simply be compared by a common standard. It also discusses the following objection:

> claiming a right to have a product made available *when the market would not otherwise have supplied it*, presents grave difficulties. It is one thing to insist that suppliers guarantee not to poison the customer; it is another to insist that companies should supply any particular range of products (my italics).[324]

This argument turns the objection against 'Hobson's choice' on its head; it presents the status quo as a world in which GM food is the standard. The Council here seems to miss the point that critics of biotechnology argue against the *introduction* of GM foods rather than for the distribution of non-GM foods. Currently consumers have the option, like they have always had, to eat non-GM food, and this is something which might be taken away.

Labelling is the topic of a conflict between the United States and Europe, with the US rejecting European labelling laws, because firstly, it is costly and creates a lot of extra work for food producers, and secondly, mandatory labelling might create a consumer bias against GM foods.[325] Biotechnology advocates claim that rather than aiming to protect consumer autonomy, countries use mandatory labelling in order to erect trade barriers.[326] Brom argues in this context that food is not only important to people because of its nutritious aspects, but also carries cultural and social significance. In international debates about food trade this latter dimension is often overlooked. This becomes clear, for example, when we take a look at the transatlantic trade dispute about the de facto moratorium since 1998 in Europe on the import of GM food. The European Union denies that a moratorium exists, but the United States, Canada, and Argentina claim that the delays on approving GM products from their countries effectively constitute a moratorium. These countries have brought the case to the WTO dispute settlement panel, which ruled in 2006 that there was an effective ban, and that this violated the WTO SPS Agreement, that this was driven by business protectionism rather than consumer and environmental protection, that it was not based on sound science, violated international trade rules and should, therefore, be lifted.[327] Green parties and environmentalists responded by saying that the United States is trying to force its unwanted products on European

---

[323] Nuffield Council on Bioethics (1999), page 9.

[324] Nuffield Council on Bioethics (1999), page 10.

[325] Brom (2004).

[326] See, for example, Australian Financial Review (2002).

[327] WTO Panel (2006).

Appendix B: Plant Biotechnology Debate 301

supermarket shelves and that the WTO ruling has shown that free trade overrides every other concern, such as health and environmental concerns, and discounts the precautionary principle.[328] Brom shows that this dispute is based on conflicting interpretations of the role of governments in food regulation. In the United States, only regulation surrounding food safety and nutrition is considered to be the state's mandate. In Europe, a broader view is taken, in which it is held that states should promote consumer autonomy – and, hence, issue labelling rules – and in which cultural and social meanings attached to food are taken into account. In Brom's eyes, there is a conflict here between the view of food as a mere commodity and food as expressing cultural and social values, and 'by ignoring this type of conflict, current international trade agreements force countries to present their dispute as a scientific disagreement on food safety', whereas Europe's resistance to GM food is not based solely on scientific arguments.[329] According to WTO rules, countries may only ban the import of certain foods if they can give sound scientific proof that human, animal, or plant health are endangered by the import, whereas other values besides safety are not sufficient. As was argued above, however, the question what constitutes sound science is in dispute.[330]

Environmental Consequences

Biotechnology advocates cite several ways in which genetic modification can help to ameliorate existing environmental problems. Because pesticides are built into plants, less chemical pesticides would have to be used and because crops are made herbicide resistant, spraying of herbicides will be more effective and, therefore, reduced. This will lead to less pollution of soils and waterways and will be beneficial to wildlife. Moreover, greenhouse gas emissions are reduced when less herbicides and pesticides will be sprayed, firstly, because less fuel is used and secondly, because genetically engineered herbicides and pesticides enable no-tillage or reduced-tillage farming, which in turn aids carbon sequestration.[331] Because of increased yields, less land will have to be exploited for agriculture. GM microorganisms and plants could be used to clean up oil spills. Furthermore, genetically engineered predators, such as mites and insects, can be used to counter crop pests.[332]

---

[328] Gillis and Blustein (8 February 2006) and Thomas (8 February 2006).

[329] Brom (2004), page 418.

[330] It can be taken to mean that rigorous, independent and interpersonally objective research has to be carried out. However, it is often considered to mean 'value-free' and 'correct' as well and this is more problematic. As was argued above, many disputes exist within the scientific community regarding the correctness of scientific findings, interpretations of research results, and even the problem definition in the first place and value judgments are involved in different stages of research.

[331] Brookes and Barfoot (2006).

[332] Goodman (1993). Understandably, concerns have been raised over this application, including concerns over the mobility of insects, and the uncontrollable appetite and behaviour of insects; they might shift their diet to nontarget species.

302   Appendix B: Plant Biotechnology Debate

Finally, plants could be genetically engineered to release viruses that could sterilize introduced animals that have become pests, such as foxes and rabbits in Australia.[333] Despite biotechnology advocates' assurances that after years of extensive growing of GM crops no environmental damage has been documented, critics raise a number of concerns relating to the environmental risks of crop biotechnology. They argue that the fact that no damage has been documented is no proof that this damage has not in fact occurred. Moreover, they point out that many field tests that have been carried out are inadequate. Some have even argued that these field trials have been carried out on a 'don't look, don't find basis'. This is because 'in many cases, adverse impacts are subtle and would almost never be registered scanning a field'.[334] In short, critics argue that biotechnology will lead to increased chemical use, which might also harm non-target species, and lead to pest resistance, that it will lead to cross hybridisation, which in turn could lead to the creation of 'superweeds', they believe it leads to a loss of biodiversity, and that it is not sustainable in the future. I will discuss each of these points in turn.

Herbicide and Pesticide Use

One of the most common claims of proponents of crop biotechnology is that it reduces the use of agro-chemicals, particularly chemical herbicides and pesticides, and thereby saves many human lives by eliminating pesticide poisoning, and decreases loss of biodiversity and poisoning of waterways.[335] Brookes and Barfoot, for instance, argue that over the last ten years of global growing of GE-crops, farmers have used 224 million kg less pesticides, primarily due to genetically engineered soy and cotton.[336] Critics argue, on the other hand, that the effects on chemical use are not so clear-cut; they argue that GM crops only serve to keep the 'herbicide and pesticide-treadmill' going.[337] They point out that herbicide tolerant crops are actually designed to be used only in combination with herbicides of the same company that provided the seeds and it is, therefore, not in the company's interest to reduce herbicide spraying. Moreover, the fact that crops are tolerant to herbicides means that farmers can spray as much as they like.[338] The industry's counter-claim is that farmers will not spray more than necessary, because they want to cut costs. Charles Benbrook argues that the scientific evidence does not support claims by the industry that GM crops have led to a decrease in herbicide and pesticide use; in

---

[333] Morell (1993).

[334] Anderson (2000), page 42.

[335] For example, Vasil (2003).

[336] Brookes and Barfoot (2006).

[337] McAfee (2003), Salleh (2000).

[338] Brookes and Coghlan (1998). Brookes and Coghlan write on page 46 that 'biotech's biggest money-spinners to date are crops that seem designed to keep farmers hooked on chemicals'...'in theory, [farmers] can spray as much as they want and not endanger their crop'.

Appendix B: Plant Biotechnology Debate 303

reality, the picture is much more complex and depends on what aspect of its use is examined, the amount of active ingredients or the number of pounds per acre sprayed.[339] Whatever measure employed, however, the numbers are not so significant that they should be cause for the optimistic claims by the industry. Moreover, Benbrook contrasts herbicide tolerant crop use with other weed management systems, and argues that farmers that invest in herbicide tolerant crops will rely more exclusively on herbicides than farmers that use 'multitactic integrated weed management systems'.[340] He does point out that despite small increases in herbicide use, plantings of Monsanto's Roundup-Ready (RR) soybeans have at least been successful commercially, because they have simplified weed management.[341] Another negative side-effect of herbicide tolerant crops that is sometimes mentioned is that these crops can actually grow in soil that has been contaminated with dangerous herbicides, leading to possible safety problems after consumption of these plants.[342] Moreover, after using Roundup the need to plough fields is reduced and this could play a role in soil erosion.[343] It is also argued that Roundup can kill a wider range of weeds than conventional herbicides, which is not such a problem in the US with its great wilderness areas that can conserve biodiversity. In Europe, however, the conservation of biodiversity is actually dependent on farmland.[344]

In the case of pesticides, applications of insecticide on transgenic corn have actually increased, whereas the effects of cotton are harder to establish. Despite the fact that according to the USDA figures, insecticide use on cotton has greatly been reduced in a number of states, the cause-effect relationship is not straightforward, because an aggressive program to eradicate exactly those insects targeted by cotton had already led to the reduction of these insects prior to the planting of transgenic cotton, and therefore, less insecticides were necessary. Similarly, in Australia, the prediction that pesticide spraying of transgenic cotton would decrease by 90% was not met, although in 1996–1997 there was a 52% decrease.[345] Benbrook cites a study by Lewis et al., based on 50 years of experience with pest management in general, and not just GM crops, that concluded that

> any toxin-based intervention will be met by "countermoves that neutralize their effectiveness".…. "The use of therapeutic tools, whether biological, chemical, or physical, as the primary means of controlling pests rather than as occasional supplements to natural

---

[339] Benbrook (2001), page 204. Official US Department of Agriculture (USDA) data based on four years of herbicide use shows that the number of active ingredients of herbicide used on a per acre-basis in the United States has slightly been reduced, while measured on the basis of pounds used per acre, herbicide use has slightly increased.

[340] Benbrook (2001), page 204. See also Salleh (2001).

[341] RR soy counters some of the problems farmers experience with low-dose herbicides, such as high costs, control difficulties, weed resistance, and crop damage (Benbrook 2001).

[342] Snow and Palma (1997).

[343] Brookes and Coghlan (1998).

[344] Brookes and Coghlan (1998).

[345] Salleh (2000). Salleh points out that this decrease did not translate into a higher yield and that some farmers complained that the advantages did not outweigh the extra costs they had made.

304 Appendix B: Plant Biotechnology Debate

regulators to bring them into acceptable bounds violates fundamental unifying principles and cannot be sustainable".[346]

In a similar vein, many people think that even if GM pesticides could be proven to be friendlier for the environment than traditional pesticides, the use of pesticides is not the right strategy in the first place.[347] In that case, traditional pesticides are not the correct standard with which to compare GM pesticides. Nevertheless, Benbrook does not conclude that transgenic crops should be abandoned altogether, but rather that they should only form a minor part of a total scheme of pest management: 'sophisticated pest management systems in the future will rely on biotech to help evoke, and sometimes strengthen, natural plant defence mechanisms'.[348]

The use of the soil bacterium *Bacillus thuringiensis* (Bt) as an inbuilt insecticide in crops such as cotton and corn is much debated. The environmentally benign Bt was originally only used sparingly by organic farmers in the form of a spray. According to biotech industry spokespersons, the advantage of building Bt into plants is that it remains effective for longer because it does not break down under the influence of UV light. Also, it is present at all times in the whole plant, so that caterpillars, the prime target of Bt, would be affected no matter what part of the plant they fed on.[349] The main drawback of the so-called 'biopesticides' is that pests can more easily become resistant. Because *Bt* is always present in much higher doses, the evolutionary pressures on pests are greater, as there are always a number of tolerant individuals that will mate and develop a new resistant strain. There will be fewer survivors if the dose is higher, but these individuals will also be tougher and hence more persistent.[350] The evolution of pest resistance is problematic for several reasons. Insect populations will be harder, if not impossible, to control and farmers will have to rely on more aggressive methods of pest control, which will be harmful for the environment. Pest problems can develop that are greater than the problems that farmers tried to solve in the first place. Pests might, for example, gain the capacity to feed on plant species that were previously unpalatable for them. Like many others, Snow and Palma, believe that pest resistance to the *Bt*-toxins is 'one of the most urgent ecological risks associated with transgenic plants'.[351] Moreover, it is argued that *Bt* resistance created by genetically modifying crops disadvantages organic farmers, because it renders useless one of the few pesticides they are allowed to use.[352] Several strategies can be employed to avoid or slow down pest resistance. By planting areas of nontransgenic crops in between transgenic crops selective pressure on the insects is varied, because they have more non-resistant insects to interbreed with. Another strategy is to make sure the pesticides are only

---

[346] Benbrook (2001). Benbrook cites Lewis et al. (1997).

[347] Jensen (2006).

[348] Benbrook (2001), page 207.

[349] Salleh (2001).

[350] Salleh (2001).

[351] Snow and Palma (1997), page 91.

[352] Hindmarsh (1991).

Appendix B: Plant Biotechnology Debate 305

expressed in some parts of the plant, such as the fruit or the young leaves. Also, if the concentration of the toxin is high enough it will kill all insects and thus no selection process will take place. A final strategy is to use a combination of pesticides, although this has been known to lead to multi-toxin resistance.[353] Generally, it is very difficult to predict the environmental results of different pest management regimes, as there are many unintended side-effects. For example, when one insect species is greatly reduced in number, this might allow a competitor to become more abundant, or the species that predated on the insect species might decline as well. Also, targeted insect species might move to other crops. However, these problems exist, perhaps even to a greater extent, with conventional broad-spectrum pesticides as well.[354] Finally, it should be noted that scientific evidence is put forward to show that *Bt* residues from *Bt*-corn can leach into soil, where it is still toxic for three weeks.[355]

Contamination

One possibly harmful effect of transgenic crops on the environment is caused by what some call 'contamination' or even 'genetic pollution'. Others consider these terms to be used rhetorically by biotech critics and the terms most scientists prefer are hybridisation or cross-fertilisation.[356] Hybridisation refers to the passing of genes from one organism to another. Two types of gene transfer can be distinguished; firstly, 'vertical gene transfer', which refers to sexual transfer through pollen or seeds. Pollen travel by wind and are carried by insects, and seeds are spilled in and around fields during harvest, and are spilled in transport.[357] The second type is 'horizontal gene transfer' which happens asexually when bacteria and viruses exchange genetic material and transfer this to plants. Some fear that this latter type of transfer will lead to transgenesis of non-target species and the unintended creation of novel genetic combinations that are potentially dangerous.[358] The 'escape' of transgenic plant genes could cause several problems: firstly, for farmers the potential danger is that the transgenic crops hybridise with wild relatives and this might make it more difficult to control these wild relatives, many of which farmers already consider to be weeds. Secondly, infestations of these weeds might occur in non-agricultural areas, such as along roadsides, in recreational areas, and

---

[353] Snow and Palma (1997).

[354] Snow and Palma (1997).

[355] However, it is not known what the effects of this are in terms of pest control or harm to non-target species (Mayer 2002, page 144).

[356] Hughes and Bryant, for example, liken the use of the words 'contamination' and 'pollution' to 'shouting 'fire' in a crowded room' (Hughes and Bryant 2002, page 134).

[357] Brookes (1998).

[358] McAfee (2003).

in nature reserves. This might lead to a decline of native species populations.[359] Thirdly, some argue that rare species of plants could be threatened with extinction if they are contaminated by genes from transgenic plants. The risk of contamination causes particular concerns for two groups. Firstly, the risk of contamination is higher in developing countries, because there are still many wild relatives and there is more uncultivated land nearby agricultural pastures.[360] According to Rosset, contamination of 'locally adapted crop varieties', especially in the megabiodiverse areas in the Third World, leads to a loss of available genetic resources and could actually threaten 'global food security'.[361] Secondly, if GM crops contaminate crops of organic farms, the latter might be robbed of their livelihood, as genetic modification is rejected by organic certifying associations.[362]

In reply, biotechnology advocates argue that there is no reason why hybridisation would be more common in the case of transgenic crops than in conventional crops.[363] Hughes and Bryant argue that cross-hybridisation between crops and their wild relatives is rare and even if this occurs, there is a good chance that the resulting hybrids will be infertile. Moreover, they think that if a weed would acquire a gene from a transgenic crop, this would usually make the weed's chances of survival slimmer, as it would make the plant 'less fit, competitive or invasive'. They think that the notion that 'superweeds' would spring up is exaggerated: many factors contribute to the spread of a species and just one new gene will not lead to an invasion by a superweed.[364] Alison Snow and Pedro Palma concur that proliferation of weeds has occurred only occasionally in the past through hybridisation between non-transgenic crops. However, they expect hybridisation to occur much more frequently after the commercial introduction of transgenic crops, because the exposition of weeds to novel traits will be greater.[365] Contrary to Hughes and Bryant's estimation, they think that there are transgenic traits that increase survival chances, such as disease-, stress-, or herbicide resistance, and these will increase the chances of contamination. It is commonly agreed by biologists that the small-scale field tests that have been done to examine the environmental risks of GM crops have not been designed

---

[359] Snow and Palma (1997).

[360] Conway (2000).

[361] Rosset (2006), page 90.

[362] According to the UK Soil Association, organic farmers are struggling after the widespread contamination of their crops, which has left them unable to sell their products on the market for organic goods. According to their report, many farmers' freedom of choice was limited, because contamination in effect forced them to proceed growing GM crops. See Meziani and Warwick (2002).

[363] It is argued that genetic modification in itself does not increase the risk of contamination; through traditional breeding, herbicide- and pest-resistant crops have already existed for decades, and this has not led to any great harms (Snow and Palma 1997, Hughes and Bryant 2002, Comstock 2000, Brookes 1998).

[364] One of the reasons 'contaminated' plants are less competitive in the wild is that they have to spend extra energy on their newly acquired trait, such as the production of herbicide-tolerance (Hughes and Bryant 2002, page 130).

[365] Snow and Palma (1997).

Appendix B: Plant Biotechnology Debate

to analyse the risks of extensive commercialisation.[366] Not only are the plots too small (<100 acres), but the duration (one or two seasons) is too short to examine effects on (often beneficial) nontarget species and to study the development of insect resistance.[367] Still, some proponents of biotechnology claim that the level of testing that GM crops are subjected to is 'far in excess of the testing to which similar plant varieties developed by breeding and selection are ever subjected ... transgenic crops/products are among the most exhaustively tested, characterized and regulated plants in history'.[368]

Loss of Biodiversity

A potential consequence of the contamination of centres of crop diversity by transgenes is a loss of biodiversity worldwide.[369] Many proponents as well as opponents of genetic engineering acknowledge that GM crops are associated with a loss of biodiversity, but many think this is due to the wider context of intensive agriculture in which genetic modification is applied and not to the technology per se.[370] An analogy is often made with the introduction and subsequent invasion of exotic plants or animals in the past, which have led to a loss of native species and the creation of new weeds.[371] Rifkin thinks that, ironically, while the biotech industry relies on genetic diversity for its raw materials, the practice of genetic engineering is likely to lead to genetic uniformity, because biotechnologists rely too much on the couple of genes that they can successfully engineer into a whole variety of plants.[372] According to Jorge Mayer, however, transgenesis does not automatically lead to the promotion of

---

[366] 'Monsanto, for example, tests engineered crops in plots the size of tennis courts before growing them commercially in fields that stretch beyond the horizon'. On small plots pollen turn out not travel further than three meters from the field's edge, whereas on commercial scale plots, pollen are known to travel up to two kilometers (Brookes 1998, page 39). Note that some critics of biotechnology, in fact, argue that field testing should not be carried out, but their reason is that these in themselves constitute an environmental release of GMOs that can have unpredictable consequences.

[367] Also, many possible effects are excluded from the field tests, while in test reports it is stated that these effects were not observed. Of course, when one does not look for certain effects, they will not be observed, but this is often taken to mean that they do in fact not occur (Snow and Palma 1997).

[368] Vasil (2003), page 850. Note that some critics of biotechnology, in fact, argue that field testing should not be carried out, but their reason is that these in themselves constitute an environmental release of GMOs that can have unpredictable consequences. See Hindmarsh (1991).

[369] Rifkin (1998).

[370] For example, Hughes and Bryant (2002).

[371] An example is the introduction of the blackberry into Australia, which quickly took over and changed whole ecosystems. See Bovenkerk et al. (2002).

[372] Rifkin (1998), page 108.

monocultures, because – as already mentioned in the context of Golden Rice – most transgenic traits can easily be bred into local plant varieties.[373]

Another possible cause of a loss of biodiversity is potential damage to non-target, often beneficial, species. This could disrupt the ecological balance between pests and beneficial insects, that often keep these pests in check.[374] A much cited study into the effects of *Bt*-corn showed that the larvae of monarch butterflies that fed on the corn were killed. These research findings were later questioned, because they were held not to reflect the conditions in the field.[375] Some interpreted the test results as showing the dangers of GM crops, whereas others take it to show the opposite. Either way, it is argued that this is just an example of unintended damage that was actually detected, while there may be many more effects on non-target species and accompanying long-term changes in ecosystems, that go undetected under current testing methods that focus on short-term harm to a limited set of organisms.[376] A problem of a loss of biodiversity is that the genetic uniformity caused by widespread planting of GM crops could in fact weaken a crop's defence mechanisms against disease and render them more prone to infestations by weeds and insect pests.[377]

Sustainability

Spokespersons from the biotech industry claim that sustainable agriculture is only possible with the use of biotechnology.[378] Critics label this hypocritical coming from an industry that uses many chemicals and heavily pollutes the environment. Moreover, they point out that biotech companies promote intensive monocultural agriculture, which environmentalists do not regard as sustainable.[379] Jack Kloppenburg and Beth Burrows argue that as current agricultural technologies have not been used in order to promote environmental goals, biotechnology cannot be expected to be deployed for this purpose either. They think that 'ecological and social sustainability follow from social arrangements, not from the technologies developed'.[380] They also argue that the biotech industry frames problems in terms of the symptom rather than of the underlying problem. For instance, Monsanto argues

---

[373] Mayer (2005).

[374] Brookes and Coghlan (1998).

[375] Even though within three metres from the pasture an increase in Monarch butterfly mortality was noted, and a small increase was present ten metres from the field edge, this represented just a small portion of the feeding area of the caterpillars (Hughes and Bryant 2002).

[376] McAfee (2003), page 208, note 3.

[377] In the summer of 1998, for example, on some farms in Missouri, half of the crops of GM soya that happened not to be resistant to mould died of mould infestation (Mack 1998, Salleh 2001).

[378] Kloppenburg and Burrows (1996). Kloppenburg and Burrows cite two top executives of Monsanto.

[379] Kloppenburg and Burrows (1996).

[380] Kloppenburg and Burrows (1996), page 62.

Appendix B: Plant Biotechnology Debate   309

that the Potato Beetle is a problem that needs to be solved with the help of biotechnology, while in their eyes the real problem is potato monoculture.[381] Critics argue
that more sustainable alternatives are available, such as integrated pest management
and organic farming, in which local peasants play a larger role.[382] One of the underlying causes of the different views on the sustainable promise of gene technology is
that different interpretations of sustainability are used. Some argue that sustainability refers to the capability to provide food security for future generations, or even
to economic sustainability, while others also look at the sustainability of a healthy
environment.

The Nuffield Council appears to take the first approach.[383] It does not expect
much gain for food security based on organic farming in developing countries,
where in a sense farmers are forced to grow crops organically because they cannot afford chemical inputs. But unlike organic farmers in the developed world, in
developing countries crop yields are too low, so leftovers cannot be used to fer-
tilise the soil, manure from livestock is either not available, or of poor quality and
is often used for fuel. Without these fertilisers the soils erode quickly. Also, poor
farmers often have no means to combat pest infestations.[384] Organic farmers, on
the other hand, argue that high-intensive monocultural farming appears to be economically effective, but in effect it is not, due to the many hidden costs of negative
side effects, such as health costs associated with chemical use, water testing and
treatment costs, salination, and blue green algae.[385] Trewavas argues, however, that
conventional farming methods are actually more sustainable and are better able
to conserve wildlife than organic farming methods, because conventional farming
uses land more efficiently, leaving more room for wildlife refuges and other nature
conserving land uses, such as willow plantations.[386] However, it is not clear what
Trewavas considers to constitute 'conventional farming' methods. He gives examples from conventional mixed farming in small plots, from traditional ley systems,
and from integrated management practices, which 'combines the best of traditional
farming with responsible use of modern technology'.[387] These are not representative of today's intensive, high input agriculture. Moreover, Trewavas approaches the
issue of the merits of conventional versus organic farming in a rationalistic scientific

---

[381] Kloppenburg and Burrows (1996).

[382] Rosset (2006), Hindmarsh (1991).

[383] Nuffield Council on Bioethics (1999).

[384] Nuffield Council on Bioethics (2003, section 4.4).

[385] Kinnear (1999).

[386] He holds that organic farming, which is based on the rejection of synthetic herbicides and pesticides, and soluble mineral inputs, leads to higher costs because the yield are lower and, therefore,
land is used less efficiently. Conventional farming methods can have the same yield as organic
farming, whilst using just 50–70% of the farmland. Moreover, as organic farmers cannot use synthetic herbicides, they often use mechanical weeding methods, which causes harm to birds, worms
and invertebrates, and uses fossil fuels. The natural pesticides they are allowed to use are often
unsafe for humans (Trewavas 2001).

[387] Trewavas (2001), page 410.

310 Appendix B: Plant Biotechnology Debate

way, whereas for the organic movement and organic consumers other issues play a more important role. Annette Mørkeberg and John Porter argue that his approach is highly positivistic, demanding strong scientific proof and eschewing anecdotal evidence, whereas proponents of organic farming 'rely on personal experiences and beliefs that make them more receptive to the idea that there is a difference between organic and conventionally produced food'.[388] They suggest that this topic should be dealt with taking into account a broader social context.

Due to the controversy over GM crops, many have also become critical of conventional intensive, chemically based, agricultural practices. On the one hand, some reject the idea that the only alternative for GM foods will be sold in health food stores, but on the other hand, many critics of GM food themselves make the case for more organic food and more funding of research into organic farming.[389] Some biotechnology advocates lament what they see as a tendency to blame biotechnology for industrializing agriculture and creating monocultures, while this process is in fact due to socio-economic forces rather than this new technology.[390] One can wonder whether this is really an accurate depiction of the critics' point of view. Most critics regard biotechnology as a continuation and an intensification of industrial agriculture, and they think that biotechnology will lead to even more monocultures, but they do not claim that industrial agriculture is *caused* by biotechnology. Some, however, argue that biotechnology and modern intensive agriculture are not necessarily co-dependent on each other. As more research is being done into tailoring biotechnological applications to local needs in the developing world, agricultural biotechnology does not necessarily seem to be a part of modern agriculture.[391]

Discussion

As should have become clear in the foregoing section, the debate about the merits of plant biotechnology is multifaceted and complex, and involves many different topics and subtle variations of opinion. Proponents focus on the benefits that biotechnology will bring about for agriculture and hold that there is no essential, morally relevant, difference between genetic modification and conventional breeding methods; only the speed and precision of genetic modification set it apart. They think that biotechnology can counter what they conceive as negative side-effects of the Green Revolution by providing a truly green form of agriculture that increases global yields and will provide food security. They also think that the environmental and safety risks of biotechnology tend to be overestimated and they propose to weigh the risk and benefits on a case-by-case basis; they argue against a blanket approach to all biotechnology applications.

---

[388] Mørkeberg and Porter (2001), page 677.

[389] Barry (2001).

[390] Nuffield Council on Bioethics (1999), page 16.

[391] Comstock (2000), page 182.

Appendix B: Plant Biotechnology Debate   311

Opponents do not regard the problems associated with the Green Revolution as mere side-effects, but as symptomatic of a misguided business-like approach to agriculture. They think that biotechnology cannot solve environmental or food insecurity problems, but will merely involve a continuation of intensive, high input, monocultural farming. The argument that no essential difference exists between conventional farming and agricultural biotechnology does not convince opponents, because even if they would accept this claim, many of them reject conventional agricultural methods in the first place. Their rejection of biotechnology should, therefore, be seen in the broader context of the relatively recent developments in agriculture. Opponents in this context also focus on the unequal distribution of benefits and burdens that is already present today, but which they believe will be intensified by agricultural biotechnology. They try to dismantle the biotech industry's claim that GM will feed the world, and argue that it will merely serve to increase the gap between rich and poor. Moreover, they are afraid that the concentration of power with private companies will lead to a monopoly that will be detrimental to diversity and consumer choice, and will disadvantage those who opt for alternative farming styles, such as organic farmers. They find it unfair that local farmers and indigenous people who have contributed to the improvement of crops, provided genetic knowledge, and preserved biodiversity are not acknowledged or compensated for their efforts. Finally, they are concerned that plant biotechnology will have unknown consequences, both for human and environmental health and they point out that risk analysis is not as objective and value-free as it is often presented.

One striking feature of the plant biotechnology debate is that many of the arguments appear to turn on issues of power and control. Both biotechnology advocates and its critics claim that self-interested or political motives lie behind their opponents' arguments; some of their accounts almost read like conspiracy theories. For example, the European moratorium on the import of transgenic crops is regarded as a 'protectionist manoeuvre' and a bowing down to certain political parties.[392] Critics of biotechnology, on their part, have argued that the public was left 'in the dark' about this new technology and its possible risks for as long as possible, in order to serve corporate interests.[393] Richard Hindmarsh has made an extensive study of the methods employed by those with 'vested interests' in biotechnology in order to push biotechnology developments forward, focusing on the Australian context. Some of the methods he describes are 'cooptation' of environmental groups onto gene technology committees; as these groups were always in the minority, their views could be said to be represented, but they would not have an effective voice in discussion outcomes. This strategy could be likened to 'repressive tolerance' as described by Herbert Marcuse.[394] Another strategy was the blocking of

---

[392] Vasil (2003).

[393] Hindmarsh (1994).

[394] See Herbert Marcuse (1969).

312   Appendix B: Plant Biotechnology Debate

certain influential persons that were critical of GM (such as Peter Singer) from committees. Further strategies were either blocking public education or public awareness campaigns by the government, or exactly the opposite, stimulating public 'indoctrination' under the guise of education.[395] As a result of these and other strategies, according to Hindmarsh, 'society is being masked to the fact that biotechnology, in particular genetic engineering, is only one of a number of possible pathways for development'.[396]

Analysis and Conclusion

What conclusions can we draw from the foregoing reconstruction of the debate about the merits of animal and plant biotechnology? First of all, it should be clear that we are dealing with a complex, multi-facetted and heated debate in which many different positions are defended, ranging from outright rejection to moderate criticism, and from a qualified support of some applications to an optimistic embrace of all forms of gene technology. In short, we are dealing with a very diverse debate which involves many disagreements on many different levels. The advent of biotechnology has not only led to a heated debate, but also to name calling, vocal protests, and acts of civil disobedience, or in the eyes of some, 'eco-sabotage'. Field tests of GM crops have been destroyed by action groups, such as Greenpeace, farmers who have discovered secret field trials of GM crops on the borders of their land are outraged, citizens' concerns have been responded to by calling the public irrational, and scientists critical of genetic modification have been ostracized.[397] It is my contention that biotechnology has given rise to such a heated controversy because a lot is at stake: not only economic interests and the health of people and their environment, but ultimately our deeply held beliefs and views about the kind of world we want to live in. Even though the debate is often constructed narrowly as being about the merits of a novel technology, the arguments on both sides apply to a wider group of ideas or concerns surrounding agriculture, socio-economic relations, nature conservation, medicalisation, and so on. Positions in this debate cannot be seen apart from broader views about what would constitute a desirable society, about what are proper relationships within society and with nonhuman nature, and even about our own nature. For example, many intrinsic objections, such as the objection to 'playing God' and the objection against the instrumentalisation of animals, refer to fundamental questions about the boundaries of scientific endeavour and our proper relationship to nonhumans. Also, one's estimation of the potential and dangers of biotechnology seem to be influenced by one's optimism or pessimism about technological developments and their perceived inevitability. Furthermore, how one

---

[395] Hindmarsh (1992, 1994).

[396] Hindmarsh (1994), page 448.

[397] See Chin (2000).

Appendix B: Plant Biotechnology Debate 313

estimates the potentials of, and even the need for, the application of gene technology to agriculture is strongly influenced by how one perceives the development of modern agriculture in general. This evaluation is supported by a Canadian study into public attitudes towards genetic engineering, which concluded that 'attitudes to genetic engineering were affected by a respondent's "core beliefs", that is, their knowledge of science and technology, attitude to nature and attitude to God'.[398]

An analysis comparing scientists' and non-scientists' stances on GM concluded that two different worldviews could be discerned.[399] The pro-GM stance was associated with a worldview that regarded natural objects as either 'devoid of purposes', but capable of being 'rendered with purposiveness by humans' or as possessing their own purposiveness, which may be legitimately changed by humans'. The anti-GM stance regarded nature as possessing its own purposiveness, which should not be interfered with by humans. Even though many of the assumptions underlying participants' views remain implicit in the debate, some participants also mention the broader context explicitly. For example, the Food Ethics Council, in its comments submitted to the Nuffield Council on Bioethics enquiry into GM crops, states that

> A key concern in the debate about the use of GM technology is the fundamental question of the kind of society that we wish to live. The past twenty years has seen an extraordinary growth in the power of recombinant DNA technology, a simultaneous massive increase in globalisation and the inter-dependence of economies, and a switch to the reliance on market forces to determine not only the provision of what were formerly public services but also the direction of public policy. We believe that this is a dangerous combination. It may lead to the marginalisation of minority and poorer communities, the short-term loss of biodiversity in agriculture, and the restriction of consumer choice.[400]

People's estimation of the merits of biotechnology, then, are dependent on their wider points of view. Several studies have concluded that very little genuine dialogue between opponents and proponents of biotechnology is possible and this contributes to the persistence of the disagreements.[401] Part of the problem seems to be that these groups argue on different levels. Geneticists for example, look at organisms on a genotypical level, while ecologists and many critics of biotechnology look at them on the phenotypical level. Similarly, Levidow and Carr argue that a lot of scientific disagreement over biotechnology arises from 'divergent cognitive frameworks': ecologists tend to view the environment as a 'fragile 'ecological balance'', whereas geneticists see it as 'resilient, capable of stabilizing itself'.[402] It appears that one major thread underlying the different viewpoints of opponents and proponents of biotechnology is whether they employ an atomistic or a holistic worldview. However, as noted before, it is not the case that all scientists working with genetic engineering are trapped, as it were, in a reductionist framework.

---

[398] Norton et al. (1998). Norton et al. cite Decima Research (1993).

[399] Deckers (2005), page 467.

[400] Lund and Mepham (1998).

[401] For example Norton (1998), Zoeteman et al. (2005).

[402] Levidow and Carr (1997), page 36.

Grice and Lawrence's study among Australian scientists found that the world-views or 'mindsets' of scientists are by no means coherent and do not fit perfectly into a reductionistic paradigm. Rather, scientists seem to work within an eclectic paradigm, sometimes displaying reductionistic elements and at other times holistic ones.[403] While we cannot say that scientists work from clear-cut atomistic or holistic worldviews, opponents and proponents tend to focus on opposite paradigms and worldviews when debating the merits of biotechnology. For example, their standpoint on the merits of biotechnology seems to depend on whether they see biotechnology as part of a larger – undesirable – development or just as one technology as such. This reconstruction of the biotechnology debate has made clear, then, that the different parties to the debate are not always on the same wavelength, so to speak. Some of the basic assumptions of the participants in the debate are contrary to each other – such as assumptions about the precision of rDNA techniques –, the participants tend to focus on different levels – for example, on an atomistic as opposed to a holistic level –, and they draw that boundary between science and ethics differently.

What this reconstruction has also aimed to make clear is that many arguments that are ostensibly limited to scientific or empirical questions in fact have moral aspects. An example of my claim is provided in the debate about the supposed unnaturality of crossing the species barrier. As I have argued, how one answers the question whether or not crossing the species barrier is natural is dependent on one's point of view. If one looks at the phenomenon on the genetic level, one sees a high level of similarity between species and indeed species boundaries seem artificial. If one regards species on the phenotypical level, species boundaries do become real and the natural crossing of species boundaries only occurs in closely related species. As I have also argued, the view that crossing species barriers is or is not natural in reality denotes the opinion that it is or is not 'good' or 'acceptable'; it is therefore not in the first place an empirical claim, but primarily a moral statement which refers to several other concerns people might have about GM. Another example of the moral aspects of ostensibly empirical discussions is that the questions whether biotechnology will further increase the gap between rich and poor and whether patenting is warranted in part depend upon moral assumptions about distributive justice. But perhaps the best example of the implicit value aspects of the biotechnology debate is provided by the debate surrounding risks. As I have pointed out, the framing of questions regarding risks tend to determine the outcomes of risk assessments and this framing carries – often implicit – value judgments. Framing determines amongst other things what types of consequences are considered, how much risk is deemed acceptable, and what benefits justify what level of risk taking. For example, does the risk of causing minor allergic reactions in some people weigh up to the benefits of genetically modifying food? The answer to this question is determined to a large extent on what one conceives to be these so-called benefits. Of course, if one does not share biotechnologists' optimism about these benefits, one is willing to accept a lot less risk, if any.

---

[403] Grice and Lawrence (2006/2007).

Similarly, the question of how many type-I errors we should allow in order to avoid type-II errors cannot be decided by empirical analysis or scientific research, but requires us to make a moral decision. The debate about risks also supports my first conclusion that broader points of view are involved: fundamental views about the status of scientific research and the level of precision of gene technology influence empirical assessments of the safety of GMO's, and these assessments are, therefore, not as straightforward as they are often presented.

Bibliography to Appendices A and B

Achterberg, Wouter. 1989. "Transgene Dieren: Ethische grenzen (Transgenic Animals: Ethical boundaries)." In *PAN-symposium "Biotechnologie met dieren: praktische, ethische en juridische problemen": De Queeste, Leusden, 11 maart 1989: teksten van de inleidingen*, ed. P.M. Schenk. Wageningen: Landbouw Universiteit Wageningen.
———. 1994. *Samenleving, Natuur en Duurzaamheld (Society, Nature and Sustainability)*. Assen: Van Gorcum.
Anderson, Luke. 2000. *Genetic Engineering, Food, and Our Environment. A Brief Guide*. Melbourne: Scribe Publications.
Attfield, Robin. 1998. "Intrinsic Value and Transgenic Animals." In *Animal Biotechnology and Ethics* (pp. 172–189), eds. A. Holland and A. Johnson. London: Chapman & Hall.
Australian Financial Review. 2002. "Sticky Labels: The New Barrier to Trade and Aid." *Australian Financial Review*, November 2.
Balzer, Phillipp, Klaus Peter Rippe, and Peter Schaber. 2000. "Two Concepts of Dignity for Humans and Non-Human Organisms in the Context of Genetic Engineering." *Journal of Agricultural and Environmental Ethics* 13:7–27.
Barinaga, Marcia. 2000. "Asilomar Revisited: Lessons for Today." *Science* 287 (5458):1584.
Barry, John. 2001. "GM Food, Biotechnology, Risk and Democracy in the UK: A Sceptical Green Perspective." In *ECPR (European Consortium for Political Research), Joint Sessions*. Grenoble.
Benbrook, Charles. 2001. "Do GM Crops Mean Less Pesticide Use?" *Pesticide Outlook*:204–7.
Bharathan, Geeta, Shanti Chandrashekaran, Tony May, and John Bryant. 2002. "Crop Biotechnology and Developing Countries." In *Bioethics for Scientists*, eds. J. Bryant, L. Baggott la Velle, and J. Searle. Chichester: Wiley.
Bovenkerk, Bernice, Frans W.A. Brom, and Babs J. Van den Bergh. 2002. "Brave New Birds. The Use of 'Animal Integrity' in Animal Ethics." *The Hastings Center Report* 32 (1):16–22.
Brom, Frans W.A. 1997a. *Onherstelbaar verbeterd. Biotechnologie bij dieren als een moreel probleem (Irrepairibly Improved. Animal Biotechnology as a Moral Problem)*. Utrecht: University of Utrecht.
———. 2003. "The Expressive-Communicative Function of Bio-Patent Legislation: The Need for Further Public Debate." In *Patente am Leben? Ethische, rechtliche und politische Aspekte der Biopatentierung*, eds. C. Baumgartner and D.H. Mieth. Paderborn, Germany: Mentis.
Brom, Frans W.A. 1997b. "Animal Welfare, Public Policy and Ethics." In *Animal Consciousness and Animal Ethics: Perspectives from the Netherlands*, eds. M. Dol, S. Kasanmoentalib, S. Lijmbach, E. Rivas, and R. Van den Bos. Assen: Van Gorcum.
———. 2004. "WTO, Public Reason and Food. Public Reasoning in the 'Trade Conflict' on GM-Food." *Ethical Theory and Moral Practice* 7 (4):417–31.
Brookes, Graham, and Peter Barfoot. 2006. "Global Impact of Biotech Crops: Socio-Economic and Environmental Effects in the First Ten Years of Commercial Use." *AgBioForum* 9 (3):139–51.
Brookes, Martin. 1998. "Running Wild." *New Scientist* 160 (2158):38–41.

318   Bibliography to Appendices A and B

Brookes, Martin, and Andy Coghlan. 1998. "Live and Let Live." *New Scientist* 160 (2158):46–9.
Bruce, Donald. 2001. "Finding a Balance over Precaution." *Journal of Agricultural and Environmental Ethics* 15 (1):7–16.
Buiatti, Marcello. 2005. "Biologies, Agricultures, Biotechnologies." *Tailoring Biotechnologies* 1 (2):9–30.
Burkhardt, Jeffrey. 2003. "The Inevitability of Animal Biotechnology? Ethics and the Scientific Attitude." In *The Animal Ethics Reader*, eds. S.J. Armstrong and R.D. Botzler. London/New York: Routledge.
Busch, Lawrence, Robin Grove-White, Sheila Jasanoff, David Winickoff, and Brian Wynne. 30 April 2004. "Amicus Curiae Brief Submitted to the Dispute Settlement Panel of the World Trade Organization in the Case of EC: Measures Affecting the Approval and Marketing of Biotech Products". http://www.lancs.ac.uk/fss/ieppp/wtoamicus/
Capron, Alexander M., and Renie Schapiro. 2001. "Remember Asilomar? Reexamining science's ethical and social responsibility." *Perspectives in Biology and Medicine* 44 (2):162.
Carman, Judy. 2004. "Is GM Food Safe to Eat?" In *Recoding Nature: Critical Perspectives on Genetic Engineering*, eds. R. Hindmarsh and G. Lawrence. Sydney: UNSW Press.
Chin, Geraldine. 2000. "The Role of Public Participation in the Genetically Modified Organisms Debate." *Environmental and Planning Law Journal* 17 (6):519.
Christie, Jean. 2001. "Enclosing the Biodiversity Commons: Bioprospecting or Biopiracy?" In *Altered Genes II. The future?*, eds. R. Hindmarsh and G. Lawrence. Melbourne: Scribe Publications.
Cohen, Phil. 1998. "Strange Fruit." *New Scientist* 160 (2158):42–5.
Comstock, Gary. 2000. *Vexing Nature? On the Ethical Case Against Agricultural Biotechnology*. Boston/Dordrecht: Kluwer.
Conway, Gordon. 2000. Crop Biotechnology: Benefits, Risks and Ownership. Paper Read at OECD Conference: The Scientific and Health Aspects of Genetically Modified Foods, 28 March, at Edinburgh, UK.
Conway, Gordon, and Gary Toenniessen. 1999. "Feeding the World in the Twenty-first Century." *Nature* 402:C55–C58.
Cooper, David E. 1998. "Intervention, Humility and Animal Integrity." In *Animal Biotechnology and Ethics*, eds. A. Holland and A. Johnson. London: Chapman & Hall.
Dawkins, Richard. 1976. *The Selfish Gene*. Oxford: Oxford University Press.
De Vries, Rob. 2006. "Genetic Engineering and the Integrity of Animals." *Journal of Agricultural and Environmental Ethics* 19:469–93.
Decima Research. 1993. *Final Report to the Canadian Institute of Biotechnology on Public Attitudes towards Biotechnology*. Ottawa: Canadian Institution of Biotechnology.
Deckers, Jan. 2005. "Are Scientists Right and Non-scientists Wrong? Reflections on discussions of GM." *Journal of Agricultural and Environmental Ethics* (18):451–78.
Douglas, Mary. 1986. *Risk Acceptability According to the Social Sciences*. New York: Russel Sage.
Eckersley, Robyn. 1992. *Environmentalism and Political Theory. Toward an Ecocentric Approach*. New York: UCL Press.
———. 2004. Biosafety and Ecological Security: Resisting the Trade in GM Food. Paper Read at International Sources of Insecurity Conference, 17–19 November, at Melbourne.
Eisenberg, Rebecca. 1996. "Patents: Help or Hindrance to Technology Transfer?" In *Biotechnology. Science, Engineering, and Ethical Challenges for the 21st Century*, eds. F.B. Rudolph and L.V. McIntire. Washington, DC: Joseph Henry Press.
European Commission. 2003. "Europeans and Biotechnology in 2002. Eurobarometer 58.0."
Fedoroff, Nina, and Nancy Marie Brown. 2004. *Mendel in the Kitchen. A Scientist's View of Genetically Modified Foods*. Washington, DC: Joseph Henry Press.
Food Ethics Council. 2003. *Engineering Nutrition. GM Crops for Global Justice?* Brighton: Food Ethics Council.

Bibliography to Appendices A and B

Fox, Michael. 1990a. "Transgenic Animals: Ethical and Animal Welfare Concerns." In *The Bio-Revolution. Cornucopia or Pandora's Box?*, eds. P. Wheale and R. McNally. London: Pluto Press.

———. 1990b. "Why BST Must Be Opposed." In *The Bio-revolution. Cornucopia or Pandora's Box?*, eds. P. Wheale and R. McNally. London: Pluto Press.

Frempong, Godfred. 2006. "Tailoring Biotechnology in Ghana: Implications for Genomics Development." *Tailoring Biotechnologies* 2 (1):51–62.

Frey, R.G. 1980. *Interests and Rights: The Case Against Animals*. Oxford: The Clarendon Press.

Gee, Henry. 2000. "Futurology: What Next?" *Guardian*, June 26, 14.

Genetic Resources Action International (GRAIN) (2001), *Grains of Delusion: Golden Rice Seen from the Ground*. This document was researched, written, and published as a joined undertaking between BIOTHAI (Thailand), CEDAC (Cambodia), DRCSC (India), MASIPAG (Philippines), PAN-Indonesia, and UBINIG (Bangladesh). Los Baños, Laguna, Philippines: MASIPAG

Gillis, Justin, and Blustein, Paul. 8 February 2006. "WTO Ruling Backs Biotech Crops." *Washington Post*. http://www.washingtonpost.com. Accessed 20 July 2006.

Goldhorn, Wolfgang. 1990. "The Welfare Implications of BST." In *The Bio-revolution. Cornucopia or Pandora's Box?*, eds. P. Wheale and R. McNally. London: Pluto Press.

Goodman, Billy. 1993. "Debating the Use of Transgenic Predators." *Science* 262 (5139):1507.

Grice, Janet, and Geoffrey Lawrence. 2006/2007. "Misreading Mindsets? Paradigms of Genetic Engineering Among Australian Scientists." *Tailoring Biotechnologies* 2 (3):87–106.

Grommers, F.J., L.J.E. Rutgers, and J.M. Wijsmuller. 1995. "Welzijn – Intrinsieke Waarde – Integriteit: Ontwikkeling in de Herwaardering van het Gedomesticeerde Dier" (Welfare – Intrinsic Value – Integrity: Developments in the revaluation of the domesticated animal). *Tijdschrift voor Diergeneeskunde* 120:490–4.

Grove-White, Robin, and Bronislaw Szerszynski. 1992. "Getting Behind Environmental Ethics." *Environmental Values* 1 (4):285–96.

Heeger, F. Robert. 1997. "Respect for Animal Integrity." In *Science, Ethics, Sustainability*, ed. A. Nordgren. Uppsala: Uppsala University.

Hindmarsh, Richard. 1991. "The Flawed "Sustainable" Promise of Genetic Engineering." *The Ecologist* 21 (5):196–205.

———. 1992. "CSIRO's Genetic Engineering Exhibition: Public Acceptance or Public Awareness?" *Search* 23 (7):212–3.

———. 1994. Power Relations, Social-Ecocentrism, and Genetic Engineering: Agro-Biotechnology in the Australian Context. Doctoral Thesis, Griffith University, Brisbane.

Hindmarsh, Richard, and Geoffrey Lawrence, eds. 2004. *Recoding Nature: Critical Perspectives on Genetic Engineering*. Sydney: University of New South Wales Press Ltd.

Hindmarsh, Sarah, and Richard Hindmarsh. 2002. *Laying the Molecular Foundations of GM Rice Across Asia*. Penang, Malaysia: Pesticide Action Network Asia and the Pacific.

Ho, Mae-Wan. 1999. *Genetic Engineering: Dream or Nightmare?* Second ed. Dublin: Gateway.

Ho, Mae-Wan. 1998. *Genetic Engineering Dream or Nightmare? The Brave New World of Bad Science and Big Business*. Bath: Gateway Books.

Holtung, N. 1996. "Is Welfare All That Matters in Our Moral Obligations to Animals?" *Acta Agriculturae Scandinavica* Sect. A, Animal Sci. Suppl. 27:16–21.

Hughes, Steve. 2002. "The Patenting of Genes for Agricultural Biotechnology." In *Bioethics for Scientists*, eds. J. Bryant, L. Baggott la Velle and J. Searle. Chichester: Wiley

Hughes, Steve, and John Bryant. 2002. "GM Crops and Food: A Scientific Perspective." In *Bioethics for Scientists*, eds. J. Bryant, L. Baggott la Velle, and J. Searle. Chichester: Wiley.

Jasanoff, Sheila. 2001. Citizens at Risk: Reflections on the US and EU. Paper Read at Third Congress of the European Society for Agricultural and Food Ethics. Food Safety, Food Quality and Food Ethics, 3–5 October, at Florence, Italy.

Jensen, Karsten Klint. 2006. "Conflict over Risks in Food Production: A Challenge for Democracy." *Journal of Agricultural and Environmental Ethics* (19):269–83.

320

Bibliography to Appendices A and B

Jochemsen, H., ed. 2000. *Toetsen en Begrenzen. Een ethische en politieke beoordeling van de moderne biotechnologie (Testing and Limiting. An Ethical and Political Assessment of Modern Biotechnology)*. Amsterdam: ChristenUnie.

Johnson, Lawrence. 1992. "Toward the Moral Considerability of Species and Ecosystems." *Environmental Ethics* 14:145–157.

Kanoute, Amadou. 2003. "GM 'Assistance' for Africa." *Nation* 277 (4):7–8.

Kinnear, Scott. 1999. "The Story So Far. Expert presentation." In *First Australian Consensus Conference on Gene Technology in the Food Chain*. Canberra.

Kloppenburg, Jack, Jr., and Beth Burrows. 1996. "Biotechnology to the Rescue? Twelve Reasons Why Biotechnology is Incompatible with Sustainable Agriculture." *The Ecologist* 26 (2):61–7.

Kloppenburg, Jack Ralph, Jr. 1988. *First the Seed. The Political Economy of Plant Biotechnology 1492–2000*. Cambridge: Cambridge University Press.

Kuiper, Harry A., Gijs A. Kleter, Hub P.J.M. Noteborn, and Esther J. Kok. 2002. "Substantial Equivalence – An Appropriate Paradigm for the Safety Assessment of Genetically Modified Foods?" *Toxicology* 181–182:427–31.

Lammerts van Bueren, Edith T., and Johannes Wirz. 1997. The Future of DNA Paper read at IfGene Conference on Presuppositions in Science and Expectations in Society, at Dornach, Switzerland.

Lebacqz, K. 2001. "Who "Owns" Cells and Tissues." *Health Care Analysis* 9: 353–67.

Levidow, Les. 2001. "The GM Crops Debate: Utilitarian Bioethics?" *A Journal of Socialist Ecology* 12:44–55.

Levidow, Les, and Susan Carr. 1997. "How Biotechnology Regulation Sets a Risk/Ethics Boundary." *Agriculture and Human Values* 14 (1):29–43.

Lewis, W.J., J.C. Van Lenteren, S.C. Phatak, and J.H. Tumlinson. 1997. "A Total System Approach to Sustainable Pest Management." *Proceedings of the National Academy of Sciences* 94 (12):243–8.

Lindhout, Pim, and Daniel Danial. 2006. "Participatory Genomics in Quinoa." *Tailoring Biotechnologies* 2 (1):31–50.

Love, Rosaleen. 2001. "Knowing Your Genes: Who Will Have the Last Laugh?" In *Altered Genes II: The Future?*, eds. R. Hindmarsh and G. Lawrence. Carlton North, VIC: Scribe Publications.

Lund, Peter, and Ben Mepham. 1998. "Comments of the Food Ethics Council Submitted to the Nuffield Council on Bioethics Enquiry into Genetically Modified Crops: Social and Ethical Issues."

Mack, Debbie. 1998. "Food for All." *New Scientist* 160 (2158):50–2.

Marcuse, Herbert. 1969. *A Critique of Pure Tolerance*. Boston: Beacon Press.

Marris, Claire. 2001. "Public Views on GMOs: Deconstructing the Myths. Stakeholders in the GMO Debate Often Describe Public Opinion as Irrational. But Do They Really Understand the Public?" *EMBO Reports* 2 (7):545–8.

Mayer, Jorge E. 2005. "The Golden Rice Controversy: Useless Science or Unfounded Criticism?" *BioScience* 55 (9):726–7.

Mayer, Sue. 2002. "Questioning GM Foods." In *Bioethics for Scientists*, eds. J. Bryant, L. Baggott la Velle, and J. Searle. Chichester: Wiley.

Mayer, Sue, and Andy Stirling. 2002. "Finding a Precautionary Approach to Technological Developments – Lessons for the Evaluation of GM Crops." *Journal of Agricultural and Environmental Ethics* 15 (1):57–71.

McAfee, Kathleen. 2003. "Neoliberalism on the Molecular Scale. Economic and Genetic Reductionism in Biotechnology Battles." *Geoforum* 34:203–19.

McHughen, Alan. 2000. *A Consumer's Guide to GM Food: From Green Genes to Red Herrings*. Oxford: Oxford University Press.

Meyer, Gitte. 2001. "Fighting Poverty with Biotechnology? Report from a Copenhagen Workshop on Biotechnology and the Third World." Copenhagen: University of Copenhagen. Centre for Bioethics and Risk Assessment.

Bibliography to Appendices A and B

Meziani, Gundula, and Hugh Warwick. 2002. *Seeds of Doubt. Executive Summary Briefing Paper.* Soil Association 2002 [cited September 17 2002].

Midgley, Mary. 1983. *Animals and Why they Matter.* Harmondsworth: Penguin Books.

———. 2000. "Biotechnology and Monstrosity." *Hastings Center Report* 30 (5):7–15.

Millar, Kate, and Sandy Tomkins. 2006. The Implications of the Use of GM in Aquaculture: Issues for international development and trade. Paper read at Ethics and the Politics of Food. European Society for Agricultural and Food Ethics, 21–24 June, at Oslo.

Miller, Henry I. 1999. "Substantial Equivalence: Its Uses and Abuses." *Nature Biotechnology* 17 (11):1042–3.

Miller, Henry I., and Douglas Gunary. 1993. "Serious Flaws in the Horizontal Approach to Biotechnology Risk." *Science* 262 (5139):1500–1.

Millstone, Erik. 2006. Can Food Safety Policy-Making be Both Scientifically and Democratically Legitimated? If So, How? Paper read at Sixth Congress of the European Society for Agricultural and Food Ethics. Ethics and the Politics of Food, June 22–24, at Oslo, Norway.

Millstone, Erik, Eric Brunner, and Sue Mayer. 1999. "Beyond 'Substantial Equivalence'." *Nature* 401:525–6.

Morell, Virginia. 1993. "Australian Pest Control by Virus Causes Concern." *Science* 261 (5122):683–4.

Mørkeberg, Annette, and John R. Porter. 2001. "Organic Movement Reveals a Shift in the Social Position of Science." *Nature* 412:677.

Norton, Janet. 1998. "Throwing up Concerns about Novel Foods." In *Altered Genes. Reconstructing Nature: The Debate*, eds. R. Hindmarsh, G. Lawrence, and J. Norton. St. Leonards, NSW: Allen & Unwin.

Norton, Janet, Geoffrey Lawrence, and Graham Wood. 1998. "Australian Public's Perception of Genetically-Engineered Foods." *Australasian Biotechnology* 8 (3):172–81.

Nuffield Council on Bioethics. 1999. *Genetically Modified Crops: The Ethical and Social Issues.* London: Nuffield Council on Bioethics.

———. 2003. "The Use of Genetically Modified Crops in Developing Countries. A follow-up discussion paper."

OECD. 1986. *Recombinant DNA Safety Considerations.* Paris: OECD.

Ortiz, Sarah Elizabeth Gavrell. 2004. "Beyond Welfare. Animal Integrity, Animal Dignity, and Genetic Engineering." *Ethics and the Environment* 9 (1):94–120.

Paula, Lino, and Frans Birrer. 2006. "Including Public Perspectives in Industrial Biotechnology and the Biobased Economy." *Journal of Agricultural and Environmental Ethics* 19:253–67.

Price, S.C. 1999. "Public and Private Plant Breeding." *Nature Biotechnology* 17:938.

Pluhar, Evelyn. 1986. "The Moral Justifiability of Genetic Manipulation." *Between the Species* 2:136–138.

Regan, Tom. 1984. *The Case for Animal Rights.* London: Routledge and Kegan Paul.

Regouin, Eric, and Frank Tillie. 2003. "Een Schaap met Vijf Poten? Stand van zaken in onderzoek en ontwikkeling van niet-biomedische toepassingen van biotechnologie bij dieren (A Five-legged Sheep? State of affairs in research and development of non-biomedical applications of animal biotechnology)." *Expertisecentrum LNV.*

Reiss, Michael J., and Roger Straughan. 1996. *Improving Nature? The Science and Ethics of Genetic Engineering.* Cambridge: Cambridge University Press.

Rifkin, Jeremy. 1998. *The Biotech Century. Harnessing the Gene and Remaking the World.* London: Victor Gollancz.

Robins, Rosemary. 2001. "Overburdening Risk: Policy Frameworks and the Public Uptake of Gene Technology." *Public Understanding of Science* 10:19–36.

Roeser, Sabine. 2006. Moral Intuitions and Risk Evaluations. Paper read at Ethical Aspects of Risk, 14–16 June, at Delft, the Netherlands.

Rogers, Nicole. 2002. "Seeds, Weeds and Greed: An Analysis of the Gene Technology Act 2000 (Cth), Its Effect on Property Rights, and the Legal and Policy Dimensions of a Constitutional Challenge." *Macquarie Law Journal* 2:1–30.

Roht-Arriaza, Naomi. 1996. "Of Seeds and Shamans: The Appropriation of the Scientific and Technical Knowledge of Indigenous and Local Communities." *Michigan Journal of International Law* 17:919–65.

Rollin, Bernard E. 1986. The Frankenstein Thing. Paper read at Genetic Engineering of Animals. An agricultural perspective. Symposium on genetic engineering of animals, September 9–12, at University of California, Davis.

———. 1995. *The Frankenstein Syndrome: Ethical and Social Issues in the Genetic Engineering of Animals*. New York: Cambridge University Press.

———. 1998. "On *Telos* and Genetic Engineering." In *Animal Biotechnology and Ethics*, eds. A. Holland and A. Johnson. London: Chapman & Hall.

Rolston, Holmes. 2002. "What Do We Mean by the Intrinsic Value and Integrity of Plants and Animals?" In *Genetic Engineering and the Intrinsic Value and Integrity of Animals and Plants*, eds. D. Heaf and J. Wirtz. Hafan: Ifgene.

Rosset, Peter M. 2006. "Genetically Modified Crops for a Hungry World: How Useful Are They Really?" *Tailoring Biotechnologies* 2 (1):79–94.

Rutgers, L.J.E., and F.R. Heeger. 1999. "Inherent Worth and Respect for Animal Integrity." In *Recognizing the Intrinsic Value of Animals: Beyond Animal Welfare*, eds. M. Dol et al. Assen: Van Gorcum.

Salleh, Ariel. 2000. "The Meta-Industrial Class and Why We Need It." *Democracy and Nature* 6 (1):27–36.

Salleh, Anna. 2001. "Wearing Out Our Genes? The Case of Transgenic Cotton " In *Altered Genes II: The Future?*, eds. R. Hindmarsh and G. Lawrence. Melbourne, Australia: Scribe Publications.

Shiva, Vandana. 2000. *Stolen Harvest. The Hijacking of the Global Food Supply*. Cambridge, MA: South End Press.

Singer, Peter. 1975. *Animal Liberation: A new ethics for our treatment of animals*. New York: New York Review.

———. 1979. *Practical Ethics*. Cambridge: Cambridge University Press.

Smith, Neil. 2003. "Farmers Risk Irrelevancy If They Fail to Involve Themselves." *Australian Farm Journal* 13 (10):16–8.

Snow, Allison A., and Pedro Moran Palma. 1997. "Commercialization of Transgenic Plants: Potential Ecological Risks." *BioScience* 47 (2):86–97.

Sterckx, Sigrid. 2004. Ethical Aspects of the Legal Regulation of Biodiversity. Paper read at EurSafe, at Leuven, Belgium.

Straughan, Roger. 1992. *Ethics, Morality and Crop Biotechnology*. UK: Reading University.

Sunstein, Cass R., and Martha C. Nussbaum, eds. 2004. *Animal Rights. Current Debates and New Directions*. Oxford: Oxford University Press.

Thomas, Leigh. 8 February 2006. "Europe Downplays WTO Ruling Genetically Modified Crops." *Farmnews*. http://www.terradaily.com. Accessed on 20 July 2006.

Thompson, P.B. 1997a. "Food Biotechnology's Challenge to Cultural Integrity and Individual Consent." *Hastings Center Report* 27 (4):34–8.

———. 1997b. "Ethics and the Genetic Engineering of Food Animals." *The Journal of Agricultural and Environmental Ethics* 10:1–23.

Thompson, Paul B. 2005. "Research Ethics for Animal Biotechnology", https://www.agralin.nl/ojs/index.php/frontis/article/download/879/445

Tilman, D. 1998. "The Greening of the Green Revolution." *Nature* 396:211–2.

Trewavas, A. 2002. "Malthus Foiled Again and Again." *Nature* 418:668–70.

Trewavas, Anthony. 1999. "Much Food, Many Problems." *Nature* 402:231–2.

———. 2001. "Urban Myths of Organic Farming. Organic Agriculture Began As an Ideology, But Can It Meet Today's Needs?" *Nature* 410:409–10.

Trias, José E. 1996. "Conflict of Interest in Basic Biomedical Research." In *Biotechnology. Science, Engineering, And Ethical Challenges for the 21st Century*, eds. F.B. Rudolph and L.V. McIntire. Washington, DC: Joseph Henry Press.

Bibliography to Appendices A and B

Turner, Gillian, and Brian Wynne. 1992. "Risk Communication: A Literature Review and Some Implications for Biotechnology." In *Biotechnology in public. A review of recent research*, ed. J. Durant, 109–141. London: Science Museum for the European Federation of Biotechnology.

Van den Belt, Henk. 2003. "Enclosing the Genetic Commons: Biopatenting on a Global Scale." In *Patente am Leben? Ethische, rechtliche und politische Aspekte der Biopatentierung*, eds. C. Baumgartner and D.H. Mieth. Paderborn, Germany: Mentis.

Van den Belt, Henk, and Bart Gremmen. 2002. "Between Precautionary Principle and "Sound Science": Distributing the Burdens of Proof." *Journal of Agricultural and Environmental Ethics* 15:103–22.

Van Staveren, G.D. 1991. *Het overschrijden van soortgrenzen. De rol van het soortbegrip binnen de discussie rond genetische manipulatie (Crossing of Species Barriers. The role of the notion of species in the discussion about genetic manipulation)*.

Van Willigenburg, Theo. 1997. "Natuurberoep in de Bio-ethiek en Verantwoordelijkheid voor Handelen (The Appeal to Nature in Bio-ethics and the Responsibility for Action)." In *Museum Aarde. Natuur: Criterium of Contructie?*, eds. J. Keulartz and M. Korthals, 166–80. Meppel: Boom.

Vandeveer, Donald. 1986. "Interspecific Justice." In *People, Penguins, and Plastic Trees. Basic Issues in Environmental Ethics*, eds. D. Vandeveer and C. Pierce, 85–99 Belmont: Wadsworth.

Vasil, Indra K. 1998. "Biotechnology and Food Security for the 21st Century: A Real-World Perspective." *Nature Biotechnology* 16:399–400.

———. 2003. "The Science and Politics of Plant Biotechnology – A Personal Perspective." *Nature Biotechnology* 21:849–51.

Verhoog, Henk. 1991. "Genetische Manipulatie van Dieren (Genetic Manipulation of Animals)." *Filosofie in Praktijk* 12 (2):87–106.

———. 1992. "The Concept of Intrinsic Value and Transgenic Animals." *Journal of Agricultural and Environmental Ethics* 5 (2):147–60.

———. 1998. "Xenotransplantatie: Het beest in de mens (Xenotransplantation: The Animal in Humans)." In *En toen was er DNA.... wat moeten we ermee? Over morele en maatschappelijke dilemma's (And then there was DNA....what to do with it? About moral and social dilemmas)*, eds. E.T. Lammerts van Bueren and J. Van der Meulen. Zeist, Netherlands: Indigo.

Vint, Robert. 2002. "Force-Feeding the World. America's 'GM or Death' Ultimatum to Africa Reveals the Depravity of Its GM Marketing Policy." *Genetic Food Alert* 1. http://www.ukabc.org/forcefeeding.pdf.

Vorstenbosch, Jan. 1993. "The Concept of Integrity. Its Significance for the Ethical Discussion on Biotechnology and Animals." *Livestock Production Science* 36:109–12.

Wambugu, Florence. 1998. 'Benefits and risks of genetically modified crops, CERES Forum on Food Products from Plant Biotechnology II, 8–9 June, Berlin, Germany.

Wambugu, Florence. 1999. "Why Africa Needs Agricultural Biotech." *Nature* 400:15–6.

Wenz, Peter. 1988. *Environmental Justice*. Albany, NY: State University of New York Press.

Wills, P.R. 2001. "Disrupting Evolution: Biotechnology's Real Result." In *Altered Genes II*, eds. R. Hindmarsh and G. Lawrence. Melbourne: Scribe Publications.

———. 2002. Biological Complexity and Genetic Engineering. Paper read at Environment, Community and Culture, at Brisbane.

Wilmut, Ian. 1995. "Modification of Farm Animals by Genetic Engineering and Immunomodulation." In *Issues in Agricultural Bioethics*, eds. T.B. Mempham, G.A. Tucker, and J. Wiseman, 229–46. Nottingham: Nottingham University Press.

WTO Panel. 2006. "European Communities – Measures Affecting the Approval and Marketing of Biotech Products (DS291, DS292, DS293)."

Wynne, B. 2001. Public Lack of Confidence in Science? Have We Understood Its Causes Correctly? Paper read at EurSafe 2001: Food Safety, Food Quality and Food Ethics, 3–5 October, at Florence, Italy.

Wynne, Brian. 1996. "May the Sheep Safely Graze? A Reflexive View of the Expert-Lay Knowledge Divide." In *Risk, Environment and Modernity: Towards a New Ecology*, eds. S. Lash, B. Szerszynski, and B. Wynne. London: Sage.

Zoeteman, Bastiaan C.J., Miranda Berendsen, and Pepijn Kuyper. 2005. *Biotechnologie en de Dialoog der Doven. Dertig jaar genetische modificatie in Nederland (Biotechnology and the Dialogue of the Deaf. Thirty years of genetic modification in the Netherlands)*. Bilthoven: Commissie Genetische Modificatie.

Websites

http://www.agresearch.co.nz (accessed on 1/5/2005)
http://www.ausbiotech.org/ (accessed on 22/5/2006)
http://www.austmus.gov.au/consensus (accessed on 17/11/2005)
http://www.gene.ch/genet/2000/Dec/msg00034.html (accessed on 25/10/2006)
http://www.isaaa.org (accessed on 22/5/2006)
http://www.niaba.nl/sa_files/niaba_biotechnologie.pdf (accessed on 1/5/2005)
http://www.terradaily.com (accessed on 20/7/2006)
http://www.washingtonpost.com (accessed on 20/7/2006)
http://filer.case.edu/~ijd3/authorship/neem.html (accessed on 16/05/2007)

Appendix C
Interview Questions

Questions Pertaining to Members of Ethics Committees

1. The use of biotechnological procedures with animals and plants has given rise to moral concerns within society. What moral (as opposed to legal, technical or practical) concerns have been discussed within your committee? In your eyes, what moral concerns have been overlooked?
2. In your eyes, which of these moral concerns have been dealt with adequately and which have not?
3. (Specifically for the Committee for Animal Biotechnology) The CAB judges the acceptability of an application for biotechnological procedures with animals according to a '5 step plan'. The first 4 steps can independently lead to a rejection of the application and in the absence of such a rejection, in the fifth step the importance of the goal of the application is weighed against the moral objections.

   a. Do you think that all the relevant moral concerns are dealt with within this step plan?
   b. The CAB has been criticized for not following this step plan but rather making an integral judgment (which means that a 'yes, if' and not a 'no, unless' policy has been followed). Do you agree with this criticism?

4. Are you satisfied with the terms of reference of your committee, as set by the Minister? In your view, what are the advantages and disadvantages of these terms of reference?
5. Your committee is composed of experts from different fields relevant to animal or plant biotechnology and ethics, who are appointed in a personal capacity.

   a. Do you think the members adequately represent the opinions present in their field of expertise?
   b. Do you think the members are implicitly understood as representing certain segments within society?
   c. Do you think the Minister knows the initial stance on biotechnology of each member prior to his or her appointment?

6. Do you think that members of the lay public should be present in the composition of your committee? Why?

7. (Specifically for the CAB) Do you consider the public consultation process as laid down in the Decree for Animal Biotechnology (Animal Health and Welfare Law, article 66) to be a satisfactory way of involving the public in the decision making procedure about biotechnological procedures?

8. (Specifically for the CAB) The CAB has been given three distinct functions. Firstly, it has to clarify and strengthen the moral position of animals in view of modern biotechnology. Secondly, it in effect acts as a licensing authority by advising the Minister of Agriculture on specific license applications for biotechnological procedures with animals. Thirdly, it has to identify and offer recommendations about morally problematic developments in the field of biotechnology at an early stage and thereby stimulate public discussion.

   a. To what extent do you think the CAB has fulfilled each of these roles?
   b. Do you consider these different roles to be compatible?

9. What role do you think ethics committees should play within public debate?

10. After the evaluation of the functioning of the CAB (in 2000) the Minister has decided to supplement the public consultation procedure with six-monthly public debates.

   a. Have these debates been held already? If yes:
   b. To your knowledge, was there major public interest or were mainly specialized interest groups present?
   c. What are your observations about the debates?

11. From research commissioned by the Rathenau Institute (Paula, 2001) it becomes clear that lay persons who are asked to review licence applications for animal biotechnology are generally more critical and more likely to give a negative recommendation than the (majority of) the CAB (In case of CAB). How do you explain this difference? (In case of other committees) Do you think this difference also exists in relation to your committee?

12. A group of scientists has put forward an alternative view of genetics, called Fluid Genomics, in which a more holistic picture of genes and their surroundings is painted than in mainstream genetics.

   a. Has this view ever been discussed within your committee?
   b. Do you consider room for expert controversy desirable within committee discussions?

13. In your opinion, has your committee adequately dealt with technological uncertainties? For example, the CAB has advised the Minister that it cannot determine whether biotechnology causes harm to the welfare or health of invertebrates and therefore has advised that experiments with invertebrates should be exempt from licensing.

14. Some people argue that in the case of technological uncertainties, policies should be based on public debate. What are your thoughts on this?

Appendix C: Interview Questions

15. In your view, is it always desirable for ethics committees to strive for consensus or do you think that majority/ minority recommendations are also valuable?
16. In some of its recommendations, the CAB points out that there are groups in society that will always have objections against biotechnological procedures with animals, but that what matters is how grave these objections are in view of the possible benefits of the procedures. What do you think about this statement?
17. In general, how do you think that ethics committees should deal with persistent moral disagreement in society?
18. How many negative compared to positive recommendations have been given?

Questions Pertaining to Representatives of Interest Groups

1. What is the name of your organization in English?
2. The use of biotechnological procedures with animals and plants has given rise to moral concerns within society. What moral concerns can you identify?
3. Do you think these moral concerns have been dealt with adequately by ethics committees, in particular by the Committee for Animal Biotechnology (CAB) and Committee Genetic Modification (COGEM) in the Netherlands and the Gene Technology Ethics Committee (GTEC) in Australia? In your eyes, what moral concerns have been overlooked?
4. (Specifically for the Animal Protection Association in the Netherlands) The CAB judges the acceptability of an application for biotechnological procedures with animals according to a '5 step plan'. The first 4 steps can independently lead to a rejection of the application and in the absence of such a rejection, in the fifth step the importance of the goal of the application is weighed against the moral objections.

   a. Do you think that all the relevant moral concerns are dealt with within this step plan?
   b. The CAB has been criticized for not following this step plan but rather making an integral judgment (which means that a 'yes, if' and not a 'no, unless' policy has been followed). Do you agree with this criticism?

5. Ethics committees regarding animal and plant biotechnology are composed of experts from different relevant fields, who are appointed in a personal capacity.

   a. In your view, is this a balanced composition?
   b. Do you think the composition of these committees should be supplemented by representatives of interest groups or lay persons?

6. Is it true that representatives of your organization were invited to take place in animal experiments committees, but declined this invitation?
7. Do you consider a public consultation process where members of the public can submit comments to draft proposals to be a satisfactory way of involving

the public in the decision making procedure about biotechnological procedures? Do you feel that your comments have been given serious review?

8. Do you think ethics committees have functioned as stimulator of public debate? More generally, what role do you think ethics committees should play within public debate?
9. In the case of majority/ minority recommendations by ethics committees, the Minister tends to adopt the view of the majority. Do you think this is justified?
10. (Specifically for the Animal Protection Association) After the evaluation of the functioning of the CAB (in 2000) the Minister has decided to supplement the public consultation procedure with six-monthly public debates.

    a. Have these debates been held already? If yes, have you attended theses?
    b. To your knowledge, was there major public interest or were mainly specialized interest groups present?
    c. What are your observations about the debates?
    d. What role do you think the outcomes of the debates will play vis-à-vis the decisions made by the Minister?

11. What moral disagreements exist between the ethics committees and your interest group? In your eyes, are these disagreements based on fundamental value conflicts?
12. What scientific or technological disagreements exist between the committees and your interest group? How do you explain these disagreements?
13. A group of scientists has put forward an alternative view of genetics, called Fluid Genomics, in which a more holistic picture of genes and their surroundings is painted than in mainstream genetics. Have you heard of this view? If so, what is your opinion of it?
14. In general, how do you think society should deal with persistent moral disagreements?

Questions Pertaining to Organizers of Public Debates

1. In general, what issues have been the subject of organized public debates?
2. What type of public debates have you organized? For instance, have you organized consensus conferences, citizens' juries, deliberative polls or other types of public debate?
3. Can you describe the different steps in the organized public debate process?
4. What role did the media play regarding the public debates you have organized?
5. What role did (scientific) experts play in the public debates you have organized? For example, was there open expert controversy?
6. How have you selected the participants of the public debates you have organized?
7. Did the participants mainly consist of lay persons, mainly of representatives of specialized interest groups, mainly of experts, or a combination of these?

Appendix C: Interview Questions

8. (Only in the Netherlands) After the evaluation of the functioning of the Committee for Animal Biotechnology (in 2000) the Minister for Agriculture has decided to supplement the public consultation procedure, as laid down in Decree for Animal Biotechnology, with six-monthly public debates.

   a. Have these debates been held already? If yes:
   b. To your knowledge, was there major public interest or were mainly specialized interest groups present?
   c. What are your observations about the debates?
   d. Do you think the six-monthly debates should supplement or replace the public consultation procedure?

9. What moral (as opposed to practical, legal, or technical) disagreements did you encounter in public debates regarding genetic engineering and cloning? In your eyes, were these disagreements based on fundamental value conflicts or on more superficial differences (such as different interpretations of the facts)?

10. What disagreements, if any, about scientific facts did you encounter? How do you explain these disagreements?

11. In general, how do you think society should deal with persistent moral disagreements?

Index

A

Ackerman, Bruce, 64, 66, 70, 72, 76, 83, 91–92, 128

Active states, 197

Activism, 14, 103, 119, 124, 133–134, 143, 236

Aggregation, 4, 12, 64, 79–86, 97, 100, 121, 224

Agonism, 112–113, 121, 124

Agonistic, 93, 111

Animal biotechnology, 21, 35–36, 47, 158–159, 164–166, 170, 177–178, 181, 184, 209, 245–247, 249–252, 256–259, 283

Animal Health and Welfare Law, 158

Animal integrity, 35–36, 166, 172, 177, 252–258

Animal welfare, 35–36, 55, 172, 249–250, 252–253, 257, 259

Australian Broadcasting Corporation, 9, 217–218

Australian Museum, 199, 201, 206, 212

B

Bargaining, 4–5, 11–13, 64, 79–81, 84, 86, 97, 100, 102, 104, 132, 134

Benhabib, Seyla, 97, 100, 104, 109–110, 128–129, 131

Biodiversity, 42, 47–48, 161, 263, 279–280, 284–285, 302–303, 307–308, 311, 313

Biopiracy, 29, 39, 179, 213, 283–287

Bohman, James, 5, 81, 85, 92, 104, 108, 140

Broadening of the debate, 105, 110, 164, 205, 209

Brom, Frans W.A., 33, 35–37, 39–40, 65, 74, 96, 159, 166–167, 171, 246, 248–249, 252–253, 255, 25, 280, 282–283, 293–294, 296, 300–301

C

Chambers, Simone, 101, 121

Citizens' jury, 192–193

Cloning, 1, 9, 19–21, 178, 189–190, 199, 202–205, 208–209, 215–216, 222, 240, 246, 329

Cohen, Joshua, 25, 29, 96, 104, 110, 126, 135, 287–288, 299

Committee for Animal Biotechnology, 9, 36, 142, 157–161, 165–166, 176, 255–256, 325, 327, 329

Comparative analysis, 5–7, 10, 92, 152–157, 190, 194–198, 223, 225–226, 236, 238–242

Connolly, William, 81

Consensual style democracy, 183

Consensus, 2–3, 6, 8–10, 12–15, 22, 40, 47, 51–54, 56, 69, 93, 104–106, 110–114, 122–127, 133–136, 138–143, 153–155, 159, 164–165, 167, 172, 183, 185, 190–194, 196–203, 205–230, 235–242, 244, 249, 269, 277, 288, 294, 327–328

Consensus conference, 9–10, 54, 114, 127, 141–142, 191–193, 197, 199–202, 206–208, 210–214, 216–222, 226, 228, 230, 241–242

Containment, 52, 127, 151–186, 227–228, 241

Conversational restraint, 12–13, 64, 66, 70–73, 81–85, 92, 128, 134, 136

Cooke, Maeve, 98, 119

Co-option, 119

Corporatist, 10, 153, 157, 183, 197, 226, 230, 241

Criterion of absence of power differences, 84

Criterion of non-exclusion, 84, 136

Criterion of open-endedness, 84, 129, 134, 136

Cross-cutting exposure, 98, 103

D

Deepening of the debate, 105, 110
Deliberative democracy, 4–8, 10–13, 15, 23, 64, 83–86, 92–94, 97, 98–99, 101–104, 107, 110–111, 113, 115–116, 119–124, 126–128, 130–138, 140–142, 152, 154, 183, 190, 195, 205, 236–237, 239, 241–244
Deliberative fora, 5–6, 8, 14, 110, 135–136, 138–141, 152, 189–190, 192–194, 236, 239, 242, 244
Deliberative mini-publics, 6, 189, 191–195, 223–224, 226, 229–230, 237, 239, 244
Deliberative poll, 133, 140, 191–194, 237, 328
Depoliticisation, 3, 10–12, 14, 53, 56, 126–127, 183, 185, 190, 216, 226–228, 238
Difference democrats, 93, 111, 115, 117–118, 120, 123, 131–133, 140, 243
Dissensus, 51–53, 56, 106, 112–114, 135–136, 236
Dryzek, John, 5, 13–14, 80, 104, 107, 110–111, 115, 117, 119, 123, 128, 130–131, 134, 139, 189, 191, 197, 242, 244

E

Eckersley, Robyn, 13, 48, 80, 98, 100–101, 282
Einsiedel, Edna, 161, 191, 199, 201, 224
Elitism, 116, 230, 241–243
Elster, John, 104
Ethics/morality distinction, 129
Exclusive states, 197–198
Expert knowledge, 6, 54–56, 95, 132, 135–138, 189–190, 212, 228–229, 238, 242
Expert/lay distinction, 230
Experts, 3, 9–10, 40–41, 51–54, 56, 95–96, 100, 113–114, 124, 126–127, 133, 135, 138, 141, 158–159, 163, 165–166, 170, 174–179, 182–184, 191–193, 196, 199–208, 211–214, 216, 218, 220–221, 223–225, 227–230, 236, 238–239, 242, 249, 273, 288, 294–296, 325, 327–328

F

Fishkin, James, 10–11, 126, 133, 192–193
Frewer, Lynn J., 184, 191–194
Fung, Archon, 5, 8, 125, 194, 218, 237, 239

G

Gene technology, 9, 24, 34, 40–44, 95, 157, 161–163, 167–168, 172–173, 175, 178–180, 190, 195, 199–202, 206–207, 210–211, 213, 217–218, 220, 257, 294, 296, 309, 311–313, 315, 327

Gene Technology Ethics Committee, 9, 157, 161–164, 327
Gene Technology Regulator, 162, 167
Genetic determinism, 33–34
Genetic engineering, 3, 19–22, 24–26, 29–30, 32–35, 37–38, 42, 44, 46, 156–157, 176, 185, 190, 203, 219–220, 247, 251, 259, 261–263, 266, 271–272, 275, 286–287, 292, 295, 307, 312–313, 329
Genetic modification, 12, 21–22, 25–26, 30–33, 35–37, 42–43, 54–55, 154, 156, 158–159, 162, 171, 179, 202, 204, 245–248, 250–251, 256–259, 261, 263–265, 272–274, 281, 287, 297, 299, 301, 306–307, 310, 312, 327
GM crops, 30, 32, 37–39, 42–43, 45, 47–49, 55, 85, 95, 155, 161, 179, 261–262, 267–271, 273–275, 281–282, 287, 290–292, 298–299, 302–303, 306–308, 310, 312–313
GMO, 23–24, 30, 40–41, 44–45, 50, 95, 113, 155, 157, 162–163, 168–169, 173–174, 179, 182, 189, 199, 201–202, 269–270, 288, 292, 298–299, 307, 315
Goi, Simona, 76, 112–113, 121
Goodin, Robert, 193, 224
Gould, Carol, 104, 106, 114, 116
Government neutrality, 3–4, 65
Group dynamics, 111, 115, 119–121, 133–134, 137, 140, 194, 210, 217, 220–221, 224, 241
Group polarisation, 120–121, 132–133, 219
Group think, 223
Grove-White, Robin, 40–41, 295–296
Gutmann, Amy, 71, 82, 97, 100–102, 104–108, 112, 128–130

H

Habermas, Jürgen, 13, 74–76, 96–97, 103–106, 111–112, 117, 126, 130, 135, 153, 196, 225, 243
Herbicide, 37–38, 41–42, 45, 181, 261, 263–264, 266, 271–274, 282, 291, 301–306, 309
Hindmarsh, Richard, 23, 25–26, 28–29, 38, 42–43, 154–156, 173, 176, 195, 263, 267, 274–275, 277, 281–282, 289, 296, 298, 304, 307, 309, 311–312
Hisschemöller, Matthijs, 2, 12, 14, 51–53, 56, 93–94, 99, 122, 126–127, 151, 165, 185, 216, 227, 236, 238
Ho, Mae-Wan, 28, 31, 33–34, 38, 247, 277, 283, 291–292

Index

I

Inclusiveness, 13–14, 92–93, 102–103, 105–106, 110–111, 118, 121–122, 124–125, 133, 135–136, 164, 167, 183, 194–195, 197, 205–206, 208–209, 230, 239, 241
Inclusive states, 197–198
Internet, 27, 163, 177, 194–195, 214, 217, 239
Intractable disagreement, 3–6, 8, 10–13, 56, 63, 65–66, 84–85, 92–94, 103, 105, 110, 112, 122, 124, 128, 134–135, 138–140, 143, 151–152, 154, 182, 184–185, 189–190, 197, 205, 229, 235–236, 238, 240, 242–244

J

Jasanoff, Sheila, 40, 294–295

K

Kleinman, Daniel Lee, 28, 30, 45–46
Kloppenburg, Jack Ralph, Jr., 28, 30, 38, 45–46, 265, 308–309
Kymlicka, Will, 66, 68, 116

L

Labelling, 41, 46, 50, 157, 181, 199, 202, 206–207, 219, 263, 269, 288, 298–301
Lay panel report, 14, 113, 140, 192, 201–202, 207, 209, 212–213, 221–222, 225, 227–229
Lay persons, 6, 40, 53, 55–56, 94–95, 100, 135–138, 142, 170, 176, 179, 192, 200, 210, 212, 216, 218, 221, 228, 239–240, 295, 326–328
Levidow, Les, 21–23, 30, 41, 44, 169, 264, 272, 293, 295–296, 313
Liberal neutrality thesis, 64–66, 81, 129
Lijphart, Arend, 153
Lobbying, 11–12, 102, 126–127, 154, 163, 184–185, 198, 206, 209, 226–228, 240, 243
Luddites, 27–28

M

Majority/minority views, 185
Mansbridge, Jane, 118
Marginalised groups, 99, 101–102, 109–110, 113, 115–120, 126, 131–132, 137, 170, 216, 221, 241
Mill, John Stuart, 64, 67, 97–98, 189, 241
Mohr, Allison, 199, 201–202, 207, 210–214, 217–220, 227
Monsanto, 23, 28, 30, 46, 201, 210–211, 219, 267, 282, 303, 307–308
Mouffe, Chantal, 112
Mutz, Diana C., 7, 93, 98–99, 102–103, 124, 128, 133, 242

N

Nagel, Thomas, 72
Norton, Janet, 24, 44, 123, 157, 313
Novel technologies, 1, 3–4, 6, 8, 11, 13, 43, 54, 56, 63, 81, 93, 96, 122, 126–127, 189, 191, 222, 241–242, 271, 298, 312
Nuffield Council on Bioethics, 30, 37, 175, 261, 264–265, 267–271, 273–274, 276–277, 279, 281, 283, 287, 297, 299–300, 309–310, 313

O

Olson, Kevin, 113, 116, 132
Opinion transformation, 13–14, 64, 78–82, 84, 92, 96–97, 105, 123, 128, 134, 136–138, 167, 174, 177, 183, 205, 209, 230, 236, 239, 244

P

Partisan mutual adjustment, 80, 97
Passive states, 74, 197, 230
Paula, Lino, 37, 52, 95, 155, 159–160, 164–167, 170–172, 176–178, 261, 326
Pettit, Philip, 11–12, 14, 53, 126–127, 184, 216, 226–227, 238
Phillips, Anne, 115, 118–119
Piercey, Robert, 73, 75–76
Pillarisation, 153
Plant biotechnology, 2, 21, 37–43, 92, 113, 138–139, 157–158, 261–315, 325, 327
Plumwood, Val, 109
Pluralism, 3, 45, 54, 56, 63–64, 66, 74, 82, 84, 86, 104, 107, 112, 114, 123, 128, 134, 226
Political culture, 6, 8, 10, 14, 127, 132–133, 139, 152–154, 163, 183–184, 195–198, 206, 210, 227–230, 238–244, 293
Political liberalism, 4–6, 13, 64, 66, 68, 70, 138, 141–142, 236
Precautionary principle, 41, 45, 55, 173, 180, 296–298, 301
Pre-structured, 211, 216, 226
Public/private distinction, 65, 67, 69, 71, 86
Purdue, Derrick, 113, 141, 212

Q

Quality of argument, 14, 84, 92, 102, 105, 118, 124, 225

R

Rawls, John, 65–70, 72–73, 75, 79, 106, 112, 215
Remoteness, 109–110
Rescher, Nicolas, 111–112, 114, 134

Rhetoric, 20–21, 27–30, 42, 117, 119, 131, 162, 175, 210–211, 213, 219, 225, 227, 305
Rio Declaration, 296–297
Risk communication, 40, 293–294
Rowe, Gene, 184, 191–194

S

Sandel, Michael, 72
Sandøe, Peter, 127, 139, 191, 196–197
Schroten, Egbert, 65, 158
Shiva, Vandana, 29–30, 264, 272, 274, 278, 286, 289
Singer, Peter, 43, 48, 248, 253–254, 259, 312
Smith, Graham, 5, 120–121, 139, 141, 192–194, 221, 262
Species barriers, 30, 32–33, 44, 175, 205, 314
Substantial equivalence, 39, 45, 172, 201, 219, 290–292
Sunstein, Cass R., 19, 91, 105, 111, 120, 123, 219, 248

T

Telos, 37, 47, 252, 254, 257, 259
Thompson, Dennis, 30, 35–37, 41, 71, 82, 97, 100–102, 104–108, 112, 128–130, 250–251, 253, 257, 288, 299

Three-track approach, 125–127, 134, 136, 209, 229, 240–241
Transgenic, 36–37, 42–43, 158, 171, 245–247, 250, 252, 258, 261–262, 266, 268, 278, 281, 288–289, 299, 303–308, 311

U

Unstructured problems, 3, 6, 12–14, 20, 50–54, 56, 64–65, 80–81, 122–126, 134, 138, 140–141, 165, 182, 185, 189, 208, 212, 216, 236, 238

V

Van den Brink, Bert, 72, 74–75, 77, 107, 112, 153
Van der Burg, Wibren, 66, 74, 135, 154

W

Wickson, Fern, 96, 169, 173–174, 179, 181
Wynne, Brian, 40, 95, 293–295, 297

X

Xenotransplantation, 20, 221, 245

Y

Young, Iris Marion, 101, 115–119, 123–124, 130–131, 133, 214, 276, 305